THE AGRICULTURAL TRACTOR: 1855-1950

R. B. Gray

American Society of Agricultural Engineers
2950 Niles Road, Saint Joseph, Michigan

Preface

A limited edition of "Development of the Agricultural Tractor in the United States" was compiled and printed in 1954 by the Agricultural Engineering Research Branch (Farm Machinery Section), Agricultural Research Service, United States Department of Agriculture. As the demand for copies exceeded supply, and since it was not practical to arrange for a reprinting by the USDA, the American Society of Agricultural Engineers sponsored a second printing, in order that this comprehensive and notable contribution to the early history of the agricultural tractor might be given wider distribution.

All previous printings of this book use the original typewritten format furnished by the USDA. During the last reprinting in original format (1974), parts one and two were consolidated under one cover and the title changed to "The Agricultural Tractor: 1855-1950".

This printing is a departure from all previous editions. The original format has been completely revised. The text, however, remains true to the original, except where altered to fit the new format. One substantial change has been made. An index has been added. It follows Part II.

Printed: 1954, 1956, 1958, 1974
Revised: 1974, 1975

THE AGRICULTURAL TRACTOR:
1855-1950

PART I
1855-1919

UNITED STATES DEPARTMENT OF AGRICULTURE
AGRICULTURAL RESEARCH SERVICE
AGRICULTURAL ENGINEERING RESEARCH BRANCH
FARM MACHINERY SECTION

INFORMATION Series No. **107** Beltsville, Md.
June, 1954

The Agricultural Tractor: 1855-1950

Development of the Agricultural Tractor in the United States

by
R. B. Gray[1]

Introduction

This publication is an attempt to assemble in chronological order the events and dates relating to the establishment, merger, discontinuance or reorganization of steam traction engine and gas tractor manufacturing companies, and to supply some brief specifications of a few of their products. Obviously many pictures of tractors produced in the early days are not available, and of those that are, considering the many makes and models, space permits only the illustration and discussion of a relatively few. It has not been possible through lack of space, to include many names of inventors and manufacturers who have contributed to the development of the industry. Dates of the first appearance of some models may be in error but every effort has been made to make them as accurate as possible. As Fred A Wirt[2] says " . . . no two individuals in the tractor industry will interpret dates and developments in the same manner. For instance which date should be used for the introduction of a new tractor or an important change in construction; the year the first model was designed and built largely by hand; the year in which the pilot run was manufactured and sold; or the year when regular production was offered to the public?"

Some tractors illustrated in Section 1 got very little beyond the experimental stage but are shown as a record of the wide and varied ideas of what early tractors should be, others are typical of the period when introduced.

The number of companies which actually manufactured tractors has varied each year from a minimum of one or two in 1895 to a peak of 186 in 1918 and to around 50 (exclusive of garden tractors) in 1950. Over 500 tractor manufacturing companies to date had entered the field since the start of the industry.

The manuscript is divided into two sections - I, which includes a brief discussion of mechanical farm power from its inception to 1920; (in this section wheel and track-type tractors are discussed separately), and - II, which includes the development of both wheel and track type tractors, except garden, in chronological order up to and including 1950.

Acknowledgment is made of the valuable assistance given by the many tractor manufacturers and individuals who made valuable suggestions as to additions to and corrections in the original manuscript.

---

[1] *R. B. Gray, formerly Head, Farm Machinery Section; and W. R. Humphries, formerly Chief Engineering Aid (now deceased), aided materially in collecting information up to 1936.*

[2] *Fred A. Wirt, former Advertising Manager, J. I. Case Co. Racine, Wis.*

I. The Beginnings of Mechanical Power

DURING the 17th century, as a result of the widespread interest in scientific research in Europe, experiments on heat engines were taking place in many countries. While the origin of the internal combustion engine is obscure, the present engine is the result of a long-continued development the early stages of which were mainly exploratory in character. For instance, Huyghens, a Dutch physicist in 1680 experimented with gun powder. Thomas Newcomen in 1705 made a practical success of an atmospheric engine. Indeed, more than twenty-one centuries ago, Hero of Alexandria (130 B.C.) described an apparatus (fig. a) that made use of heated air for opening and closing temple doors. This is the first air engine of record and the first device known to do mechanical work with expanded air. (See Section IV - "Early Internal Combustion Engines and Tractors," for additional early developments.)

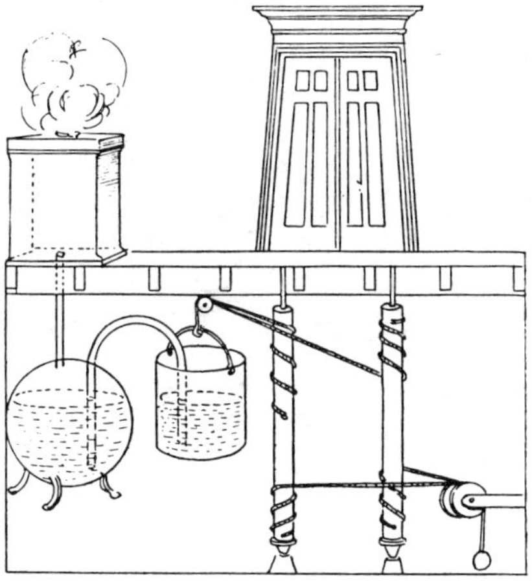

Figure a. Hero's air engine (130 B.C.)

A fire on the altar heated the air in a hollow beneath it so that it forced water from a bottle into an open pail through a siphon, the weight of the water opening the doors against counterweights. When the air cooled, the water siphoned into the bottle and as its weight was removed from the pail the counterweights closed the doors.

In 1769 James Watt took out a patent on a steam engine making use of a separate condenser and produced steam engines from 1774 to 1800, when the patents expired. The contributions made by Watt were of such significance that he is frequently referred to as the inventor of the steam engine.

The Cugnot steam road wagon built in France in 1769 is said to be the first self-propelled vehicle constructed for road use. It was used for hauling artillery.

In 1784 William Murdock, a Scotchman and close associate of Watts, built a small working model of a high-pressure locomotive engine (fig. b) which performed satisfactorily. About 1800 Richard Trevithick in England and Oliver Evans in America started experiments in the use of steam for railway carriages (fig. c). Steam turbines for marine use and steam engines for stationary and motive power were in the experimental stage during this period. In 1810 Major Pratt of England attempted to use steam power for plowing, but many years passed before significant progress was made in the practical application of steam power to agriculture.

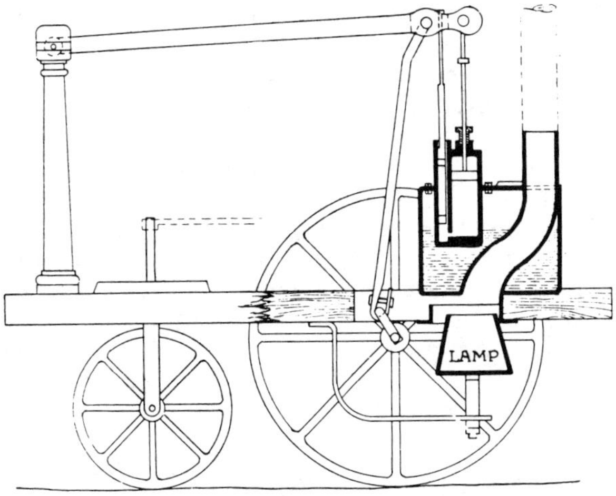

Figure b. Murdock's model of locomotive (1784)

A lamp located directly under the boiler generated the steam. The piston of the vertical cylinder was directly connected to the overhead walking beam. From this beam a long connecting-rod attached to the axle crank effected propulsion.

Figure c. Trevithick's steam carriage (1803)

One of the first practical steam vehicles was produced by Richard Trevithick, 1803. He made a number of journeys in his vehicle averaging 8 or 9 miles per hour for distances of 4 or 5 miles.

The "steam plow", or traction engine, was the forerunner of the gasoline tractor and the first step of importance in mechanical power farming in this and other countries. In addition to the demand for steam power for plowing, new farm machines invented in the first half of the 19th century stimulated a need for mechanical power. In 1831 McCormick invented his reaper. This was soon to create a demand for belt power with which to thresh the mechanically harvested grain crops. By 1847 the Pitts pattern threshing machine was being built by many shops from Maine to Mississippi and by 1860 over 50 shops were building threshers under license from Pitts Brothers. Shortly thereafter, steel plows, mowers, shellers, fodder cutters, and other machines were being offered to the farmer. With the invention of the reaper and the development of threshers, the flail with its limited threshing capacity, became obsolete. Work animals, through sweeps and tread mills (page 2), were inefficient sources of power for operating the threshers and other belt driven machines. Sources of mechanical power greater than the power horses could produce were required, hence thresher manufacturers and others undertook the production of portable steam engines.

One of the first of these produced in the United States, "The Forty-Niner", was built by A. L. Archambault in Philadelphia, Pa. in 1849 (page 2). Hoard and Bradford of Watertown, N. Y., built an engine in 1850, followed by Gaar Scott in 1852, Robinson and Co. in 1860, Rumely in 1863, and Case in 1869 (page 3).

About 1880 Baker and Hamilton marketed a portable threshing engine (page 3). This unit, equipped with an Ames engine and Laufenburg boiler, was built by the Ames Iron Works of Oswego, N. Y. Henry Ames of this company was one of the early builders and advocates of steam power on the farm and his portable engines date from 1854 when the works were founded by him.

With the introduction of portable steam engines, the familiar scene of horses powering threshing machines, through treadmills and rotary sweeps, began to disappear.

II. Self-propelled Steam Engine (Traction Engine)

The next step in the evolution of farm power was the conversion of the portable steam engine into a self-propelled steam traction engine.

In 1855 Obed Hussey of Baltimore, Md., invented and put into operation a "steam plow". In 1858 a more successful steam plowing outfit was produced by J. W. Fawkes of Christiana, Pa. (page 3). It was awarded the gold medal of the United States Society at the Chicago Fair in 1859. In the same year, the Fowler (English) cable machine (page 3) was awarded the prize for steam plows at the trial held at Warkick, England, by the Royal Society of Agriculture.

While the origin of the cable system is commonly credited to England the 1869 Report of the U. S. Commissioner of Agriculture, referring to this system, says "this same apparatus was first invented and patented in the United States by E C. Bellinger of South Carolina, Nov. 19, 1833 and was improved upon by John Fowler of England in 1854, twenty one years after Bellinger had created the infant Hercules".

Because of his faith in the development of suitable mechanical power for farming, President Abraham Lincoln in his address before the Wisconsin State Agricultural Society at Milwaukee, Sept. 30, 1859 said:

> "The successful application of steam-power to farm work is a desideratum — especially a steam plow. It is not enough that a machine operated by steam will really plow. To be successful, it must, all things considered, plow better than can be done with animal power. it must do all the work as well, and cheaper, or more rapidly, so as to get through more perfectly in season; or in some way afford an advantage over plowing with animals, else it is no success."

In 1868 the Standish steam rotary plow, "Mayflower," was built by Philander Standish at Pacheco (near Martinez), Calif. (page 3). This 12-horsepower outfit was later offered for sale in several sizes, ranging from 10 to 60 horsepower. Operating speed was from 1.7 to 3.4 miles per hour and plowing rate up to 5 acres an hour.

Also in 1868 Owen Redmond of Rochester, N. Y., patented a steam plow, (page 3) and according to the 1870 Report of the U. S. Commissioner of Agriculture, "a gang of six plows, designed to go with the engine, has since been constructed, intended to be operated by one man, who also might be the fireman."

The manufacture of farm machines was greatly accelerated with the opening of the western prairie land (Homesteading Era) following the war between the States. The J. I. Case thresher factory at Racine, Wis., the McCormick's harvester factory at Chicago, and a number of other establishments became beehives of industry. D. M. Osborne of Auburn, N. Y., whose firm was established in 1856, was beginning reaper production in quantity. "Bill" Whitely and his reaper had already given Springfield, Ohio the name of Reaper City.

By 1870 farming was becoming an industry in which more and more power was needed. In that year more than 2,000 establishments were manufacturing farm machines valued at $48,000,000. Twenty-one years

Farm "Horsepower," above "treadmill," below "sweep." Common threshing methods antedating the portable "threshing engine."

"The Forty-Niner", one of the first portable steam "farm engines" produced in the United States. A. L. Archambault, Philadelphia, Pa., 1849. This was built in 4, 10, and 30 horsepower sizes. Engine shown is 4 horsepower weighing about 4000 pounds.

Case Steam Engine No. 1 built about 1870 and now in the Smithsonian Institution, Washington, D.C.

Baker and Hamilton threshing engine, marketed about 1880. The boiler was provided with a jacket of 2-inch staves held in place by brass bands, and was capable of burning wood, coal, or straw.

One of the first successful "steam plows" produced in the United States, by J. W. Fawkes, 1858. The main frame of iron was 8 feet wide by 12 feet long and rested on the axle of a roller (driver) 6 feet in diameter and 6 wide. One cylinder 9 inches in diameter with 15-inch stroke was provided on each side of the boiler and the engine was so geared that the drive roller made one revolution to six strokes of the piston.

Steam Ploughing. One phase of early plowing embodied the use of two steam tractors which alternately pulled the plow across the field by means of cable. Known as "Cable" or "English," the system was tried out in this country on some of the large sugar beet ranches in the West but was soon replaced by more efficient and less costly equipment (1859).

The Standish Rotary Steam Plow "Mayflower". Built at Pacheco (near Martinez) Calif., 1868. The vertical boiler and a pair of horizontal engines were mounted on a frame carried on two main wheels two feet wide, and two guide or steering wheels. The frame at the rear held two, three or four vertical shafts upon which the rotary "plow wheels" with the cutter blades were secured. Drivers, 6 feet in diameter on the small outfits and 10 feet on the large ones, were covered with soft wood.

Redmond's Steam Plow (1868). Weighed nearly 5000 pounds and was about 6 feet wide. Tender carried water tank of two barrel capacity upon which was mounted engineer's seat. Drive wheels fitted with protruding lugs automatically pushed through rim.

Who said rubber tires were new? This "Thompson's Rubber Tire Steamer" is reproduced from the Pacific Rural Press of Jan. 14, 1871. It hauled through soft ground, pulled a gang of seven plows, and could speed along the road 10 miles per hour.

Atlas Engine (1881) Atlas Engine Works, Indianapolis, Indiana. Showing chain drive — first reduction, and horse steering.

Peerless Engine and Steam Lift Plow (1886) Geiser Manufacturing Company, Waynesboro, Pennsylvania.

previously (1849), when many experiments were being conducted to apply steam to agriculture, about 1,300 small farm-implement establishments, employing fewer than 7,000 persons, were producing farm machines valued only at about $6,000,000. (In 1950, the value of farm machines, including tractors, produced was approximately $1,800,000,000).

Prior to about 1870, thresher manufacturers had not encouraged the manufacture of portable steam engines. By that date, however, the production of portable steam engines had become an important part of their business, and became more important as the demand increased for self-propelled steam engines to pull threshers from place to place.

The Thomson's Rubber Tired Steamer (page 4), which was built in England, made its first appearance in the United States in 1871, with wheels fitted with rubber cleats, when it was entered in the California State and San Joaquin Valley Fairs. Thomson, the inventor, is said to have previously entered his steamer fitted with pneumatic tires in tests in England and France, but the pneumatics performed unsuccessfully. This machine probably marks the first attempt to use rubber on tractor wheels.

While the main efforts in providing self-propulsion systems for steam tractors seemed to center largely around the use of wheel propulsion, many inventors were at work devising methods for providing better traction through the application of tracks and other devices, and many unique ideas were worked out. (See Track-Type Tractors Sec. VII).

Between 1870 and 1880, farm machinery inventors devised a suitable gearing for the rear wheels of portable steam engines, and also a chain or belt drive from the engine flywheel to a countershaft of this gearing to provide self propulsion (page 4). The bevel gear and inclined shaft developed by C. and G. Cooper of Mt. Vernon, Ohio was also a popular method of drive at this time. By its use the farmer could convert his portable steam engine into a traction engine in the field. In 1880 a United States patent was issued for a steering device, although English tractors were fitted with steering gears as early as 1863. This was followed by the introduction of a clutch and gear train between the engine and rear drive wheels. The steering gears on these early "steamers" were not at first considered as entirely reliable by some manufacturers, and operators were cautioned about their use on public highways. The suggestion was made that it might be safer to guide the machine with a team (page 5). Others considered that horses were not frightened when meeting a traction engine preceeded by a steering team, and still others felt that the additional horsepower provided by this team was advantageous — both reasons for horse steering remaining for a while.

Farmers started buying self-propelled steam engines in the late 1870's in considerable numbers. In 1890 approximately 3000 steam tractors and 2661 steam threshers were built. In 1894 several plow manufacturers advertised multiple-bottom steam tractor plows or gangs. By 1900 more than 30 firms were manufacturing some 5,000 large steam-traction engines a year. These tractors were greatly improved over earlier models. The gearing, shafting, and other wearing parts were built to withstand the immense strains imposed upon them in pulling large threshers, and plowing many furrows at one time. Big wheat farms and ranches in the Dakotas, Colorado, Montana, Nebraska, Kansas, California, and Western Canada were using steam traction engines. About this time the Geiser and Friede Companies both of Waynesboro, Pennsylvania offered to the trade steam-lifts for engine gangs (page 4). This development indicated, even this early, that thought was being given to cutting down labor requirements in plowing; also to lightening the burden of manually lifting the plows by levers which were continually sticking or giving trouble and resulting in the use of very strong language by the operator.

"Clouds of smoke and hissing steam; a broad prairie stretching for miles and miles without a break . . ." are the initial words in Ellis and Rumely's "Power and the Plow" and heralds an era which, through mechanical power, the lifting of the farmer's burden had made a significant start.

Through the years before the final abandonment of the "steamer" and the universal acceptance of the gasoline tractor many improvements had been made and the performance of these huge self-propelled power plants had been and still is the pride of those traction engine engineers who pioneered in the ultimate placing of power in the hands of the 6 million farmers in this country.

While most of the traction engines have passed on, the spirit still lingers with those numberless enthusiasts whose privilege it was to operate these machines in the early days. This is evidenced by the "picnics" held in various parts of the country, and to which thousands of men and their families flock each year. The annual reunion of the Rough and Tumble Engineers Historical Association held each year at Kinzers, Pa. is an example of such an event.

E. R. Potter[3] of Saskatoon furnishes a statement, typical of the performance of many of these steam power vehicles, — "Wm. Rowse of Hanley, Saskatoon, broke 33 acres in first half days breaking of prairie sod using a Reeves 40 H.P. engine and 16-14 in. plows. In 1909 Swanson of Semans, Saskatoon using the same outfit used to be credited with breaking as much as 160 a. in 24 hrs.

"An Alberta farmer in a 10 hr. shift turned over 75 a. of back setting (plowing again in the fall after breaking prairie land in the spring), 8 in. deep using one engine in 1910.

"At the 1909 Brandon Manitoba Fair a Gaar-Scott steamer of 25 H.P. and 14-14 in. plows plowed for exhibition at the rate of an acre in 7 min. 58 seconds."

The trend in the development of steam tractors from 1885 to 1919 is shown on pages 5-9.

Some Steam Tractors*

John Fowler & Co., (Leeds) Ltd., Leeds, England (1877)

[3]*Mr. Potter also furnished many pictures and offered many valuable suggestions for improvement of the text. He is connected with the Western Development Museum of Saskatchewan at Saskatoon (Canada).*

**Most of these photographs are through the courtesy of Arthur S. Young, Kinzers, Pennsylvania.*

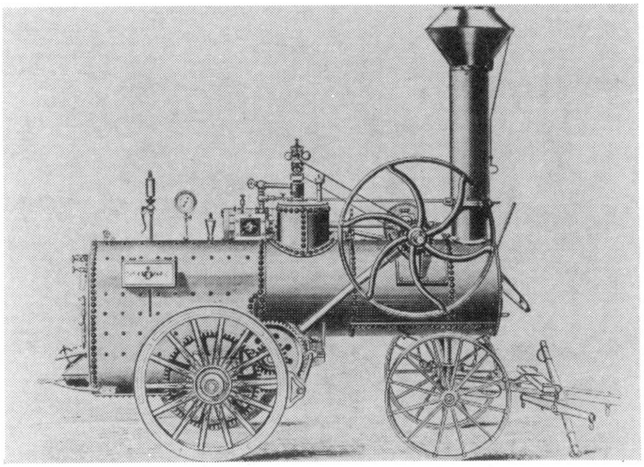

Nichols & Shepard Co., Battle Creek, Mich. (1879)

Gaar-Scott & Co., Richmond, Ind. (1880)

Atlas Engine Works, Indianapolis, Ind. (1881) (chain drive)

Buffalo-Pitts Co., Buffalo, N.Y. (1882)

Frick Co., Waynesboro, Pa. (1885) "Daniel Boone" (eclipse engine)

Geiser Mfg. Co., Waynesboro, Pa. (1893) ("Peerless" Model "R")

J. I. Case T. M. Co., Racine, Wis. (1885)

D. June & Co., Fremont, Ohio (1896) ("Champion")

O.S. Kelly Co., Springfield, Ohio (1890) ("Springfield" engine)

Port Huron Engine & Thresher Co., Port Huron, Mich. (1899)

Minnesota Thresher Mfg. Co., Stillwater, Minn. (1891 or 1893)

M. Rumely Co., La Porte, Ind. (1899) (Straw burner)

Aultman Co., Canton, Ohio (1901) ("Mogul")

Advance Thresher Co., Battle Creek, Mich. (1902) (Coal and wood burner)

Northwest Thresher Co., Stillwater, Minn. (1903)

Minneapolis Threshing Machine Co., Minneapolis, Minn. (1904) (Two cylinder engine)

Russell and Co., Massillon, Ohio (1905) (One cylinder engine)

M. Rumely Co., - 15 and 20 HP, La Porte, Ind. (1905) (Two cylinder straw burner)

Huber Mfg. Co., - 18 HP, Marion, Ohio (1906) (Return flue, one or two cylinder engine)

Gaar-Scott & Co., - 22 and 25 HP, Richmond, Ind. (1910) (Simple two cylinder engine)

Reeves and Co., - 13 HP, Columbus, Ind. (1912) (Simple two cylinder engine)

Wood Bros., Des Moines, Iowa (1913)

Avery Co., - 40 HP, Peoria, Ill. (1913) (Two cylinder locomotive style, undermounted)

Birdsall Engine Co., Newark, N. Y. (1914) (Two cylinder engine)

Nichols & Shepard Co., Battle Creek, Mich. (1915)

J. I. Case Threshing Machine Co., Racine, Wis. (1915)

Geiser, (Emerson-Brantingham), Waynesboro, Pa. (1916) ("Anderson" type)

Reeves (Emerson-Brantingham), Rockford, Ill. (1916) ("Canadian Special")

Avery Company, Peoria, Ill. (1917)

Huber Manufacturing Co., Marion, Ohio (1918)

Banting Manufacturing Co., Toledo, Ohio (1919)

III. Functional Units of Steam Traction Engines

Both the horizontal-tube and vertical-tube boilers were used in early steam traction engines. The former was decidedly the more popular. While the two types were entirely different in form they had many operative points in common.

Horizontal-Type Boilers

The horizontal-type boiler was constructed (a) with direct flue, (b) with return flue, or (c) with firebox return flue. All three were used extensively and were quite satisfactory.

The direct-flue boiler (page 9) was known as the locomotive firebox, straight-flue boiler.

With the return-flue boiler (page 9), there was one main flue which carried the products of combustion from the firebox to the front end of the boiler. From here they passed back through the numerous small flues to the smokebox in the rear and out of the stack.

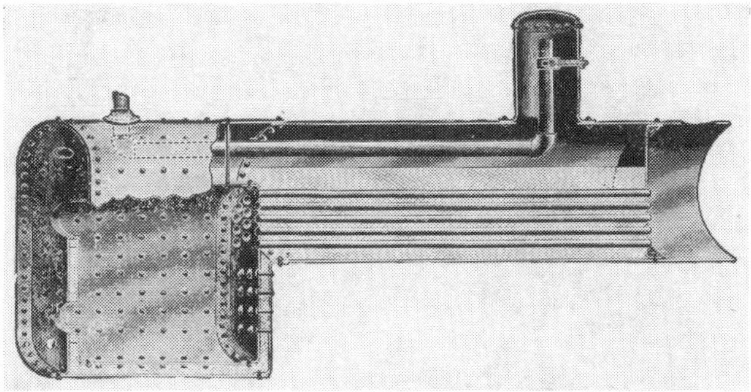

Typical firebox direct flue steam boiler. The smoke and products of combustion passed from the firebox through numerous small tubes into the smokebox at the front end of the boiler, thence into the chimney and out into the air.

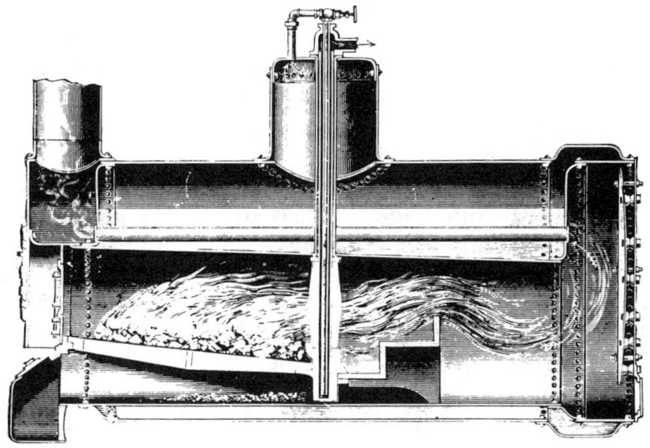

Typical return flue steam boiler. The products of combustion travel twice the length of the boiler before reaching the outer air. The grates were placed in the rear end of the large flue, the latter serving the purpose of a firebox. Little space was provided the grates to catch ashes and cinders and there was the constant danger of burning out grates.

Little space was provided under the grates of types (a) and (b) types to catch ashes and cinders, and there was the constant danger of burning out grates. This danger was largely overcome in the firebox return-flue boiler. With this type, heated surfaces were entirely surrounded by water, the grate area was comparatively large, and the boiler contained a maximum of heating surface.

The return flue and firebox return-flue boilers were soon to be discarded and practically universal use was then made of the open-bottom boiler with an ash pan.

Vertical-Type Boilers

Boilers of the vertical type had a cylindrical shell provided with a firebox at the lower end, either with horizontal water tubes placed in courses, so that each course was at right angles to the course next below and next above (page 10), or with fire tubes extending vertically from the flue sheet above the fire to the top of the boiler (page 10).

Engine and Boiler Mounting

Two principal approaches were used in constructing

Westinghouse vertical boiler (horizontal water tubes), 1886.

C. L. Best Steam Tractor (vertical fire tube boiler), 1885. Space between firebox and outer shell formed the water leg and made place for mud to collect.

the steam traction engine. One was to make the boiler the central structure, and attach all other parts thereto, such as engine, drive gears, steering gear, and main truck. The other method was to provide a separate framework on which to mount the boiler and attach all of the parts.

In order to prevent heavy shocks and jars to the engine in traveling over rough roads, heavy coil springs were frequently placed between the boiler and the axles, both front and rear. Springs were also placed in the steering gear to prevent breakage when the front wheels hit an obstruction.

In most cases the early engine was mounted on the boiler, called topmounted, and the boiler was mounted upon the truck. In some cases of the locomotive type, the engines were mounted under the boiler on a separate chassis and were known as under-mounted engines.

There were several methods of mounting boilers. One common method, known as "side mounting," was to attach stub axles of the drivers to brackets placed at about the middle of the sides of the firebox. In another type, known as "rear mounting," one continuous axle located back of the firebox was used. On return-flue boilers a continuous axle was often mounted under the firebox. This was known as the "under mounted boiler."

On page 11 is a typical steam traction engine of about 1917, with the parts named.

Transmission — Wheels

The power of the steam traction engine was usually transmitted to the traction wheels by a simple train of spure gears generally made of cast iron. A typical arrangement consisted of a friction clutch with attached driving pinion which engaged an intermediate gear, and which, in turn, engaged a large compensating gear on the countershaft. Pinions on either end of the countershaft drove the large master gears which were fastened in the drive wheels by either rigid or spring connections. At first, traction engines were geared with one forward speed and made from 2 or 3 miles per hour on the road; later, some were geared with two forward speeds, one "slow" and one "fast." These were especially useful in hilly country.

The front or steering wheels were often of the built-up type (steel) tire to which the outer ends of the spokes were riveted, but were more frequently made of cast iron rims. A flange or collar around the middle of the front wheels tended to prevent them from lateral slippage. Steering was accomplished by guiding the front wheels with a chain, winding shaft or roller, worm gear, and hand wheel. Occasionally, steering was accomplished by power from the engine.

The rear traction or drive wheels were mainly of the built-up steel tire type, although, previous to 1900, many manufacturers used cast iron drive wheels. To these were attached round or flat spokes which were fastened to a cast iron hub. Cleats of steel or malleable iron were mounted diagonally on the outside of the tire rims to increase traction. Some rims were cast, in which case the cleats were a part of the cast. The cleat faces were chilled to resist wear. The spokes were set into the rim and hub when the rim was cast.

Two of the illustrations on page 12 show extreme attempts to provide adequate traction in soft land. The problem of supporting weight under such conditions, no doubt, inspired the idea of substituting tracks for wheels.

IV. Early Internal Combustion Engines and Tractors

The origin of the internal combustion engine might be considered as dating back to 1678 when Abbe′ Hautefeuille, a Frenchman, credited with the first one, "conceived the idea of burning a small quantity of gunpowder in a chamber connected with a reservoir of water. After combustion, the gases were cooled resulting in a partial vacuum, and the water was raised from the reservoir by atmospheric pressure. A few years later he described another engine in which the actual explosive energy of the powder was used to raise water through a system of pipes and check valves. In this type the water acted as a piston, so that this is the first record of a direct-acting engine."[4]

In 1680 Huyghens, a Dutch physicist, conceived the idea of securing power from explosions. This was the first time that the explosive energy of powder had been harnessed. A working model, employing a piston and a cylinder, was constructed with a small receptacle screwed to the bottom of the cylinder in which the powder could be ignited (page 12). The uncertainty of the functioning of this device caused it to be abandoned.

---

[4]*Page 25 - Gas Power - Hirshfeld and Ulbricht (1913).*

Case Steam Traction Engine

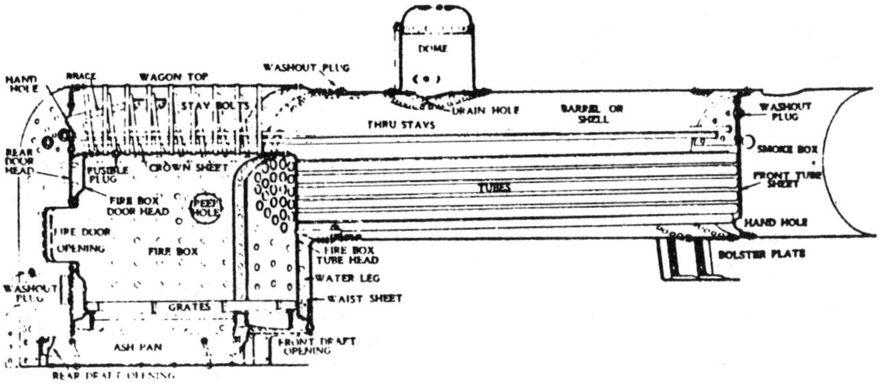

FIG. 1. SECTIONAL VIEW OF BOILER

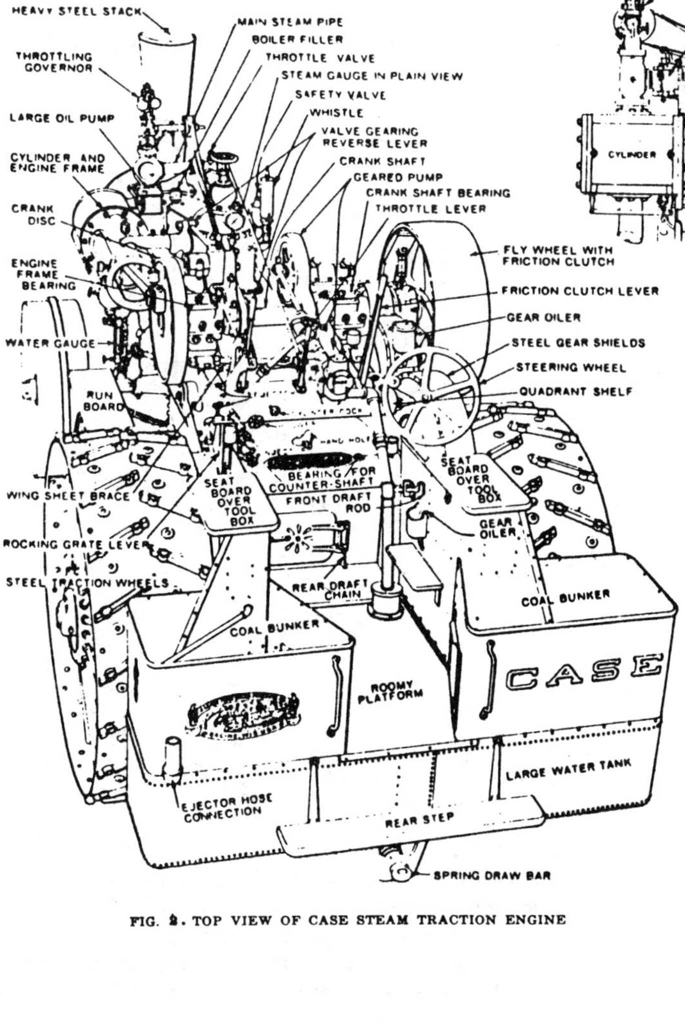

FIG. 2. TOP VIEW OF CASE STEAM TRACTION ENGINE

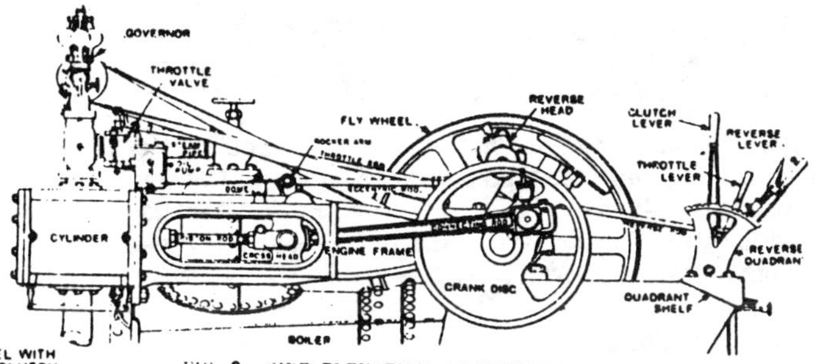

FIG. 3. SIDE ELEVATION OF ENGINE PROPER

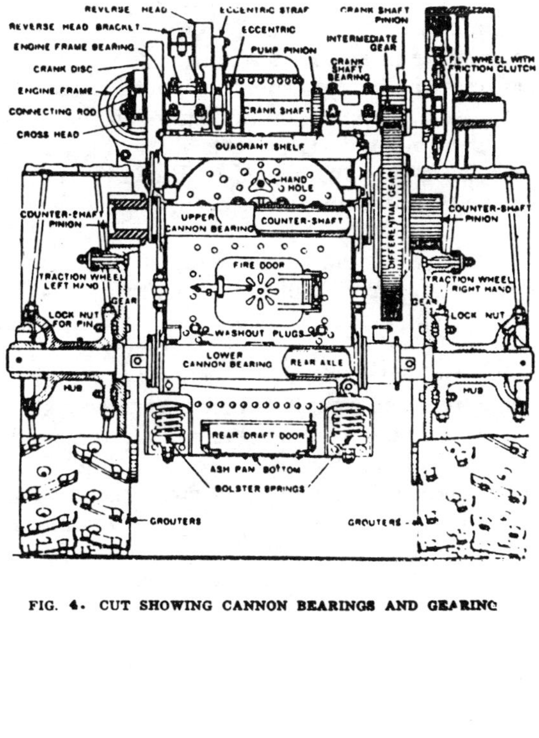

FIG. 4. CUT SHOWING CANNON BEARINGS AND GEARING

A big-wheel outfit built by the Best Mfg. Co. in 1900 for the Middle River Farming Company, Stockton, Calif. The wood-covered drivers were 15 feet wide and 9 feet in diameter. Total weight of outfit was 41 tons.

Holt Steam Traction Engine, built about 1903 specially for Charles Moreing, a large land owner, in the San Joaquin Valley, Calif. Each wheel was 7-1/2 feet in diameter and 6 feet wide. The wheels were mounted in groups of three thereby making a total width of 18 feet on each side. Shown working a 44-foot strip of soft peat land in the San Joaquin Valley, California.

In 1690, Papen, another French physicist, experimented along this line (page 12) but as his efforts were unsuccessful he directed his attention to steam.

Because of the many difficulties encountered in the early days, such as poor workmanship and the unavailability of suitable material, efforts along this line seem to have been largely abandoned and for over a century developments in steam power occupied the inventive mind.

It was during this period that Watson discovered the distillation of gas from coal which, with the experience of steam engineers, aided John Barber, an Englishman, in 1790 to build the first engine to employ gas as a fuel. The device employed a paddle-wheel (gas-turbine principle) that revolved rapidly when a continuous flame from a gas exploder was directed against the paddles.

In 1794 Robert Street, also an Englishman, "patented the first real engine. It contained a cylinder in which worked a piston connected to a lever and operated a pump. Turpentine as the fuel, was introduced and heated, and the air drawn in produced the combustible mixture."[5]

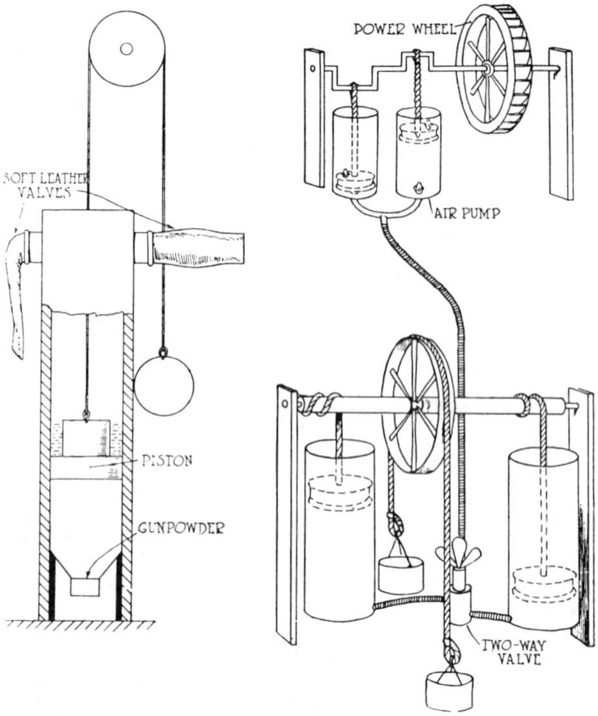

Left: Huygen's gunpowder engine of 1680 in which the flame and gas of the explosion threw the piston upward, the leather exhaust tubes flattening as the piston dropped and prevented air entering until the piston passed them. A rope pulley and weight communicated power to the pulley shaft.
Right: Papen's engine of 1690 which was somewhat similar but steam was used instead of powder. Papin's engine was used in raising any sort of weight out of a deep mine.

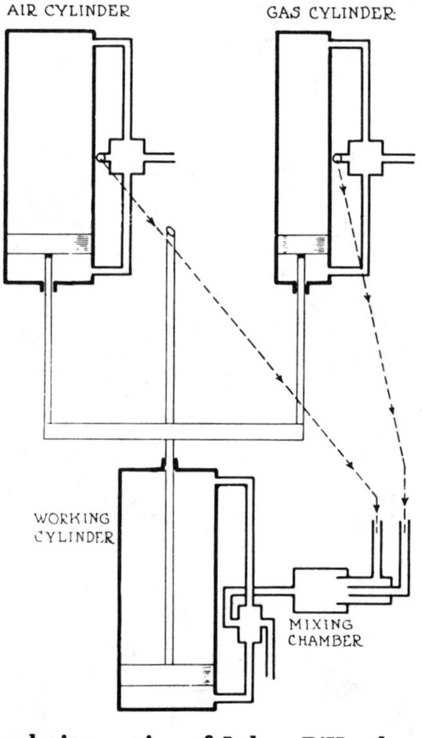

The 1801 explosion engine of Lebon D'Humbersin was designed with three cylinders, the two upper ones serving for storing air and gas. The contents of the two cylinders were brought into a mixing chamber and thence into a combustion chamber where they burned and expanded, finally reaching the working cylinder where useful work was accomplished.

[5] *Page 1 - Internal Combustion Engines by Wallace L. Lind (1920).*

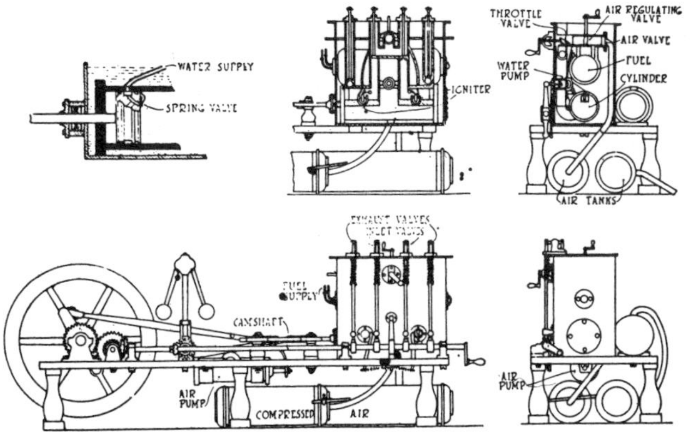

Stuart Perry (1844-46) engine fitted with valves in the cylinder head, water jacket, cylinder head and valve parts. A double-acting water pump (when the piston registered with the water supply pipe) delivered water into a piston groove, and, by means of a spring valve in the piston, part of the water was injected into the cylinder to aid in lubrication and to add its expansive force.

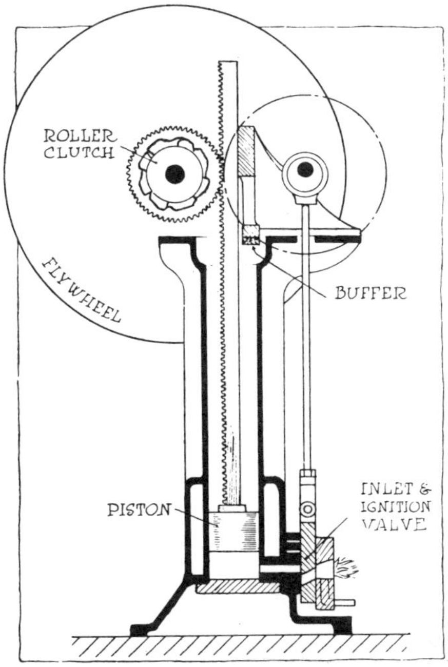

Otto and Langen's Free-Piston engine (1876). The explosion threw the piston to the top of the stroke, the rack revolving a free wheel clutch. Atmospheric pressure then drove the piston down again, the clutch gripping and turning the shaft.

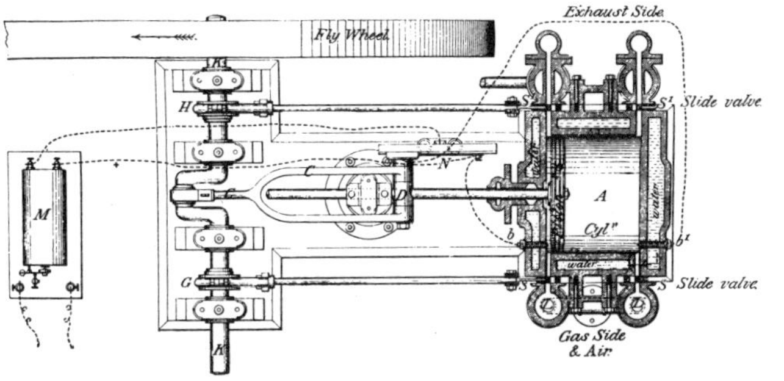

Lenoir horizontal gas engine (1860). The inlet valve was closed at the half stroke and the charge was immediately fired by means of electric spark; the last half of the piston stroke was allotted to power impulse.

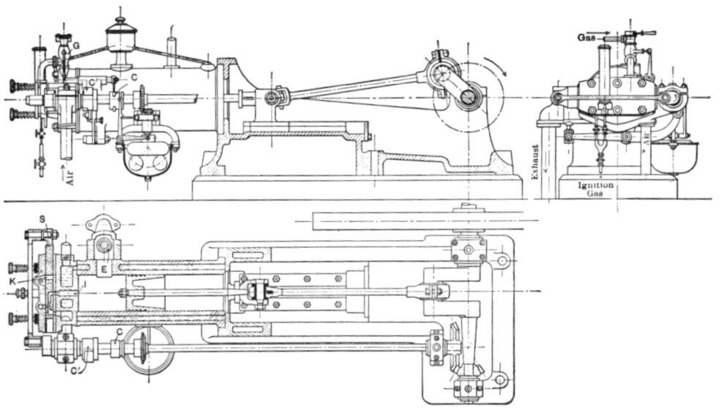

Otto's engine (1884). At the beginning of the suction stroke the poppett exhaust valve E is closed. As the piston moves outward only air is drawn through port I, but a moment later valve G opens and the mixture is admitted for the rest of the stroke. At the outer dead-center the slide valve S has closed the port I in the head, and compression begins. When the piston reaches the inner dead center the ignition cavity K is exposed and the charge is ignited. The piston moves outward again with the expansion of the gases and on the next (return) stroke the gases are exhausted through the valve E which has been opened by the cam C′.

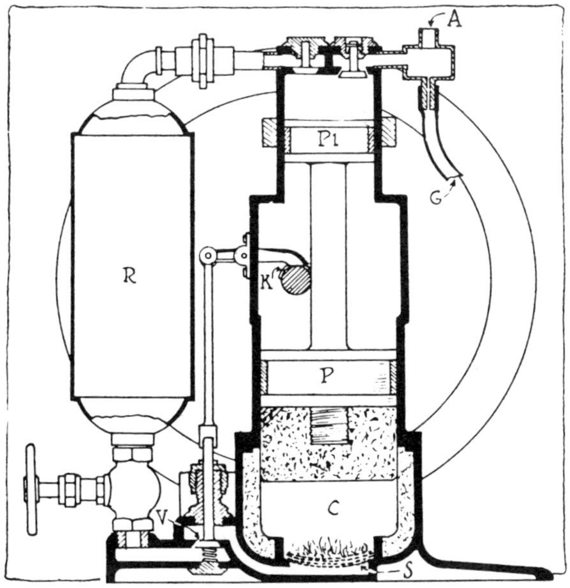

Brayton's two-cycle engine (1872). The upper portion of the compound piston P1, pumped an explosive mixture into the larger reservoir, R, where it was stored under pressure. The explosive mixture, under valve control, passing through several thicknesses of wire mesh at the bottom of the cylinder, was kept burning in constant flame above the wire mesh, driving piston upward by expansion due to heat rather than from explosion.

An explosion-type engine was patented in 1801 by Lebon D'Humbersin, a Frenchman, who is classed by some as the father of the present day internal combustion engine. In his gas motor, both the gas and the air were compressed in external chambers previous to ignition (page 12).

In this connection the efforts and contributions of Stuart Perry of New York, 1844-46, should not be overlooked. Perry's engine was double-acting, being built much as a steam engine, an explosion taking place at either side of the piston. Perry used an air starter and his patent drawings show no attempt at attaching a starting crank. However, there was a miniature crank which was used to put the camshaft in proper position for starting; that is, to open the intake valve to admit the explosive mixture when it was ready and then to close the valve and revolve the cover of the platinum ignition tube so as to let the heat of the tube explode the mixture (page 13).

In 1860 Lenoir, (page 13) another Frenchman, developed a gas engine of a non-compression type, built along the lines of a double acting steam engine. Electric ignition was used. Several hundred of these engines were built from 1860 to 1865 but they proved very uneconomical because of excessive fuel consumption (from 80 to 110 cubic feet of gas per horsepower hour); hence they were unsuccessful. While Hugon, a Frenchman, was credited with experimenting with an engine of the Lenoir type before Lenoir, he did not bring his out until 1864. Ignition was accomplished by flame. Twenty-five years later, however, Lenoir brought out a single acting four-cycle engine which may well be classified with our modern engines. Electric ignition was used.

Beau de Rochas, a French engineer, about this time formulated the theory of the cycle of operations to utilize more efficiently the heat supplied. He laid down conditions from which resulted the four-cycle principle embodying compression of the charge before ignition.

Brayton's two-cycle engine of 1872 proved to be a popular one in the 70's (page 13). It was considered one of the best explosive engines of the time. Compression was the principle that made the gas engine ultimately successful — and it seemed to be left to N. A. Otto and Eugene Langen, two German inventors, to make practical application of it.

The internal combustion engine as a practical power unit began to assume importance with the Otto engine in 1876. This was not Otto's first effort, however, for in 1861 he attempted to improve on Lenoir's engine by giving it a full power stroke, and, later with Langen, Otto invented an entirely new type known as the free-piston engine (page 13).

In 1878 Otto exhibited a still newer type of engine of recognized merit at the Paris Exposition, and in 1884 brought out a model the fundamentals of which were, in modified form, used many years thereafter. (see page 13) Economy tests on early Otto engines gave thermal efficiencies on the brake of from 9 to 12 percent, while one experimenter obtained a value of 15.5 percent.

Otto's four-stroke cycle engine was thought to be at a serious disadvantage because of only one power stroke for every two revolutions of the crankshaft. Experimenters accordingly turned their attention to the development of an engine operating on a two-stroke cycle in which there is a power stroke each revolution of the crankshaft.

A two-stroke cycle engine was produced by Dugald Clerk, an English engineer, in 1879, but owing to a number of difficulties encountered it was given up and it was not until some years later that a commercial two-stroke cycle machine was successfully developed.

The patents of the Otto cycle were so basic in character that during the years of their life the general manufacture and sale of this improved form of generating power was greatly restricted. With the expiration of these patents in 1890, however, many concerns entered the gas-engine field. Within a few years, many practical improvements were made, and the practicability of the internal combustion engine as a power source for self-propulsion was established.

In 1899 there were approximately100 firms engaged in the manufacture of internal combustion engines in the United States; in 1911, 400 firms, and in 1914, 549 (not including the automobile manufacturers).

In 1893 Rudolph Diesel, a German scientist and inventor, born in Paris 1858 of Bavarian parentage, described a new form of internal combustion engine which we now know as the Diesel engine. Diesel's first engine was patented in 1892, the inventor claiming to have originated a compression-ignition engine to burn coal dust blown into the engine cylinder by compressed air, although it was further claimed any kind of fuel could be used.

Coal dust — a useless by-product in mining areas — was the first fuel considered because of its economical possibilities. Its use, however, was unsuccessful and after a serious explosion further experiments with the dust fuel were discontinued. Later a successful oil-burning compression-ignition engine was developed by Diesel and a number of European manufacturers were licensed to build similar engines.

This engine depends on the high temperatures of the highly compressed air (500 to 600 pounds per square inch) to ignite the fuel oil charge as it is injected into the combustion chamber during the early part of the working stroke; the amount injected depending on the load, and controlled by the governor.

The first Diesel engine (stationary) in the United States was built by a company organized by Adolph Busch, of St. Louis, Mo., who had purchased the first American Diesel license in 1896. The engine was a 2-cylinder 60 horsepower unit weighing about 400 pounds per horsepower. The first Diesel tractor in the United States was built in 1931 by the Caterpillar Tractor Co., Peoria, Ill. It was of the track-type powered with a 4-cylinder Diesel engine and made use of a 2-cylinder gas engine for starting.

It is difficult to state definitely who first made use of petroleum fuel in an internal combustion engine. The discovery of its potentialities for this purpose, however, paved the way for the development, improvement and increasing use of the tractor and automobile — dependent mainly on the availability of reliable sources of petroleum fuels. (In the case of the automobile, particularly, a contributing factor toward its wide acceptance, was of course, the development and extension of all-weather roads.)

It seems to have fallen to the lot of one E. L. Drake, who had been a conductor on a New England railroad, to start the real commercial development of the oil industry by having a well drilled on August 29, 1859 at Titusville, Pa. (Oil Creek). Mr. (Colonel) Drake has since been considered the "Founder of the Petroleum Industry."[6]

As has been stated the early experimenters used as fuels gun powder, turpentine and then gas — natural or artificial — but it was not until the discovery that gasoline was a fuel to be found in quantity that the gasoline engine development made headway.

When candles were the main source of house lighting in the olden days there was a clamor for something which would furnish more light for the long dark nights

[6]*Gasoline, What We Should Know About It, by Tad Boyd. (1925)*

— especially in winter. The kerosene oil or coal oil lamp appeared. When it was found that petroleum as it came from the ground would burn, it was also found to contain a fraction which burned too readily for use in the lamps — it was dangerous. This was gasoline which had to be distilled off and which became a drug on the market. This lighter fraction was also distilled off as a waste in the production of lube oils. It has to be thrown away as it was useless, it was poured into streams until laws were enacted to prevent it — once ignited it would burn on the surface endangering villages, boats or anything that happened to be along the stream. The Dutch East Indies refiners dumped it into the ocean and burned it.

Now millions of barrels of liquid fuels are produced and used annually and without it the wheels of industry would cease to turn.

Heldt in The Gasoline Automobile (Vol. IV-1925) says, "For ten years after Drake's discovery of oil at Oil Creek the petroleum industry remained confined to Pennsylvania but from 1870 on it spread to other parts of the United States."

V. Early Tractor Development

The early development of the tractor followed a pattern similar to that of the steam traction engine — the stationary engine was mounted on skids or wheels to make it portable and then a drive was devised to make it self-propelled. The trend continued for many years but gradually absorbed certain automotive principles until today (1954) we have such features as the automobile type steering and transmission, high speed engine, and a much lighter and more efficient tractor unit. Indeed, the trend went further and added such items as belt pulley for belt work, drawbar for pulling machines, power take-off (p.t.o.) for driving mounted or pulled machines and finally hydraulic controls for controlling many drawn and pulled machines — furnishing four different methods of applying power.

1889-1901

In 1889 the Charter Gas Engine Co. built six gasoline tractors which were shipped to farms in the Northwest (page 16). The power units were of the stationary type of engine of the day and were all built on Rumely steam tractor trucks.

On February 7, 1890, George Taylor, Vancouver, B. C., patented a walking type motor plow using a "petroleum engine" (page 16). One claim showed a worm or screw in contact with the ground. E. R. Potter hints that this unit may have been the first garden tractor.

In 1891 William Deering and Co. built an engine with two parallel vertical cylinders which developed 6 hosepower. This was mounted on a New Ideal mower to make it a self-propelled unit. Later this Company designed and built 12- and 16-horsepower engines. One of these engines was used on a self-propelled vehicle from 1892 to 1895; another on an experimental corn picker; and still another was mounted on a mower which was demonstrated at the Paris Exposition in 1900 and at the St. Louis Exposition in 1903.

In 1892 John Froelich mounted a gasoline engine, built by the Van Duzen Gas and Gasoline Engine Co. of Cincinnati, on a Robinson running gear equipped with a traction arrangement of his own manufacture (page 16). This was probably the first gasoline tractor of record that was an operating success. It completed a 50-day threshing run belted to a Case 40 x 58 thresher, pulled the thresher over difficult terrain, and operated in temperatures from -3° to 100°F. The Froelich was the forerunner of the Waterloo Boy and the John Deere line of tractors.

In 1892 the Case Threshing Machine Co. built an experimental tractor of comparatively light weight which developed from 16 to 20 horsepower (page 16). This followed the design of William Patterson. Due to carburetion and ignition difficulties the machine was discarded after several years of experimentation. D. P. Davies, for many years vice president in charge of engineering, was at that time a young draftsman in the experimental department of the company.

In 1892 the C. H. Dissinger and Bros. Co. built their first gasoline traction engine, the "Capital," at Wrightsville, Pa. These brothers, as boys, had been apprenticed with Schliger, Schumm & Co. of Philadelphia who had brought the first Otto engine to this country in 1886. This firm later incorporated as the Otto Gas Engine Co., and sold the first internal combustion engines in the United States. Soon the Dissinger Bros. were to withdraw and organize the company bearing their name. The "Capital" traction engine was not put to practical use however until 1899.

The Hocket (Sterling) tractor (page 17) was brought out in 1893 by J. A. Hocket and a group of farmers near Sterling, Kans., after 2 years of experimentation. Three years later Hocket ran out of funds and discontinued manufacture. He published one of the first advertisements for gasoline tractors.

In 1893 John Froelich incorporated the Waterloo Gasoline Traction Engine Co. of Waterloo, Iowa, to manufacture the Froelich gas traction engine.

In 1894 Van Duzen (page 17) built a tractor similar to the Froelich. In the same year, the Otto Gas Engine Co. of Philadelphia, which already had an excellent stationary engine, manufactured its first gasoline tractor (page 17). Its second model was built in 1896 (page 17). Also in 1894 the Lambert Gas and Gasoline Co. of Anderson, Ind. started building tractors (page 17).

In 1895 the Waterloo Gasoline Traction Engine Co. (organized in 1893) was reorganized as the Waterloo Gasoline Engine Co. Having achieved considerable success in building gasoline engines, this concern continued its program of building experimental tractors. Almost 20 years passed, however, before this Company began production of its kerosene-burning tractor known as the "Waterloo Boy."

Also in 1895 the Davis Gasoline Traction Co. and the Webster Engine Co., both of Waterloo, were first listed as tractor builders.

In 1897 the McCormick Co. built its first gasoline engine. The following year its second engine was developed and installed in a chassis with two forward speeds and a reverse gear, and a friction transmission. The engine was of a 2-cylinder opposed type, with air-cooled cylinders, water-cooled heads, and rotary valves.

In 1898 several new companies entered the gasoline engine field. The most important was the Huber Mfg. Co. of Marion, Ohio, which had been manufacturing steam tractors. In 1898 this Company acquired the patent rights and interests of the Van Duzen Co. of Cincinnati. Huber built 30 single-cylinder tractors the first year, using a motor which very closely followed the lines of the Van Duzen motor, while the principles of transmission resembled those used on the Huber traction engine, which had already met with considerable success (page 18).

In 1899 S. S. Morton built his first tractor which was a combination machine to take the place of a portable engine (page 18). He continued his experimental work and obtained patents on a traction track mounted with a gasoline engine in 1902 and 1903. These were soon purchased by the Ohio Mfg. Co., Upper Sandusky, Ohio. After a demonstration for International Harvester Co., the latter bought several machines in 1905. For several years tractors for the International Harvester Co. were built there. The International Harvester Co. line, therefore, finds its tractor business beginning with the first efforts of Morton in 1899.

In 1897 Kinnard-Haines built a gasoline tractor using a steam-tractor running gear and a single-cylinder 12 x 18 engine which it had been manufacturing since 1894. One of these machines was produced in 1897, four in 1898, and 28 in 1899 when they were called "Flour City" (on page 18 is an illustration of the 1901 model).

In 1901 the Waterous Engine Works of St. Paul, Minn., started production on one tractor model.

Some Early Gasoline Tractors (1889-1901)

1889 Charter

10 to 20 H.P. A Charter single-cylinder engine was mounted on a steam-tractor truck. Grease cups or wick-filled oil pockets provided main-bearing lubrication, and sight-feed oilers were used for piston and connecting rod bearings. Gears were entirely exposed.

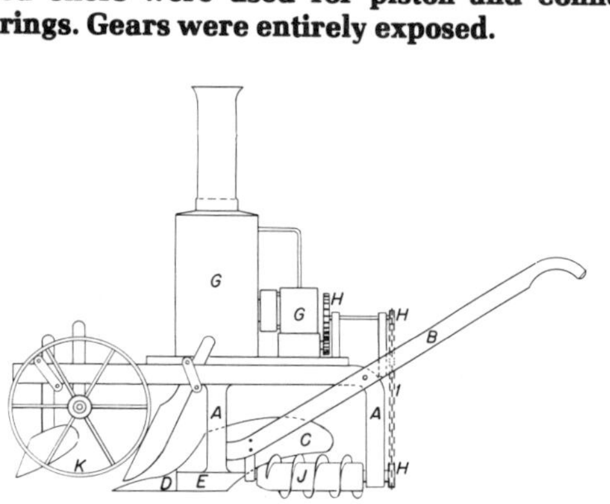

1890 Taylor worm drive

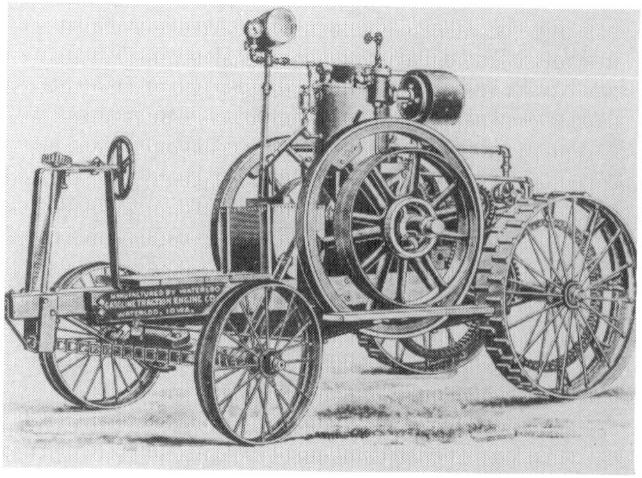

1892 Froelich

20 H.P. (14-inch bore x 14-inch stroke) single-cylinder vertical type engine. Electric ignition: battery and contact points; flyball governor, and exhaust held open and intake closed when above speed. It was argued that the vertical-type engine overcame the end motion caused by the pistons in horizontal engines. Steering by hand-wheel in front, which through shafting and gears, actuated a sprocket meshing with a tight hook-link chain attached to each end of the front axle just inside the front wheels.

1892 Paterson (Case)

16 to 20 H.P. engine. Large tank on front end of tractor is carburetor. Atomization of the gasoline was obtained by drawing air through the fuel to pick up gasoline vapor before entering engine. Engine was 2-cylinder horizontal balanced type, with single-throw crankshaft which connects directly to the piston in the open end of near cylinder by connecting rod and far piston (with cylinder opening in opposite direction) through pitman, rocker arm and short connecting rod. Ignition: make-and-break, bolt in piston head contacted stationary insulated electrode on compression stroke, and spark occurred just after dead center on outward stroke.

"STERLING" Gasoline Traction

1893 Hockett (Sterling)

Hockett exposed transmission gears to internal gears in drive wheels. It has been said that the familiar outline of the steam tractor was followed to avoid frightening farm horses. Had whistle, 2-shoe clutch chain-and-roller steering; all transmission gears exposed.

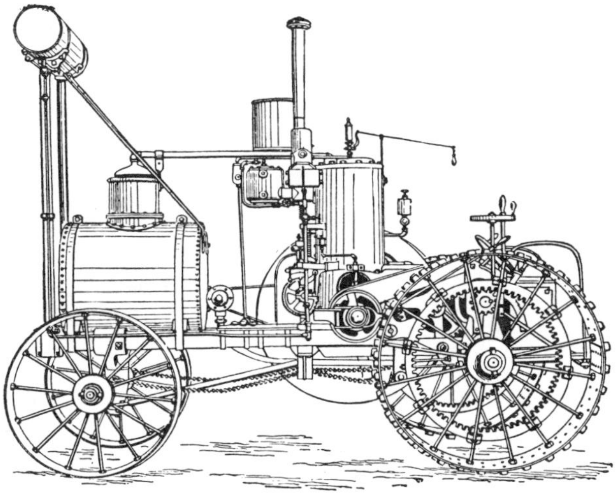

1894 Van Duzen

1-cylinder engine. Original ignition was to start by blank cartridge and then depend on platinum tube which retained heat, to ignite next charge. Later a parlor match, ignited by the release of a spring, replaced the blank cartridge. Internal gear in drive wheels and chain steering. This tractor somewhat resembled a steam outfit and even had a whistle.

1894 Otto

15 H.P. Bevel gears and drive shaft in first reduction.

1894 Lambert

1-cylinder horizontal engine; lubrication by sight feed oilers; open crankcase.

1896 Otto

15 H.P.

1898 Huber

1-cylinder vertical engine with "huge flywheels to carry it over the compression stroke". Spray nozzle arranged in intake to partially vaporize the fuel. Ignition by heated platinum wire.

1899 Morton

Single-cylinder (horizontal) hopper-cooled engine was mounted on a light, short wheel-base chassis. It was gear driven except for the friction drive from engine to countershaft.

1901 Kinnard-Haines (Flour City)

30 H. P., 200 to 250 r.p.m., 1 cylinder horizontal engine. Chain drive from flywheel to secondary shaft on which were mounted clutch and drive sprocket; final drive: bull gear and pinion. First Flour City model was built in 1897.

1902-1908

C. W. Hart and C. H. Parr in Charles City, Iowa, built their first tractor model in 1902 (page 23). They designed their tractor for draw-bar work rather than for belt work as evidenced by the ruggedness of its transmission which well withstood the heavy strains imposed when plowing. In 1903 they built their second model (page 24). These men had started experimental work on gasoline engines in 1895, and by 1905 had established the first business in the U. S. devoted exclusively to the manufacture of tractors.

Also in 1902 the Massey-Harris Co. of Racine, Wis., built a tractor powered with a 4-cylinder engine capable of pulling ten 14-inch plows. This was known as the Wallis "Bear."

In 1903 S. P. and J. Lauson of Milwaukee, Wis., built a tractor from patterns originally obtained from the Otto Co. of Philadelphia. In the same year, the Morton Traction Co. of York, Pa., built 12 tractors with friction transmission. The Foos Gas Engine Co. Springfield, Ohio, Fairbanks Morse, Co. of Chicago, Ill., Alamo Engine Co., Hillsdale, Michigan, Columbus Co. of Columbus, Ohio, Stover Engine Co. of Freeport, Ill., St. Mary's Machinery Co., St. Mary's, Ohio, and Springfield Boiler Mfg. Co. of Springfield, Ill., were other stationary engine manufacturers who mounted their engines on Morton trucks with a view to developing tractor trade.

In 1904 C. H. A. Dissinger and Brother Co., Wrightsville, Pa., started producing the Dissinger tractor (page 24) and the Electric Wheel Co., Quincy, Ill., began building experimental tractors (page 24).

In 1905, 12 new companies entered the field. The International Harvester Co. started building a three-wheel tractor, with a single wide wheel in front at its Keystone Works in Rock Falls, Ill., under the direction of E. A. Johnston. The Ohio Manfacturing Co. (page 25) started building the Morton combination machine (traction truck) for the International Harvester Co. and continued it for several years.

A name which was destined to play an important role in tractor development, according to Frank Kranick[7], is that of C. M. Eason. In Mr. Eason's first attempt he, with Ansel Wysong built their first tractor in 1905 at Topeka, Kans. (page 24). This had a single rear drive wheel and two front steering wheels. Tests proved that the single drive wheel had inadequate traction so the job was rebuilt with 2 rear drive wheels and two front wheels, 4 inches wide mounted side by side, making in effect a single front steering wheel 8 inches wide - a forerunner of the so-called tricycle tractor which appeared many years later. This unit plowed with a 2 bottom 14-inch plow, disced, planted and harvested 80 acres of wheat on Wysong's farm in Meade County, Kansas. Later a Brenen engine replaced the Cushman and their machine then christened "Autotractor" pulled 3 - 14-inch bottoms in buffalo sod. On a trial run on a cold November day the crankshaft broke. They then attempted to make a big tractor along the same general lines.

In 1912 Eason joined the Wallis Tractor Co. as assistant to Hendrickson, the chief engineer, and was one of the co-inventors of the frameless or "unit type"

---

[7] *Mr. Kranick, formerly with the J. I. Case Co., also furnished a number of pictures and information helpful in making the text authentic.*

Table showing Results of Light Agricultural Motor Competition

at Winnipeg Industrial Exhibition, 1908.

DESCRIPTION.						HAULING																	PLOWING						
		Size of Cylinder				Weight—Lbs			Supplies								1000 Ft. Lbs.		1000 Ft. Lbs Per Pint		Water Gallons						Performance		
									Gasoline Pints		Water Pints																		
Competitor's Number	Maker	Number	Diameter	Stroke	Horse Power	Motor	Load	Total	Start	Used	Start	Used	Length of Course	No. of Rounds	Total distance run—Feet	Time—Minutes	Load	Total with Motor	Load	Total	Taken	Used	No of Furrows	Duration of test—Min.	Acres plowed	Gasoline used—Pints	Pints per Acre	Minutes per Acre	Points gained of Possible 155
4	International Harvester Co.	1	9	15	20	11520	9150	20670	176	55	260	224	2999 Feet	11	32989	121	320988.7	701022	5836.1	12746	32.5	28	3	118	2.042	50	24.4	57.8	101.5
7	Transit Thresher Co.	4	6	8	35	13350	18040	31390	316	64	896	0		5	14995	124	279897.9	430081	4373.4	7501	112	0	6	118	2.957	64	21.6	39.9	103.3
2	I. H. C	1	8	14	15	9920	9 50	19070	172	31.5	260	128		10	29990	116	274408.5	571909	8711.3	18155	32.5	16	3	121	1.795	33	18.4	67.3	117
8	Kinnard-Haines Co.	4	6½	7	30	13530	18040	31570	193	75.5	1312	48		9	26991	120	486917.6	852106	6449.2	11286	164	6	6	121	3.246	65	20	37.3	117.6
5	Marshall Sons & Co., Ltd.	2	6¾	7	30	10680	9790	20470	162	58	560	144		11	32989	121	322962.3	675285	5568.3	11643	70	18	3	119	1.473	56	38	80.7	108.3
3	I H. C.	3	7	9	40	12910	9150	22060	240	53.5	352	256		11	32989	116	301849.3	425888	5642	7965	44	32	5	120	2.847	59	20.7	42.1	107.7
6	Universal	4	6	6	20	6900	9540	16440	164	....	400	...		Withdrawn owing to breakage					.....	.....	....	..	..	...	.....	..	....	....	....

RESULTS OF AGRICULTURAL MOTOR CONTEST AT THE WINNIPEG EXHIBITION, JULY 7—17, 1909

Class	Makers	Cylinders: Number	Cylinders: Diameter	Cylinders: Stroke	Rated Revs per Minute: Engine (from)	Engine (to)	Pulley (from)	Pulley (to)	Horse Power: Nominal	Horse Power: Specified Brake	Pulley: Diameter	Pulley: Face	Drivers: Diameter	Drivers: Face	Total Width over Drivers	Speeds	Miles per Hour	Capacity of Fuel Tank	Capacity of Water Tank	Weights: Total	Weights: On Drivers	Cash Price F.O.B. Winnipeg
			in.	in.	from	to	from	to			in.	in.	in.	in.	ft. in.							$
A Internal Combustion 20 h p. and under	International Harvester Co.	1	×	14	250	300	250	300	15	19 5	36	9½	56	22	7 6	1.75	2½	25	58	9,900	7,100	1,700
	Avery Co.	4	4¾	5	1,000	1,500	1,000	1,500	12	36	9½	9	41	18	7 3	1.5	15	20	75	5,000	2,300	2,500
	Marshall, Sons & Co.	2	7	7	700	800	350	400	12	30	24	7	78	18	9 2	2.25	3½	30	54	14,500	....	2,500
B Internal Combustion 20-30 h.p.	Russell & Co.	3	8	10	......	350	......	350	20	44	28	10½	52	36	9 0	.....	.....	42	71	17,150	12,690	2,400
	International Harvester Co.	1	8¾	15	240	290	240	290	20	28	40	10½	70	20	8 5	1.75	2.5	25	58	13,500	9,600	2,200
	Kinnard Haines Co.	4	6½	7	.....	650	.....	325	30	45	30	9	34	18	8 1	.....	2¼	26	190	13,875	....	2,250
	International Harvester Co.	1	8¾	15	240	290	240	290	20	28	38	10½	70	18	6 8	1.9	2.4	25	54	12,500	9,000	2,200
	International Harvester Co.	1	8¾	15	240	290	240	290	20	28	38	10½	54	22	8 6	1.75	2.5	25	58	12,600	9,400	2,200
C Internal Combustion over 30 h.p	International Harvester Co.	2	8¾	14	300	325	300	325	35	45	32	10½	75	18	......	2.25	3.5	51	97	20,000	14,000	2,700
	Gas Traction Co	4	6	8	500	800	500	800	25	35	18½	9	96	18	......	.....	.....	37	...	13,600	9,000	2,750
	Marshall Sons & Co.	4	7	7	700	800	350	400	25	60	23¾	10	73	24	10 3	.....	.....	61	79	22,000	16,570	3,400
	Kinnard Haines Co.	4	7½	8	......	500	......	315	40	60	30	10	96	24	9 3	.....	.....	40	290	18,500	....	2,700
	International Harvester Co.	1	10	15	240	290	240	290	25	30	40	10½	72	18	......	2	2.5	37	71	17,000	12,400	2,400
D Steam Engines	J. I. Case Co.	1	12	12	......	230	......	230	32	110	43½	16	84	36	12 10	.....	2.37	lbs 2,500	305	40,260	32,360	4,050
	M. Rumely Co.	2	7¾	14	......	235	.....	235	36	120	40	12	84	40	12 0	.....	2.33	1,400	400	37,000	....	3,974
	Marshall Sons & Co.	1	9½	12		230	......	230	25	60	37	10	75	30	......	.....	2.25	350	200	26,000	17,000	3,250
	Russell & Co.	2	7½ / 11	12	......	230	......	230	30	60	42	14	84	36	11 2	.....	2.5	875	450	38,432	19,320	3,400
	Avery Co	2	7	10	250	300	250	300	30	90	40	12	80	26	11 8	1.9 / 2.82	2.62 / 3.15	2,700	320	40,860	31,120	3,650

Class	Makers	Brake Tests: H. P. Devel'd	Revs. Engine Pulley	Water Used Gals.	$ Water Used	Fuel Used	$ Fuel Used	H. P. Hours per Unit of Fuel	Running Time	Mean Effective Pressure	Water Evap'd per lb. of Coal	Lbs. of Water per H. P. Hour
A	International Harvester Co.	19.7	249	4.18	7.2	1.47	5.9	13.4	60	89	....	....
	Avery Co.	17.9	865	7.5	100	4.3	21.5	8.3	120	46.3	....	....
	Marshall, Sons & Co.	....	....	....	....	....	....	....	...	....	....	....
B	Russell & Co.	26.2	311	7.5	10.5	6.4	15.2	5.5	81	44.2	....	....
	International Harvester Co.	22.0	254	4.22	9	3.7	14.8	11.89	120	76	....	....
	Kinnard Haines Co.	29.7	260	8.3	4.8	7.3	28.0	8.0	118½	52.6	....	....
	International Harvester Co.	24.2	215	9.2	17.0	3.9	15.6	12.4	120	90.4	....	....
	International Harvester Co.	21.9	260	14.5	25.0	3.6	14.4	13.8	120	84	....	....
C	International Harvester Co.	46.0	292	21	21.6	12.1	22.4	7.6	120	73.8	....	....
	Gas Traction Co	Score not computed because test was not completed										
	Marshall Sons & Co.	58.5	365	21.3	30	15.4	25.2	7.6	120	58.8	....	....
	Kinnard Haines Co.	45.6	295	6.4	2.2	10.9	27.2	8.4	120	54.4	....	....
	International Harvester Co.	31.6	253	12	18.3	5.6	15.1	11.3	120	84	....	....
D	J. I. Case Co.	97.5	229	580	190.0	700	28	30	126	62	8.3	23.3
	M. Rumely Co.	105.5	238	771	193.0	1034	74	20.4	120	66	7.4	36.5
	Marshall Sons & Co.											
	Russell & Co.	66.8	211	435	97.0	442	50	30.2	120	.....	9.8	32.5
	Avery Co	73.6	254	520	162	815	30	18.1	120	74.5	6.4	35.3

Class	Makers	Hauling Tests: Load Hauled	Average Draw Bar Pull	Distance Travelled	Fuel Used	Time Hauling Minutes	1000 Ft. lbs. per Unit of Fuel	Draw Bar H. P.	Draw Bar H. P. / Brake H. P.	Water Used	1000 ft. lbs. Hauled per gal. W'r
A	International Harvester Co.	13370	1350	26800	3.64	105½	9940	10.4	.528	8.7	....
	Avery Co.	7604	818	45560	5.43	117	6867	9.7	.542	2.7	....
	Marshall, Sons & Co.	13870	1650	21440	10.21	111½	3465	9.7	......	28	....
B	Russell & Co.	20800	2236	18760	6.18	79¼	6787	16.1	.815	12.5	....
	International Harvester Co.	29970	2650	18760	4.95	124	10050	12.25	.557	8.3	....
	Kinnard Haines Co.	19480	2395	24120	6.93	98	8336	17.8	.600	12	....
	International Harvester Co.	17000	1775	26800	6.63	127½	7383	11.3	.467	12	....
	International Harvester Co.	18200	1750	26800	5.86	118½	8007	12.0	.482	17.7	....
C	International Harvester Co.	WITHDRAWN									
	Gas Traction Co	21090	2750	24120	11.0	98	6030	20.9	......	4.6	....
	Marshall Sons & Co.	20000	4550	21410	12.14	108.5	8035	27.2	.465	38	....
	Kinnard Haines Co.	27745	3050	24120	8.21	113.5	8960	19.7	.432	10	....
	International Harvester Co.	13500	3250	21440	7.14	113	9759	18.7	.581	16	....
D	J. I. Case Co.	37000	3960	21620	613	105	14000	24.7	.253	496	172
	M. Rumely Co.	40260	4860	24120	666	127	17580	28	.265	536	218
	Marshall Sons & Co.	WITHDRAWN									
	Russell & Co.	34000	4200	21400	664	119	13500	22.3	.341	376	239
	Avery Co	34000	4200	8760	500	95	15750	25.1	.341	396	199

Class	Makers	Plowing Tests: Plows Number	Plows Width	Acres Plowed	Time Plowing (Mins.)	Average Draw Bar Pull	Water Used	Fuel Used	Fuel per Acre	Engine No	Total number of points scored out of possible 148	Prizes Awarded
			in.									
A	International Harvester Co.	3	12	1.09	75½	1510	8.12	1.64	1.5	5	115.4	1st
	Avery Co.	2	14	1.06	85	1380	.35	3.57	3.38	15	106	2nd
	Marshall, Sons & Co.	3	14	1.59	103	2950	.....	7.86	4.9	12	100	3rd
B	Russell & Co.	4	14	1.7	121½	2200	......	7.93	4.66	6	106.8	2nd
	International Harvester Co.	4	14	2.17	117	2850	10.6	5.21	2.4	7	112.1	1st
	Kinnard Haines Co.	6	14	2.55	101	3625	6.5	6.14	2.41	13	105.8	......
	International Harvester Co.	4	14	1.7	84.5	2700	......	4.57	2.69	17	104.1	......
	International Harvester Co.	4	14	1.7	91.5	3100	10	4.29	2.52	21	106.5	3rd
C	International Harvester Co.									2	......	...
	Gas Traction Co	6	14	1.27	44½	5250	......	4.79	3.77	3	......	.....
	Marshall Sons & Co.	8	14	2.55	71	5500	......	9.61	3.77	8	102	2nd
	Kinnard Haines Co.	8	14	3.4	99	4180	......	8.43	2.48	16	109	1st
	International Harvester Co.	WITHDRAWN								19	100	3rd
D	J. I. Case Co.	12½ / 11½	14	3.6	62	9300	348	442	123	1	121.3	1st
	M. Rumely Co.	14	14	4.29	75	9860	......	580	135	9	106.2	....
	Marshall Sons & Co.									11	......	......
	Russell & Co.	10	14	3.18	76	7400	470	454	143	14	118.5	2nd
	Avery Co	8⅞ / 10⅞	14	4.0	72	7900	......	519	127	20	115.7	3rd

Note:—Fuel Consumption is stated in Imperial Gallons (277.27 Cu. Ins.) or 100 Lbs. of Coal. Length of Ground Plowed, 1980 Ft. Length of Hauling Course, 2680 Ft.

Judges' Tabulated Record of The 1910 Winnipeg Motor Contest, verified by Profs. A. R. Greig and J. L. Smith, Engineers in charge

ENGINE DATA

Class	Entry Number	Maker's Name	Cylinders Number	Cylinders Diameter	Cylinders Stroke	Rated R. P. M. Engine	Rated R. P. M. Pulley	H. P. Nominal	H. P. Specified Brake	Pulley Diameter	Pulley Face	Drivers Diameter	Drivers Face	Speeds in M. per Hr. High	Speeds in M. per Hr. Low	Capacity of Fuel Tank Imp. Gals.	Capacity of Water Tanks Imp. Gals.	Weight, lbs. Total
A. Internal Combustion 20 B. H.P. and under.	1	International Har. Co.	1	8	14	250–300	250–300	15	18	36	9½	56	22	2.10	1.75	23.7	30	10,500
	5	Avery Co.	4	4¼	5	1000	1000	12	36	8¼	10	40	6	10	to 2	13	7.5	6,000
B. Internal Combustion 21 to 30 B. H.P.	2	Avery Co.	1	12	18	350	350	25	25	40	11	63	20	3	¼	41	47.5	12,000
	3	Goold, Shapley & Muir Co.	2	7¼	10	350–400	350–400	20	28	18	9	70	22	2.5	2	41	37	11,000
	4	International Har. Co.	1	8¾	15	240–300	240–300	20	23	40	10½	70	20	2.10	1.75	30	44	14,200
	6	Birrell Motor Plow Co.	4	8½	10	300–350	240	22	45	34	9	78	24	2.5		39.5	130	22,000
C Internal Combustion Over 30 B. H.P.	7	Gas Traction Co. Wpg.	4	6	8	550–650	550–650	25	45	18¼	9	96	[illegible]	2	1½	45.6	90	......
	8	Goold, Shapley & Muir Co.	2	9½	13	325–375	325–375	30	45	26	9	84	26	2½	2	49	102	17 730
	9	International Har. Co.	2	9	14	320–340	320–340	45	55	28	10½	75	24	2	2.1	39	112.5	20,990
	10	Kinnard Haines.	4	7½	8	500	300	40	60	34	10	96	24	2.25		39	277	19,000
	18	Gas Traction Co. Minn.	4	6½	8	550	450	20	60	24	9	96	24	3	2	51	81	17,500
	11	Rumely Co.	2	10	12	375	375	25	50	36	11	64	26	1.9		80 ker. .67 gas.	112	26,700
D—Steam. 60 B.H.P. & under	12	J. I. Case, 36 B. H.P.	1	8½	10	250	250	...	36	40	10.5	66	20	2.4		18 cu.ft.	143	17,475
E Steam 60 to 90 Brake H. P.	13	Avery Co. 60 B. H.P.*	2	6	10	250	250	20	60	40	11	65	20	1.5	4	26 cu.ft.	303	26,000
	14	J. I. Case, 75 B. H.P.*	1	11	11	250	250	...	75	40	12	72	24	2.6		25.4 cu.ft.	217	25,800
F Steam 91 and above Brake H.P.	15	Avery Co., 90 B. H.P.	2	7	10	250	250	30	90	40	12	80	26	1.5	4.5	13 cu.ft.	371	36,000
	16	J. I. Case, 110 B. H.P.	1	12	12	230	230	..	110	43.5	16	84	36	2.37		39.4 cu.ft.	305	40,460
	17	Rumely Co., 120 B. H.P.	2	7¾	14	250	250	36	120	41	12	84	30	2.03		28.12 cu.ft.	500	46,480

ENGINE DATA — TWO HOUR BRAKE TEST

Entry Number	Maker's Name	Price F. O. B. Winnipeg	Piston Rod Diam.	Piston Speed feet per min.	Area of Piston Sq. Ft.	Piston displacement Cub. Ft. per min.	Circum'ce Engine Pulley	Authorized Steam Pressure	H.P. Developed	R. P. M. Eng. Pulley	Gals., Water Used	Per cent. Water used	Fuel Used, Lbs.	Per cent Fuel Used	H.P. Hrs. per Unit Fuel. ‡	Total running time	Lbs. water evaporated per lb. Fuel	H.P. Hrs. per 100 gal. water.	Mean Effective Press'e in Cylinder	H.P. 2 Hr. Test / Max. Test	Ave. Steam Pressure
1	International Har. Co.	$1,720.00	....	583.3–700	.349	203.6–244.3	114 "	...	15.28	249	7.8	26	21.5	90.7	9.94	2 hrs	3.03	391	69.6	.99	...
5	Avery Co.	2,500.00	....	833.3	.123	102.5	27¾"	...	14.16	1127	4.3	57.3	35.5	26.7	5.58	2 hrs	1.21	659	28.08	74	
2	Avery Co.	2,000.00	....	583.3	.442	257.8	127 11–16"		20.69	254	4.5	13.5	35.5	17.6	5.71	1.24m	27	644	31.7	....	...
3	Goold, Shapley & Muir Co.	2,300.00	....	583.3–666.7	.307	179.1–204.7	69¾"	...	28.13	350	1.1	3.05	40.5	4.5	9.48	1.57m	.27	401	72	.82	...
4	International Har. Co.	2,300.00	....	600–750	.418	250.8–313.5	125½"	.	22.11	235	10.[illegible]	22.7	28.[illegible]	[illegible]2	10.86	2 hrs.	.51	442	82.6	.98	
6	Birrell Motor Plow Co.	3,200.00	....	500–583	.394	197 –229.7	106¾"		27.21	203	3.1	2.4	65	23.5	5.86	2 hrs	.48	1755	32.5	.93	.
7	Gas Traction Co. Wpg.	3,000 cash	....	733.3–866.7	.196	143.7–169.9	59¼"	...	34.66	597	0	0	70	22.0	6.93	2 hrs.	0		50.8	.99	...
8	Goold, Shapley & Muir Co.	2 775.00	....	704.2–812.5	.492	346.5–390.7	92¾"		34.86	304	4.4	4.3	77	22.5	6.34	2 hrs.	.57	1584	49.3	.90	...
9	International Har. Co.	2,700.00	....	700	.492	344.4	89¾"	...	46.49	344	15.0	13.3	55¼	20.3	11.78	2 hrs.	.71	620	60.06	.97	...
10	Kinnard Haines.	3,400.00	....	666.7	.307	204.7	104⅜"	...	49.76	297	0	0	64	23.4	10.88	2 hrs.	0	......	56.32	99	...
18	Gas Traction Co. Minn.	3,500 cash	....	733.3	.230	168.7	74¾"	...	51.84	493	0	0	67	18.8	10.83	2 hrs.	0		64.2	.96	...
11	Rumely Co.	3,400.00	....	750	.545	408.7	113½"		46.8	378	8.6	7.67	106k 1½ g	132.5	6.88	2 hrs.	.80		52.0	.95	...
12	J. I. Case, 36 B. H.P.	1,812.50	1 5-16"	416.7	.367	152.9	124¾"	140	44.25	264	252.3	176	350	34.1	25.28	2 hrs.	7.2	35.1	62.2	.74	134
13	Avery Co. 60 B. H.P.*	3,200.00	1¼"	416.7	.193	....	106½"	175	66.79	256	398.7	162	556	39.6	24.02	2 hrs.	.17	33.5	93.03	.69	109
14	J. I. Case, 75 B. H.P.*	2,593.75	1 9-16"	458.3	.652	....	125 "	140	61.69	265	364.5	168	443	32.3	27.85	2 hrs.	.22	33.8	44.6	.65	136
15	Avery Co., 90 B. H.P.	4,000.00	1¼"	416.7	.264	....	125½"	200	97.08	265	611.2	165	701¾	100	27.66	2 hrs.	.71	31.7	95.54	.78	199
16	J. I. Case, 110 B. H.P.	3,744.00	1 15-16"	460	.775	....	137 "	162	98.61	233	604	198	606	28.4	32.54	2 hrs.	.97	32.6	62.6	.76	156
17	Rumely Co., 120 B. H.P.	4,300.00	1¼"	583.3	.261	....	129 15-16"	185	105.86	247	623	125	861	56.6	24.59	2 hrs.	.24	34.0	65.1	.78	175

HALF HOUR MAXIMUM BRAKE TEST

Class	Entry Number	Maker's Name	H.P. Developed	R. P. M. Eng. Pulley	Gal. Water Used	Fuel Used, Lbs.	H.P. Hrs. per Unit Fuel	Total running time	M. E. Pressure	Ave. Steam Pressure	Plows Number	Plows Width	Plows Make	Miles travelled	Acres plowed
A. Internal Combustion 20 B. H.P. and under.	1	International Har. Co.	15.5	247	2.75	6½	8.67	30 min	70.7	......	2	14"	Oliver	12.[illegible]	[illegible]61
	5	Avery Co.	19.2	1173	1.15	11.75	5.7	30 min	36.6	......	3	14"	P. & O.	7.1[illegible]	[illegible]20
B. Internal Combustion 21 to 30 B. H.P.	2	Avery Co.		No test on maximum or plowing											
	3	Goold, Shapley & Muir Co.*	34.15	363	1	7	7.9	14 min	84.2	......	6	12"	Verity	8.[illegible]	[illegible]65
	4	International Har. Co.	22.6	240	2.75	9.5	8.32	30 min	82.7	......	4	14"	Oliver	8.[illegible]	[illegible]00
	6	Birrell Motor Plow Co.	29.1	199	2.7	17.5	5.8	30 min	50.9	......	....	14"	Moline	3.[illegible]	[illegible]73
C Internal Combustion Over 30 B. H.P.	7	Gas Traction Co. Wpg.	35.1	616	0	23	5.3	30 min	50.0	......	6	14"	Cockshutt	12.[illegible]	[illegible]
	8	Goold, Shapley & Muir Co.	38.8	335	1.2	24.5	5.5	30 min	49.8	......	6	14"	Cockshutt	10.7[illegible]	[illegible]82
	9	International Har. Co	47.8	345	4.0	19.25	8.7	30 min	61.6	......	10	14	P & O.	12.[illegible]	[illegible]59
	10	Kinnard Haines.	50.2	302	12.3	12.0	14.6	30 min	55.9	......	6	....	Jno. Deere		
	18	Gas Traction Co Minn.	53.9	489	0	20.5	9.2	30 min	67.28	......	7	14"	Jno Deere	12.[illegible]	[illegible]43
	11	Rumely Co.	49.2	372	3.4	34k 1½ g	5.49	30 min	55.6	......	8	14"	Rumely	10.[illegible]	[illegible]01
D—Steam. 60 B.H.P. and under	12	J. I. Case, 36 H. H.P.	60.0	268	99.7	171	17.5	30 min	83.9	131	4	14"	Cockshutt	16.[illegible]	[illegible]98
E Steam 60 to 90 Brake H. P.	13	Avery Co. 60 B. H.P.*	96.6	276	149.5	238	20.3	30 min	124.8	167	8	14"	Cockshutt	5.[illegible]	[illegible]06
	14	J. I. Case 75 B. H.P.*	94.3	255	143.3	192	24.6	30 min	70.8	135	8	14"	Cockshutt	17.[illegible]	[illegible]17
F Steam 91 and above Brake H.P.	15	Avery Co. 90 B. H.P.	124.4	240	275.4	517	12.0	30 min	135.2	189	12	14"	Cockshutt	7.[illegible]	[illegible]16
	16	J. I. Case 110 B. H.P.	129.0	225	227.4	278	23.4	30 min	84.8	157	12	14"	Cockshutt	21.[illegible]	[illegible]08
	17	Rumely Co. 120 B. H.P.	132.07 135.8	243	320.8	511	17.6	10 min 30 min	84.8	176	14	14"	Jno. Deere	12.[illegible]	[illegible]07

PLOWING TEST

Entry Number	Time, minutes	Acres per Hr.	Average Draw bar Pull, lbs.	Fuel used, lbs.	Fuel used lbs. per acre	Water used, gal.	Water used per acre	D. B. Pull per 1 inch width of furrow	D. B. H. hrs. per acre plowed	Fuel Used, gals. per acre	Drawbar H.P.	D. B. H.P. / Brake H.P. %	D.B. H.P. / Max. B. H.P. in %	D. B. H.P. Hrs. per unit Fuel	Per cent water used per acre	Acres plowed per Max. B. HP. Hr.	Per cent Fuel per mile	Total points allowed in Judging Engines.	Medals Awarded
1	299	0.72	1705	81.5	22.58	27.9	7.72	60.8	15.72	3.23	11.39	75	73.5	4.87	25.7	.046	3.93	306.7	Gold Medal
5	255	0.75	1980	75.5	23.59	3.6	1.12	47.1	11.78	3.37	8.87	62.5	46.1	3.49	15	.039	7.95	275.3	Silver Medal
2																			
3	309	1.08	3250	152.5	22.92	5.4	.81	45.1	11.63	3.28	12.58	44.7	36.9	3.55	.22	.031	5.95	292.2	
4	222	1.35	2450	76.5	15.3	30.0	6.00	43.8	11.66	2.19	15.75	71.1	69.5	5.33	14	.058	4.08	329.3	Gold Medal
6	159	1.03	.....	93	34.00	10.0	3.66	....	.....	4.86	....	....	.....	....	....	.035	....	.....	
7	275	1.70	4550	179.5	16.93		0	54.16	14.3	2.42	24.24	69.8	69.0	5.91	0	.048	4.5	334.0	Bronze Medal
8	390	1.36	4200	240.5	27.26	5.3	.60	50.0	13.58	3.89	18.44	52.9	47.6	3.49	.06	.035	.655	274.6	
9	416	2.54	7350	260	14.78	119	6.76	52.5	13.9	2.11	35.30	75.8	73.8	6.59	6	.053	7.63	350.1	Silver Medal
10			Test not completed																
18	408	1.83	5400	199.5	16.05	0	.....	55.1	14.46	2.20	26.41	51.1	49.2	6.32	0	.034	4.47	354.4	Gold Medal
11	356	2.02	5500	329k 3g	27.64	27.7	2.3	49.1	13.06	3.49	26.45	56.5	53.8	3.71	2.	.041	4.88	not scored	
12	412	1.31	3580	1420	158.1	864	96.21	63.9	17.06	....	22.31	50.5	37.2	10.78	67.	.022	9.09	267.6	Gold Medal
13	170	2.14	6900	892	147.2	828.6	136.7	61.6	16.24	....	34.74	52.01	36	11.04	45.	.022	11.87	269.3	—
14	413	2.93	6850	2432	120.6	1869	92.6	61.2	16.15	...	47.34	76.6	50.1	13.4	42.7	.031	9.93	297	
15	201	3.63	11500	1820	149.6	1308	107.6	68.4	18.01	....	65.36	67.3	52.5	12.03	29.	.029	36.28	291.9	Silver Medal
16	497½	3.99	10860	3282	99.20	2713	82.01	64.6	18.78	....	74.92	75.9	58.0	18.92	27.	.031	7.19	356.1	Gold Medal
17	381	3.79	10700	2910	120.8	2244	93.22	54.6	14.79	...	56.08	53.0	41.3	12.23	14.6	.027	15.33	290.8	Bronze Medal

* These engines developed more than their allotted horse power, which threw them out of their class, consequently they were disqualified.

‡ Unit of fuel, 7 lbs. gasolene, and 100 lbs. coal; 7.9 lbs. kerosene.

Wt. of imp. gal. gasolene, 7 lbs. Wt. of imp. gal. kerosene used, 7.9 lbs.

Records of Maximum Brake Test and Plowing Test, at the Agricultural Motor Competition, held at Winnipeg, July 5th to 22nd, 1911, by the Canadian Industrial Exhibition Association

ENTRY NUMBER	MAKER'S NAME	HALF HOUR MAXIMUM BRAKE TEST: Total time running	Time lost due to engine	Average brake horse power developed	Revol. per minute of pulley	Revol per minute of engine	Fuel used in pounds	Horse power hours per unit of fuel	Water used	Steam pressure	PLOWING TEST: Number of Plows	Width of Plow	Maker of Plow	Miles Travelled (not including turns)	Length of Furrow in feet	No. Times across Field	Average Width o Plowed Land	Acres Plowed	Average Depth
	CLASS A—Gasoline																		
1	International Harvester Co.	30	0	23.65	228.9	228.9	8.5	1.39	gals. 2.37		4	14	Oliver	4.4	3864.4	6	27.83	2.47	
	CLASS B—Gasoline																		
2	Kinnard Haines	29	0	29.4	375.5	851.0	10.5	1.4	2.59		4	14	Deere	7.3	"	10	44.9	3.98	
3	Avery Co.	30	0	23.85	1006.2	1005.2	10.45	1.14	0		2	14	P. & O.	7.3	"	10	35.9	3.18	
4	I. H. C.	30	0	31.74	264.0	264.0	10.4	1.53	2.18		5	14	P. & O.	4.4	"	6	34.5	3.06	
5	M. Rumely Co.	30	0	33.58	378.2	373.2	12.45	1.35	1.56		6	14	Rumely	4.4	"	6	40.2	3.57	
6	Goold Shapley & Muir	30	0	26.2	353.8	353.8	12.7	1.03	0		4	14	Cockshutt	4.4	"	6	26.6	2.36	
7	American Abell Co.	30	0	32.71	487.8	487.7	20.4	0.08	0		4	14	Deere	8.8	"	12	55.1	4.89	
8	Canadian American																		
	CLASS C—Gasoline																		
9	Kinnard Haines	30	0	46.34	306.7	612.3	14.3	1.62	3.4		6	14	Deere	7.3	"	10	66.1	5.86	
10	Sawyer-Massey Co.	25	0	51.85	288.0	576.0	14.6	1.78	5.2		6	14	Verity	7.3	"	10	68.2	6.05	
11	I. H. C.	30	0	58.4	327.4	327.4	20.5	1.42	6.7		10	14	P. & O.	4.4	"	6	67.6	6.00	
12	Gas Traction Co.	30	0	58.1	479.7	639.6	22.3	1.30	0		8	14	Deere	5.9	"	8	72.5	6.43	
13	I. H. C.	30	0	65.62	356.9	356.9	25.0	1.31	6.0		10	14	P. & O.	4.4	"	6	68.0	6.03	
14	Goold Shapley & Muir	30	0	39.53	280.6	280.6	16.2	1.22	0		5-6	14	Cockshutt	5.9	"	8	52.0	4.61	
15	Aultman-Taylor Machinery Co.	30	0	58.51	469.7	469.7	18.5	1.58	5.0		8	14	Deere	7.3	"	10	97.6	8.66	
16	Gaar Scott Co.	25	0	77.79	292.6	585.2	34.2	1.14	0		8-10	14	D. & M.	8.6	3782	12	109.4	9.49	
17	Buffalo Pitts Co.	30	0	54.18	320.8	481.2	22.6	1.2	1.72										3½ inches
	CLASS D—Kerosene																		
18	I. H. C.	30	0	24.11	261.5	261.5	0.1g 14.4k	0.83	3.22		3	14	Oliver	4.4	3864.4	6	21.6	1.91	
19	I. H. C.	30	0	29.87	364.9	364.9	0.1g 20.6k	0.72	2.52		4.6	14	"	4.4	"	6	35.1	3.11	
20	I. H. C.	30	0	30.53	250.8	250.8	0.2g 14.2k	1.06	3.82		4	14	"	4.4	"	6	37.1	2.40	
21	M. Rumely Co.	30	0	32.19	365.7	365.7	0.8g 6.2k	0.947	1.4		4.5	14	"	5.9	"	8	40.6	3.61	
22	I. H. C.	30	0	64.53	343.7	343.7	0.3g 25.5k	1.24	4.93		10	14	"	4.4	"	6	68.0	6.03	
23	Gas Traction	30	0	49.58	507.2	676.3	1.2g 21.0k	1.12	0.82		6	14	Deere	7.3	"	10	67.6	6.00	
24	Kinnard Haines	30	0	57.16	312.4	515.2	1.5g 21.85k	1.224	2.12		8	14	"	7.3	"	10	93.0	8.16	
25	M. Rumely Co.	30	0	67.99	358.2	358.2	0.7g 33.05k	1.0072	21.5		8-10	14	"	5.7	3751	8	87.0	7.49	
	CLASS F—Steam																		
26	Buffalo Pitts Co	30	0	97.72	277.9	277.9	358.5	13.63	239.0	147.3									
27	American Abell Co.	30	0	183.31	238.6	238.6	308.5	16.74	251.6	156.14	6-8	14	"	4.3	3814	6	32.7	4.61	
	CLASS G—Steam																		
28	Avery Co.	30	0	159.34	281.0	281.0	543.5	14.66	346.2	179.0	8-10½	14	Cockshutt	7.2	3818.4	10	114.7	10.0	
29	Gaar Scott Co.	30	0	150.72	285.9	285.9	335.0	22.5	246.2	151.2	4-10	14	Deere	4.3	3784	6	37.3	4.95	
30	Sawyer Massey Co.	30	0	144.59	245.8	245.8	420.0	17.21	308.8	163.4									
31	Sawyer-Massey Co.	30	0	118.59	267.0	267.0	279.0	21.25	185.6	167.14	6-10	14	M. and D.	4.0	1338	16	168	5.17	

ENTRY NUMBER	PLOWING TEST: Total Time Required Min.	Time Lost Due to Plows. Min.	Time Lost Due to Engine. Min.	Time Required Turning. Min.	Actual Time Plowing Min.	Acres Plowed per Hour	Miles Travelled per Acre Plowed	Average Draw-Bar Pull	Average Draw-Bar Horse Power	Fuel used, including turns	Draw-Bar Horse Power per unit of Fuel	Fuel used per Acre Plowed	Water Used	Water Used per acre Plowed	Average Draw-Bar Pull per 14-in. Plow	Draw-Bar Horse Power per 1 Acre per Hr.	Possible Miles Travelled Without Replenishing Fuel	Possible Acres Plowed Without Replenishing Fuel	Cost of Fuel per Acre Plowed $	Water Per D.B.H.P. Hr., lbs.	Awards
1	181	1'	1.5'	4	144	1.03	1.78	2660	12.93	40.1	7.74	lbs. 16.23	lbs. 99.9	40.5	662.5	12.55	22.5	12.6	.4638	3.22	1st
2	208	1.5	0	7	180.5	1.25	1.83	2985	18.75	63.3	9.40	15.92	179.6	45.1	762.5	15.0	16.4	8.95	.4548	3.02	1st
3	242.5	26	2	10.5	195	.98	2.3	1800	10.81	69.8	5.02	21.95	13.1	4.12	600	11.04	15.7	6.86	.6271	3.73	
4	2-16'	1	.5	5	125.5	1.46	1.44	3150	17.64	33.7	10.95	11.01	178.8	58.4	787.5	12.08	28.5	19.8	.315	4.85	2nd
5	159	0	0	6	145	1.47	1.23	3600	18.42	68.4	6.51	19.16	78.1	21.8	633.3	12.53	11.2	9.1	.5474	1.76	3rd
6	167.5	0	.5	6	157	.90	1.87	3000	13.43	53	6.63	22.45	91.8	38.9	750	14.92	17.6	9.4	.6414	2.61	
7	276	0	.5	11	262	1.11	1.80	3000	16.09	112.3	6.26	22.96	0	0	750	14.5	24.2	3.4	.6560	0	
8																					
9	215	0	4	7	204	1.72	1.20	4600	25.84	96.5	9.10	16.47	275	46.76	750	15.01	13.1	10.5	.4705	[illegible]	3rd
10	278.5	0	11	11	253	1.43	1.17	3400	25.00	119.8	8.80	19.80	647.3	107	900	17.48	19.6	16.5	.5657	6.14	
11	150	0	0	7	139	2.59	.73	6400	32.36	74.8	10.21	12.47	306	51	640	12.49	16.1	22	.3563	4.08	
12	174	0	0	7	167	2.30	.92	5800	32.54	94.4	9.59	14.68	0	0	675	14.15	26.3	28.6	.4194	0	1st
13	149	1	1.5	6	140.5	2.58	.73	6650	33.27	81.9	9.51	13.58	345.7	57.3	665	12.90	17.2	23.6	.388	1.44	
14	222.5	0	1.5	7	212.5	1.30	1.28	3400	15.00	116.7	4.55	25.31	99.4	21.57	680	11.54	19.6	15.3	.7231	1.87	
15	210	1	2	8	195	2.66	.825	5450	32.73	117.2	9.07	13.53	513.9	50.34	681	12.30	21.8	25.8	.3866	4.83	2nd
16	328	13	2.3	11	257	2.21	.91	5450	29.17	328.7	3.8	34.63	244.8	25.79	681	13.2	16.8	18.6	.9894	1.96	
17																					
18	150.5	7	0	4	139	.82	2.30	2275	11.50	1.1G 45.7K	5.69	24.5	140	73.3	758	14.02	22.4	9.7	.3724	5.25	
19	217	0	5	5	172	1.08	1.41	4000	16.35	.6G 91.4K	5.04	29.90	175	56.27	784	15.14	12	8.5	.4545	3.74	
20	146	16	5	5	120	1.2	1.83	2700	15.91	1.1G 72.8K	4.28	30.79	110	45.8	675	13.17	14.9	8.15	.4680	3.48	
21	202	0	1	7	177.5	1.22	1.63	3400	17.95	.6G 95.K	5.95	26.48	137.8	38.17	900	14.76	12.4	7.6	.4025	2.60	
22	142.5	1	0	5	132	2.74	.73	6300	33.54	5G 115.8K	6.34	19.28	279	46.27	630	12.24	12.2	16.7	.2931	3.78	
23	201.5	0	0	9	192.5	1.97	1.46	4975	30.27	3.5G 128.3K	7.37	21.99	81.3	13.55	829	16.19	29	23.8	.3339	.837	2nd
24	234.5	0	6	8	218	2.24	.86	5900	31.70	12.5G 128.2K	8.19	17.24	220	26.96	737.5	14.15	17.4	19.5	.2824	1.91	1st
25	225	0	0	7	192	2.34	.76	6500	30.78	1G 177.3K	5.52	23.80	187.5	25.03	684	13.15	17	22.3	.3618	1.90	3rd
26																					
27	201	7	10	6	132	2.09	.935	5906	31.05	1088.5	6.273	236.	5942	1289	787	14.85	4.5	4.8	1.003	86.98	1st
28	269	8	14	16	208	2.88	.72	6600	36.70	1892	6.723	189.2	14330	1433	673	12.74	2	2.8	.804	112.7	1st
29	274.5	5.5	9	10.5	91	3.26	.87	4750	35.73	1029	5.267	208.0	7360	1487	655	10.96	6.3	7.26	.884	135.8	3rd
30																					
31	284.5	2	0	24	128	2.42	.775	7250	38.75	1088	7.338	206.6	6798	1315.0	753	15.18	3	3.86	.878	86.72	2nd

Score Card

Following are the points upon which the awards were made:

	Gas	Steam
Brake Test (150).		
Horse power hours per unit of fuel used	100	100
Water used per hour, gals. per cent. of capacity	15	10
Efficiency, as taken from mean effective pressure	15	10
Steadiness of running, vibration, condition of engine	20	10
Horse power hours per 100 gals. of water		20
	150	150
Maximum Test (50).		
Economical load compared with maximum	20	20
Condition of engine	30	30
	50	50
Plowing Test (200).		
Fuel used per draw bar horse power hour	100	90
Water used per drawbar horse power hour	20	30
Acres plowed per hour per brake horse power	20	20
Quality of plowing	20	20
Distance travelled without replenishing fuel	15	15
Condition of engine, stops, etc	25	25
	200	200
Design and Construction (100).		
Protection of parts	20	20
Accessibility	20	20
Variation of speed	20	20
Easy manipulation	20	20
Design, materials	20	20
	100	100

AGRICULTURAL MOTOR COMPETITION—CANADIAN INDUSTRIAL EXHIBITION, WINNIPEG, JULY, 1912

SCORE SHEET

CLASS 76

SECTION	Entry Number	MAKER'S NAME	ECONOMY BRAKE TEST: H.P. Hours per Unit of Fuel	H.P. Hours per Unit of Water	Steadiness, Vibration, Condition	Total	MAXIMUM BRAKE TEST: H.P. Maximum / H.P. Economy	Evenness of Load	Condition of Engine	Total
		Highest Possible No. of Points — Internal Comb.	145	15	30	190	35	20	10	65
		Highest Possible No. of Points — Steam	140	35	15	190	35	20	10	65
B—Gaso 'ne	2	Canadia Heer	125.5	15	20 4	160.9	17	14	9	40
	2	Avery	124	7.2	23.6	154.8	25	18	10	53
	4	J. I. Case	128	14.2	28.9	166.1	14	18	10	42
	5	Goold, Shapley, Muir	105.7	15	18.3	138.9	15	17	10	42
	6	International Harvester Co.	132.8	6.8	18.8	158.4	23	15	9	47
C—Gasoline	7	Sawyer Masse	7.8	14	26 4	138.2	23	18	10	51
	8	International Harvester Co	34 1		.5		8 5	19	10	47.5
	9	International Harvester Co.	122	6.6	21 5	151	23 5	7	10	50 5
	10	Goold Shapley Muir	Withdrawn							
	11	Canadian Holt	108.8	15	25	148.3	20.5	17	9	46 5
	12	Aultman Taylor	145	5.9	24.4	175.3	23	18	10	51
	13	J. I. Case	104 3	6.4	22.6	133.3	22.5	19	10	51.5
	15	Diamond Iron Works	88.7	14.5	25	128.2	18	17	10	45
D—Kerosene	1	Canadian Heer	114.2	14.2	20	148.4	17.5	14	9	40.5
	16	Rumely	92.1	7.3	20.3	119.7	29.5	19	10	58.5
	18	International Harvester Co.	124.6	6.1	21.3	152	18.5	19	10	47.5
	19	Avery	135.5	11.2	21	167.7	14	17	8	39
E—Kerosene	21	Interna ional Harvester Co.	131.4	6.5	23	160.9	21	19	10	50
	22	Aultman-Taylor	138	6.4	27.7	172.1	18.5	18	10	46.5
	23	Rumely	45	9.9	20.9	175.8	31	16	10	57
	24	J. I. Case	86.9	5.9	23.9	116.7	25.5	17	9	51.5
F—Steam	25	J. I. Case	100.3	29.6	13.5	143.9	33	15	10	58
G—Steam	26	J. I. Case	111 3	33.5	14.1	158.9	21	18	10	49
H—Steam	27	J. I. Case	140	15	7 45	167.45	27	18	10	55
	29	Sawyer-Massey	99 8	34.8	8.95	143.55	23	14	10	47

Entry Number	MAKER'S NAME	PLOWING TEST: Drawbar H.P. Hours per Unit of Fuel	Drawbar H.P. Hours per Unit of Water	Acres Plowed per Hour per Econ. Brake H.P.	Quality of Plowing	Distance Travelled Without Replenishing Fuel	Distance Travelled Without Replenishing Water	Condition of Engine	Total	DESIGN & CONSTRUCTION: Accessibility	Protection of Working Parts	Ease of Manipulation	Total	Points Penalized	Total Points	Rank	Entry No.
	Highest Possible — Internal Comb.	140	15	25	15	10	—	10	215	15	10	5	30				
	Highest Possible — Steam	120	35	25	15	5	5	10	215	15	10	5	30				
2	Canadia Heer	82.4	14.5	23.4	9	4.4		10	143.7	10.5	6	4.25	[illegible]		365	5	
2	Avery	92.8	11.5	24.7	10	4.5		10	158	8.25	6.25	4.25	18 75		37	3	
4	J. I. Case	122	14.5	19.1	9.75	10		10	185.85	6.75	6.75	4.75	18.25		411.7	1	4
5	Goold, Shapley, Muir	121.3	14.5	17.8	9.5	7.6		10	180.7	12.75	6	4	22.75		38 35	2	5
6	International Harvester Co.	87.6	5	20.2	8.5	7.8		10	138.9	11	6.75	4.25	22		366.3	4	6
7	Sawyer Masse	97 4	14.5	19.7	8.5	4.9		10	155	8 75	6 75	4.5	20		364 2	5	
8	International Harvester Co	120.2	7.5	21.8	9.7	6 3		10	175.55	11	5.75	4.25	21		403.75	2	
9	International Harvester Co.	88	13.5	22.9	7 75	4.1		10	146.25	9.5	7.25	4.75	21.5		369.35	4	9
10	Goold Shapley Muir																10
11	Canadian Holt	129.8	15	23	10.25	6.5		10	194 55	11 25	5.25	3.75	20.25	29	380.6	3	11
12	Aultman Taylor	140	8	25	9.25	6.9		10	199.15	10.75	7	4.5	22.35		447.7	1	12
13	J. I. Case	95 2	7.5	21.8	10.35	8.2		10	152.95	10.75	8.5	4.25	23.5		361 25	6	13
15	Diamond Iron Works	90.4	15	22.9	9.25	3.9		10	151.45	8.5	8.75	3.5	20.75		345.4	7	15
1	Canadian Heer	Withdrawn								10.5	6	4.25	20.75				1
16	Rumely	93.1	10	24.6	10.5	4.6		10	152.8	11.5	7	4.5	23		354	2	16
18	International Harvester Co.	118.6	7.5	19.8	10.25	10		10	176.15	10.5	7.25	4.75	22.5		398 15	1	18
19	Avery	70.6	13	20.4	9	3.5		10	126.5	8.25	6.25	4.25	13.75	15	336.95	3	19
21	Interna ional Harvester Co.	100.5	11.5	22.8	10.5	4.2		10	159.6	9 5	7.25	4.75	21.5		391 9	3	21
22	Aultman-Taylor	122.2	7.2	20.8	7.5	6.9		10	174.6	10.75	7	4.5	22.25		415 45	2	22
23	Rumely	140	12	20.4	11.25	9.8		10	203.45	10.25	6 75	4.5	21.5	8	449.75	1	23
24	J. I. Case	75.8	4.8	19.9	9	7.1		10	126.6	10.75	8.5	4.25	23.5		318 3	4	24
25	J. I. Case	81.5	29.5	22.7	9.5	3.1	4.3	10	160.6	13.5	8	4	26.5		339	1	25
26	J. I. Case	120	35	19.6	10.5	3.6	4.5	10	203.1	13.5	8	4	26.5		437	1	26
27	J. I. Case	116.7	29.6	17.1	8.5	5	4.5	10	191.4	15	8	5	29		442.85	1	27
29	Sawyer-Massey	76 9	28.5	18	11	1 5	5	10	150.9	14	8	4	27		364 15	2	29

CAUSE OF PENALTIES: No. 11—For withdrawing from brake test to adjust engine. No. 19—For Company's representative talking to Operator during the brake test. No. 23—For setting plows deeper when dynamometer readings were being taken.

of construction which set the pattern for our modern tractors today. This was incorporated in the Wallis Cub which was marketed a few years later. In 1913 he became manager of the Hyatt Roller Bearings Company's Tractor Division. In this capacity his influence became apparent in the design of the Fordson tractor about 1917.

In 1906 the Transit Thresher Co. was formed by D. M. Hartsough and Son and P. L. Lyons based on the possible success of their experimental four-cylinder high-wheel gas tractor which would draw a thresher through a field and instead of hauling the bundles to the thresher they would haul the thresher to the bundles. The idea did not materialize but 26 larger tractors were built by the Diamond Iron Works of Minneapolis under the direction of the above men in the winter of 1906-1907 for plowing and other agricultural pupproses, and the idea of threshing in transit was given up. One of these tractors was bought by Fred Glover in Glover, N. Dak., who overhauled it. With faith in the overhauled machine, Glover made a deal with Hartsough and Lyons to reorganize the Transit Thresher Co. and formed in 1908 the Gas Traction Co. in Minneapolis, Minn. The output of this Company was later called the "Big Four" (Transit, page 26). In 1912 the Gas Traction Co. was purchased by the Emerson-Brantingham Co. of Rockford, Ill.

Another model of the Waterous tractor (page 25) was produced in 1906 and the Rockford Engine Works bought one tractor and the original Otto patterns from G. P. Lauson and built tractors with an 18 horsepower engine, single-cylinder, hopper cooled. The Rockford Engine Works was also interested in the Morton tractor, using its own regular portable engine with the Morton transmission. The Rockford Engine Works built tractors with 20- and 25-horsepower engines.

Early in 1906 the International Harvester Co. built 200 single cylinder tractors with friction drive (page 25). Later in the year the design was changed to include a sliding gear transmission system with two forward speeds and one reverse, a friction clutch, and a final drive gear. In 1907 the Ford Motor Co. of Detroit, Mich. produced an experimental tractor using some of the parts from a Ford car and a binder (page 25). Also in 1907 Hart-Parr produced their Model "60" (page 25) and in 1908 their 40-80 horsepower model (page 26). While this latter model weighed 34,000 pounds the use of the so-called tricycle arrangement of wheels was made use of. This tractor appears to have been in little demand.

The origin of the word "tractor" is credited to the year 1906 and popular establishment of the name "tractor" to replace the longer expression "gasoline traction engine" is commonly attributed to Chas. W. Hart and Chas. H. Parr of Charles City, Iowa, who are credited with have built the first successful internal combustion engine tractor and founding the gasoline tractor industry; or to W. H. Williams, Sales Manager of the Company who "sat puzzling" over an advertisement he was writing. The words 'gasoline traction engine' were too cumbersome. In his mind came visions of a new word — 'tractor.' He acted on the impulse and wrote the word into the ad." However, it develops that the word had been coined by 1890 when it first appeared on record in patent No. 425,600 issued on a tractor invented by Geo. H. Edwards of Chicago.

In 1906 the first tractor school in the U. S. was held at the Minnesota State Fair grounds in St. Paul, under the direction of D. D. Mayne, Principal, School of Agriculture, University of Minnesota, and B. B. Clarke, Editor, American Thresherman. Chief instructors were William Boss of the University of Minnesota and Philip S. Rose of the North Dakota Agricultural College. H. B. White, also of the University of Minnesota, was in charge of practice work. This was the beginning of the school work which has since been carried on in practically all agricultural colleges and by tractor manufacturers.

Although tractors with internal combustion engines, burning gasoline, had been manufactured since the early '90's, it was not until the first Winnipeg trial in 1908 (under the auspices of the Winnipeg Industrial Exhibition) that the public had an opportunity to compare field operations of steam and gas tractors. In this and succeeding demonstrations held in 1908, 1909, 1910, 1911, and 1912, representatives of many countries witnessed the competition of internal combustion tractors plowing in the same fields with the steam tractors (pages 19-22 - Score Sheets, Winnipeg Motor Contest - courtesy E. R. Potter). As late as 1920, 15 companies were manufacturing steam tractors, and in that year they produced 1,766 machines.

Some 1902-1908 Models

1902 Hart Parr No. 1

2-cylinder 4 cycle 9 x 13 inch twin horizontal engine, 17-30 H.P. at 250 r.p.m., make-and-break ignition — dry cells for starting and low-tension generator for operating. Enclosed oil cooling system using a bank of commercial cast-iron heating radiator units equipped with 5 expansion "bulbs." The chassis followed the conventional lines of the steam tractor. Capacity, 4 to 5 plows.

1903 Hart Parr No. 2

2-cylinder, twin horizontal engine. Built in two sizes — 17-30 and 22-45 at 280 r.p.m. which were later enlarged and improved and called 22-45 and 30-60. Cooled with oil circulated by the centrifugal pump, homemade radiator with draft induced by exhaust. Capacity of larger engine, 8 — 14-inch plows. Power from engine transmitted to traction gears through a friction clutch for forward motion and through planetary gears for reverse.

1904 Dissinger

1-cylinder horizontal engine five sizes, 16 and 20 H.P. at 250 r.p.m., 24 H.P. at 225 r.p.m., and 35 and 45 at 200 r.p.m. Valves in separate cages. Ignition-battery make-and-break; lubrication sight-feed oiler; cooling - water forced by centrifugal pump from cylinder to coolers on top of canopy, thence to tank through inverted funnel; transmission by chains and sprockets (one spring mounted) to countershaft, bull pinion and gear final drive.

1904 Electric Wheel

12-25 H.P., 4 cylinder vertical mounted crosswise. Automatic intake valve; make-and-break ignition, battery start, magneto run; two shoe friction clutch; automotive steering; channel-iron frame.

1905 Eason-Wysong Auto-Tractor

Morton-patented traction truck mounted with gasoline engine. Truck had friction driven transmission and was so designed that almost any stationary gasoline engine could be mounted on it. This arrangement was popular around 1900.

1905 Ohio

15 H.P., 2-cylinder, 2 cycle vertical Cushman engine —later rebuilt using a 30 H.P. Brenen 2-cylinder opposed engine.

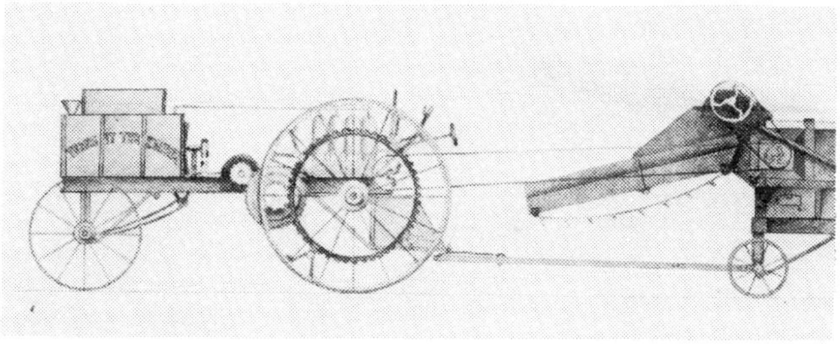

1906 Transit (Diagrammatic)

Tractor powered by 4-cylinder vertical engine mounted longitudinally. Dry-cell ignition. Thresher belted and power take-off shaft.

1906 Waterous

25 BHP, 1-cylinder horizontal water-cooled engine; make-and break ignition; lubrication by sight-feed oiler; expanding two-shoe clutch; spur-gear drive from clutch pinion to rear drive wheels mounted on floating axle; forward speed, 2 to 3 miles per hour.

1906 Ford

24 H.P., 4-cylinder vertical engine with copper water jackets; vibrator-coil, high-tension ignition. Rear wheels supplied from old grain binder; front wheels and axle, steering post and radiator from Model "K" Ford car; spur-gear drive; reverse and forward speeds.

1906 International Harvester

20 H.P., 240 r.p.m., single-cylinder engine with open crankcase and speed regulated by hit-and-miss governor; sight-feed drip oiler to lubricate engine; cooling; water circulated with plunger pump from tank through cylinder jacket thence to screen. Battery and coil make-and-break ignition; forward and reverse motion effected by shifting engine on frame rolls — friction drive from crankshaft pulley to intermediate pulley for forward motion, and by toggle operated idler friction pulley for reverse.

1907 Hart Parr 60 "Old Reliable"

30-60 H.P., 300 r.p.m., 2-cylinder twin horizontal engine; dual jump spark ignition with low-tension magneto, two coils, and dry cells — latter for starting only; oil cooling, circulation by centrifugal pump, and induced draft by exhaust; kerosene for fuel, gasoline for starting; hit-and-miss governor; mechanical oiler; forward motion through friction clutch and pinion, intermediate gear, counter-shaft, and pinion to bull gear and differential in left driver; planetary gear reverse, one forward speed, 2.3 m.p.h. and one reverse.

1907 Transit "Big Four"

35-70 H.P., 4-cylinder vertical engine. Drive wheels 8 feet high and 18 inches wide. Produced by the Gas Traction Co., of Minneapolis, Minn. resulting from the reorganization of the former Transit Thresher Co.

1908 Hart-Parr 40/80

40-80 H.P., 400 r.p.m. oil cooled with pump circulation and induced draft from the exhaust. Engine started on gasoline and operated on kerosene with water injection. Drive wheels 98 in. in diameter. Speeds forward 2-3, and 3-4 m.p. hr.; planetary gear reverse.

1909

Two important companies entered the field-Avery Co., Peoria, Ill., and Russell and Co., Massillon, Ohio. The Avery Co. was a steam tractor builder in a big way and entered the 1909 Winnipeg Motor Contest with a 12-36 tractor-truck (page 27). This was a general purpose machine powerful enough for traction work, capable of hauling on roads and suitable for stationary power. The Russell 20-40 Tractor appearing about this time was a development of a tractor purchased at the London (England) Agricultural Show in 1905 (page 27). The Olds Gas Power Co. of Lansing, Mich., also started tractor production (page 26) and the Hart-Parr Co. offered a Model 15-30 two-cylinder, threshing tractor (page 27) and a 22-45 plowing unit. The McVicker Engineering Co. designed a large tractor for the Joy-Wilson Sales Co. of Denver, Colo. (page 27). The tractors following this design were built in the shops of the Minneapolis Steel and Machinery Co. of Minneapolis, Minn. under the name of Joy-McVicker. The Geiser Co. of Waynesboro, Pa., entered the tractor field (page 26) and Rumely's built their first "Oilpull". The Hart-Parr Co. claimed to have built this — and could have built the motor as they were equipped for this class of work. First Oilpulls had 180 degree cranks, as did Hart-Parr. George Morris of Racine, Wis. built a large tractor about this time and the Sylvester Auto-Thresher appeared. From this outfit the thresher could be removed leaving the tractor ready for plowing and other field work.

Some 1909 Models

1909 Olds

Olds Gas Company, Lansing, Mich.—20 H.P., 400 r.p.m., 1-cylinder horizontal engine. Machine was an early attempt to transmit power from engine to drive wheels entirely with gearing. Weight 12,000 pounds. Forward speeds, 1, 2, and 3 miles per hour and one reverse.

1909 Geiser

25-50, 600 r.p.m., 4-cylinder vertical engine. Cooling, 2 enclosed radiators, pump, and induced draft by exhaust; multiple-feed mechanical oiler; high tension ignition with K-W magneto and coil; compressed-air starter; whistle regular equipment; disc clutch with hardwood blocks; enclosed spur-gear drive to rear wheels; two forward speeds, 2-1/4 and 3-1/4 m.p.h.

1909 Hart-Parr 15/30

15-30 at 500 r.p.m., 2-cylinder opposed engine. (8″ x 9″ bore and stroke.) Tricycle, front wheel steering and rear drivers 73″ x 20″. Final drive, spur gear pinion and internal bull gears. Speeds 2.5 and 4.75 m.p.hr.

1909 Joy-McVicker

50-140 H.P., 250 r.p.m., 4-cylinder vertical engine. All valves mechanically operated and fitted with removable valve cages. Ignition: optional, make-and-break, or jump spark, although jump spark was generally recommended. Channel iron frame. Spur-gear drive, clutch to countershaft, thence chains to rear wheels. Built also in 30-70 and 15-40 H.P. sizes.

1909 Russell 20/40

20-40 H.P., at 350 r.p.m., 3-cylinder (8 x 10) vertical engine. Tank cooling system, capacity 71 gallons, fuel tank 42 gallons. Rated 20-44 H.P. at Winnipeg trials in 1909.

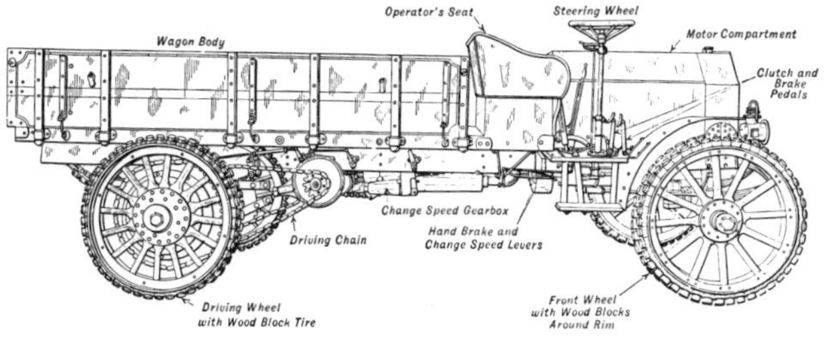

1909 Avery farm truck

Tractor Truck—12-36 H.P. at 1,000 to 1,500 r.p.m., 4-cylinder vertical motor. Equipped with belt pulley and drawbar. Capable of pulling 3-14-inch plows or carrying load of 3 tons on wagon bed. When used for plowing ballast was placed over drive wheels. Wheels were provided with wooden plugs to adapt tractor to hard roads. Extension rims with lugs were supplied for traction on soft ground. Vehicle was spring-mounted and adapted to speeds from 2 to 15 m.p.hr.

1910

There was rather bitter competition during this period between the steam and gas-tractor manufacturers. The president of the Kinnard-Haines Co. stated that when he first went into the business, the steam tractor manufacturers refused to load their machines on the same freight cars with his gasoline tractors. The International Harvester Co. started production on the 45-horsepower "Mogul" (page 28) at the Chicago Tractor Works.

The Minneapolis Steel and Machinery Co., in 1908, had built a number of large tractors for a sales company, and being encouraged by the venture developed their own "Twin City Forty" (page 29) and placed it on the market in 1910. The Gas Traction Company produced their Big Four "30" model (page 000 The Buffalo-Pitts Co., Buffalo, N. Y., introduced their first tricycle type tractor (page 30).

The Imperial Machine Co., of Minneapolis, Minn., introduced their 40-70 H.P. model (page 28) the Birrell Motor Plow Co., Auburn, N.Y. (page 28) and the Goold, Shapely and Muir Co. of Brantford, Ont., entered the tractor field. One of the smallest tractors of the time was the "York" made by the Flinchbaugh Manufacturing Co., York, Pa. (page 29) which made 9 sizes ranging from 2 to 40 horsepower. The Heider Manufacturing Co., Carroll, Iowa - later Rock Island, Ill. brought out their friction drive, (page 29) the Heer Engine Co., Portsmouth, Ohio, their Heer Four Wheel Drive, (page 29) and the Rumely Co. started building for the trade a large, heavy, high-powered machine which burned kerosene and was sometimes referred to as "Kerosene Annie" (page 29). The Adams Husker Co., of Marysville, Ohio, introduced its "Little Traction Gear" (page 30), which was an attempt to adapt mechanical power to small farms, and the Aultman-Taylor Co. Mansfield, Ohio, added gasoline tractors to their line (page 30).

The Universal Tractor Co. of Stillwater, Minn. offered a tractor by the years end. According to E. R. Potter this appears to be the same outfit sold by the Minneapolis Tractor Manufacturing Co. as "Universal" and by Rumely as "Gaspull." This year's model was a 2-cylinder 18 H. P. size.

The first "American" put out by the American Tractor Co. of Minneapolis was tested out in North Dakota in the spring of this year. It had a 4 cylinder vertical engine and crankshaft driving from both ends. Transmission gears were enclosed and running in oil. The operator's seat was placed in front as designers claimed that for speed as high as 3-1/2 m.p.hr. that was where the operator should be.

In 1910 in the United States, the combined animal and mechanical power on farms was about 28 million horsepower, of which mechanical power (almost entirely steam) represented approximately 6-1/2 million horsepower, or 24 percent of the total.

With few exceptions, the gasoline tractor engines manufactured through 1910 were of the four-stroke cycle (four cycle), had automatic intake valves, hit-and-miss governors for controlling their speed, and make-and-break ignition systems. As a rule, electric current for ignition was supplied by dry batteries for starting, and a low voltage DC magneto or generator (auto sparker) furnished the current thereafter. In some cases a low-voltage oscillating magneto furnished the spark required for starting and running. The frames of the tractors were built up of channel irons to which engine and other parts were bolted. Most large transmission gears were of cast iron, exposed to the weather, and the drive wheels often turned on a one-piece "dead" or floating axle. Selective type of transmission was most commonly used, although friction drive and planetary gear transmissions were not uncommon, and a wide variety of clutches was provided.

Some 1910 Models

1910 Imperial

First built in 1908, 40-70 H.P., 400 r.p.m., 4-cylinder double opposed engine. Vertical valves, exhaust underneath operated by rocker arms. Automatic intake above. Ignition by high tension magneto with impulse starter. Lubrication by mechanical oiler. Internal expanding-type two-shoe clutch. Drive wheel diameter, 8 feet. Steering by roller and chain, latter going around idler under each side of frame and attached to steering sector on opposite side. Two forward speeds.

1910 Mogul

45 H.P., 345 r.p.m., 2-cylinder horizontal-opposed engine. Open-tower cooling with belt-driven fan in upper right side, and rotary pump for water circulation. Hit-and-miss governor; make-and-break ignition (battery start and Motsinger auto sparker run); spur-gear transmission; forward speed, 2 m.p.h. and friction reverse; 1911 Mogul 30-60 H.P., similar specifications but larger engine, throttle governor; open-tower cooling with induced draft exhaust and chain-driven centrifugal pump.

1910 Big Four "30"

4 cylinder vertical 30 H.P. engine. Eight-foot drive wheels. Attachment on front of tractor is a self-guide device. The small wheel in front, which is attached by suitable rods to front axle drops into last furrow cut and keeps tractor in proper relation to furrow without assistance from operator.

1910 Birrell

22-45 H.P., 300 to 350 r.p.m., 4-cylinder vertical engine. Cooling; tubular radiator and fan. Rear wheel driver, 78" x 24". Speed, 2.5 m.p.h. Plows mounted underneath tractor frame.

1910 Twin City

40-65 H.P., 500 r.p.m., 4-cylinder vertical engine mounted lengthwise; cooling, pump, fan and barrel type radiator with horizontal flues; mechanical force feed and splash lubrication; high-tension magneto ignition; band-type clutch in flywheel; spur-gear transmission, and spur-gear final drive—one forward speed, 2¼ m.p.h.

1910 Rumely (Kerosene Annie)

Carburetor built from Secor patents which covered the combustion of fuels. Claimed to be first oil-burning tractor equipped with throttle governor and to use water injection under automatic control of governor. Oil cooled.

1910 Heider

12-24 H.P., 300 to 1,000 r.p.m., 4-cylinder vertical engine. Open-tower cooling system; friction transmission; chain and sprocket final drive. Two forward speeds, 2-1/2 and 4 m.p.h.

1910 York

2-1/2 H.P., 1-cylinder horizontal engine with open crankcase; thermosyphon cooling; sight-feed drip oiler lubrication. Engine could be operated in either direction by shifting exhaust valve cams; clutch, expanding shoe, centrifugally operated; selective sliding gear transmission. Final drive, spur gears from jack-shaft to bull gears of drive wheels turning on dead axle. Two forward speeds and one reverse; speed varied through hand control of hit-and-miss governor.

1910 Heer

2-cylinder horizontal opposed (also opposed crankshaft) engine in two sizes - 25 and 30 B.H.P., both at 450 r.p.m., 4-wheel drive. Jump spark ignition - batteries and magneto; throttle governor; force-feed oiler; automatic intake. Steering by roller and chain for both axles which are pivoted. At top and bottom of differential case where idlers in which chain ran when making turns. Transmission enclosed, sliding gear with open-link chains driving sheave wheels carrying the differential on both front and rear axles. Speeds 1 to 6 m.p.h.

1910 Little Traction Gear

The "gear" was designed to make use of 5- to 12-horsepower engines which a farmer might have and provide a low-cost tractor.

1910 Buffalo-Pitts

35-70 H.P., 650 r.p.m., 3-cylinder vertical engine mounted crosswise, crank throws 120 degrees apart; cooling, through pump and large radiator with induced draft from exhaust; lubrication, combination force feed and splash; high-tension magneto ignition; clutch, friction band in flywheel; transmission, spur gear; final drive, chain and sprocket for each rear wheel; one forward speed, 2.3 m.p.h.

1910 Aultman-Taylor

30-60 H.P., 500 r.p.m., 4-cylinder twin horizontal engine. Removable crankcase cover to permit tightening of bearings or removal of crank, pistons, etc. Mechanical oiler; ignition by high tension magneto. Started with batteries and would often start on "spark" without cranking. Engine started with compressed air from high-pressure tank supplied by hand pump. Spur-gear drive from two-shoe expanding clutch. Road speed, 2.2 m.p.h.

1911

The Thompson Breese (page 31), Fairbanks-Morse (page 31), and Case Companies (page 31) began building tractors, the latter bringing out their 30-60 horsepower model. The Fairbanks-Morse Co. offered their 15-25 H.P. model to the United States trade this year, it having been introduced in Canada the previous year.

The Morgan Engine Works of Alliance, Ohio, which ultimately became the property of Wallis Tractor Co., was organized to build tractors in Cleveland. This Company's first tractor proved to be too small for the trade, and was replaced by a larger one the following year.

The Dauch Mfg. Co., Sandusky, Ohio placed the 15-35 H.P. "Sandusky" tractor on the market (page 32). The Avery 20-35 was offered to the trade as was also the Fairbanks-Morse 30-60, and the Minneapolis Tractor Mfg. Co. models 20-40 and 25-50.

The Hackney Mfg. Co., St. Paul, Minn. (page 31) and Huber Mfg. Co., Marion, Ohio, were building tractors and P. E. Shirk (page 33) of Blue Ball, Pa. was furnishing Shirk Traction gears on which to mount engines in sizes from 4 to 15 H.P. The Bates Tractor Co., Lansing, Mich. introduced their 17-30 model (page 32) and the Holt Co. built a wheel-type tractor (page 31). Dent Parrett designed and built his first tractor at Ottawa, Ill., but later moved to Chicago Heights. The Minneapolis Steel and Machine Co. brought out a 25 H.P. tractor similar in construction to the "TC 40" of the preceding year and the "TC 60" of similar appearance but with a 6-cylinder engine. The Orchard (Gasport) (page 32) Gasport, N.Y., the Russell three-wheeler (page 32) and the 30-60 Rumely (page 32) appeared on the market. The International Harvester Co. added the 45 H.P. "Titan" to their line (page 33).

The Waterloo Foundry Co., Waterloo, Iowa completed construction of their "Big Chief" Model L (page 33 - courtesy E. R. Potter) with attached plow. Information does not seem to be available to indicate whether or not this Company had any connection with the Waterloo Gasoline Traction Engine Co. incorporated by Froelich in 1893.

The Minneapolis Threshing Machine Co., Minneapolis, Minn. placed on the market two models with different type engines, the Minneapolis Universal (20-40 H.P.) 2-cylinder opposed (page 33) and the Minneapolis (25 H.P.) 4-cylinder vertical (page 33). The Minneapolis Steel and Mach. Co. also of the same city appeared with the Twin City 40, (page 33) the first of a standard type.

The Gaar-Scott & Co. started production on the Tiger Pull, a 40-70 H.P. model, following their experiments in 1910 when they built some pilot models.

The Opsata Motor Plow Co., Eau Claire, Wis. brought out their Culti-Plow (page 34) in two sizes, one- and two-plow, both of the same general type of construction — plow undermounted, front wheel drive and single rear wheel for steering.

The first tractor demonstration in the United States was held at Omaha, Nebr., in 1911 as a part of an agricultural exhibit, and became an annual affair for 5 years.

1911 Fairbanks-Morse

15-25 H.P., 200 r.p.m., 1-cylinder horizontal open crankcase engine. Ignition: make-and-break (low tension) match starter; cooling system; pump and open-screen cooling tower. Foot-brake lever actuated brake band on drum of differential for stopping; brake shoe on belt pulley.

Some 1911 Models

1911 Holt

45 H.P. at 550 r.p.m., 4 cylinder vertical motor mounted lengthwise; mechanical oiler; high tension ignition; fly ball governor; radiator and fan cooling - auxiliary water tank. Bull gear and pinion final drive.

1911 Case

30-60 H. P., 365 r.p.m., 2 cylinder twin-horizontal engine; cooling system, pump and cooling tank with induced draft from exhaust; Sumter magneto high-tension ignition; mechanical oiler; expanding shoe clutch; final drive, bull gear and pinion; one forward speed, 2 m.p.h.

1911 Thompson-Breese

30 H.P. Cooling effected by water from engine delivered to cylindrical open tower where it was cooled by being discharged through brass wire screens, after which it was delivered into tanks below ready for another cooling cycle.

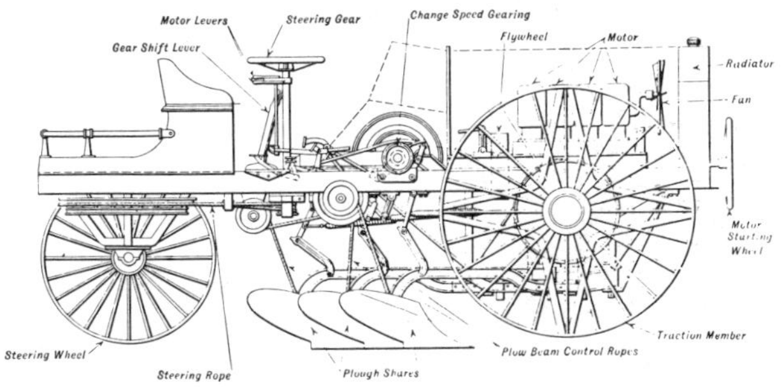

1911 Hackney "Auto-Plow"

40 H.P., 800 r.p.m., 4 cylinder vertical engine. Two drive wheels in front and steered by single rear wheel through steering attachment. Internal expanding clutch. Kingston carburetor. High-tension ignition with Remy magneto and dry cells for starting. Plows and other tillage tools mounted under chassis and controlled by hand lever and power lift from driver's seat.

1911 Sandusky

15-35 H.P., 750 r.p.m. 4 cylinder vertical engine. Cooling system: pump fan and radiator; high-tension ignition; force-feed and splash lubrications. Transmission, sliding gear. Final drive: bull gear and pinion for each drive wheel. Forward speed, 2 to 5-1/2 m.p.h.

1911 Orchard (Gasport)

20 H. P., 450 to 600 r.p.m., 2-cylinder vertical engine. (In 1910 rated 25 H.P.) K W high-tension ignition; enclosed radiator cooling system; mechanical oiler; three speed sliding-gear transmission enclosed; forward speeds: 1-3/4, 2-1/4 and 3-1/2 m.p.h.; final drive, roller chain.

1911 Russell

30-60 H.P., 650 r.p.m., 4-cylinder engine. High tension, battery ignition; force-feed oiler. Final drive, bull gear and pinion to each rear wheel: sliding-gear transmission; two forward speeds, 1-3/4 and 4-1/2 m.p.h. Single front steering wheel.

1911 Rumely Oil Pull

30-60 H.P., 375 r.p.m., 2 cylinder twin horizontal engine, with cylinders sloping and offset from center line of crank to reduce angularity of connecting rod on forward stroke. Secor-Higgins kerosene carburetor with water injection; make-and-break ignition with Bosch magneto; automatic throttle governor; mechanical oiler; oil cooling circulated by centrifugal pump and draft induced through radiator by exhaust; expanding shoe-type clutch; and spur gear final drive. Forward speed, 1.9 m.p.h. Compressed air starter.

1911 Bates

17-30-H.P., 400 r.p.m., 2-cylinder horizontal opposed engine. High-tension magneto; mechanical oiler; channel-steel frame; enclosed transmission running in oil. Spur-gear drive from transmission differential gear on countershaft. Countershaft pinions, and bull gears mounted in rear wheels turning on dead axle. Brake on drum attached to differential gear spider.

1911 Shirk

4, 6, 8, 10, 12, and 15 H.P., single-cylinder, horizontal, hopper-cooled engines. Lubrication: sight-feed oiler and screw grease cups. Spur-gear drive to differential gear mounted on live rear axle. Adapted for hauling, mowing, plowing, harrowing.

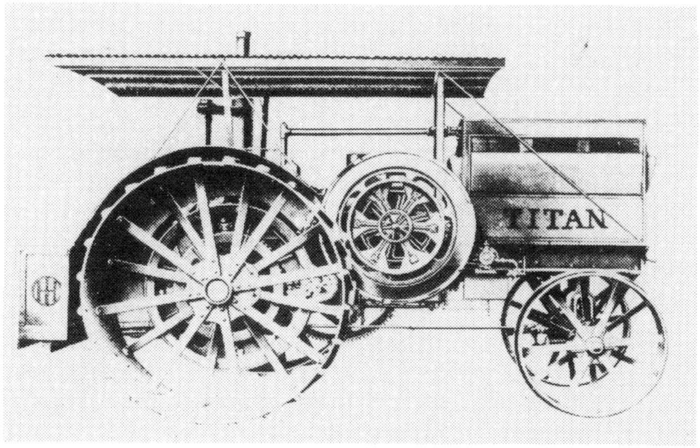

1911 Titan

45 H.P., 375 r.p.m., (later 30-60 H.P.), 2 cylinder twin horizontal; force feed oiling; make-and-break ignition; hit-and-miss governor; pump and tank cooling system. Transmission, spur gear; final drive, live axle, bull gear and differential in right rear driver. One speed forward.

1911 Big Chief, Model L 8-15 (Waterloo Foundry Co.)

8-15 H.P., 2 cylinder horizontal opposed engine. Pump, fan, and radiator cooling; make-and-break ignition with oscillating magneto mounted on each cylinder. Sliding gear transmission, enclosed, running in oil. Spur gear and pinion final drive. One speed forward and 1 reverse. Attached or integral plow, manually controlled from driver's seat.

1911 Minneapolis Universal 20-40

Minneapolis Threshing Machine Co. 20-40-H.P., 2 cylinder, opposed engines.

1911 Twin City 40 40-65

Minneapolis Steel and Machine Co. 40-65 H.P., 4 cylinder vertical engines.

1911 Minneapolis 25

Minneapolis Threshing Machine Co. 4 cylinder vertical motor. Jump spark ignition.

1911 Tiger Pull 40-70

Gaar-Scott & Co. 40-70 H.P., (600 r.p.m.) 4 cylinder T type, vertical engine. Jump spark ignition. Cooling, pump and removable-section radiator with exhaust induced draft. Double disc clutch and exposed bull gear and pinion final drive. (First models built in 1910).

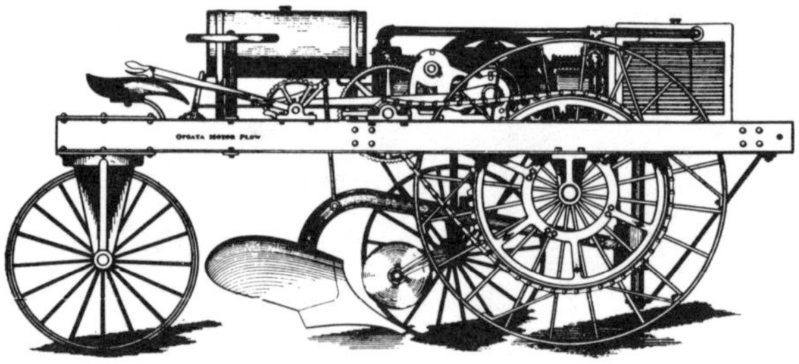

1911 Opsata Motor Plow 15

Two sizes 15 and 30 H.P., 2 cylinder opposed engine. Combination pressure and splash oiling system; pump and radiator cooling system; high tension magneto ignition; final drive, chain and sprocket to each front (driving) wheel, steering by single rear wheel. Plows underslung and controlled from driver's seat.

1912

In 1912 tractors were being delivered by the trainload from various manufacturing centers. Hart-Parr shipped 3 trainloads of tractors to Western Canada in the spring of 1911, and Rumely shipped two trainloads of oil pull tractors to Winnipeg, Canada, May 1912. Twenty-two models of gasoline tractors and about half as many steam tractors were shown at the Minnesota State Fair that year. The Emerson interests at Rockford, Ill., formed a 40 million dollar combination with Geiser, Reeves, the Gas Tractor Co. and the Rockford Engine Works. Rumely at La Porte, Ind. absorbed Gaar-Scott and Co. of Richmond, Ind. and Advance Thresher Co. of Battle Creek, Mich. and took the related production of a score of small companies. During this year the fifth Winnipeg tractor trial was held. The Hart-Parr "40" was produced, (page 35). The Case 20-40 was offered to the trade and performed exceptionally well at this trial.

The International Harvester Co. began production of the 15-30 H.P. Mogul (page 36) — a one-cylinder edition of the larger 30-60 H.P. Mogul.

The Emerson-Brantingham Co., Rockford, Ill. bought out the Gas Tractor Co. of Minneapolis, Minn., and by 1916 had several tractors on the market, including the Big-4 models of 20-35 H.P. and a 30-55 H.P. and a Big-6. The Nevada Truck and Tractor Co., Nevada, Iowa, increased its stock and started making the Nevada tractor (page 35).

The Waterloo Gasoline Engine Co. put out the first "Waterloo Boy" (Model "R") kerosene-burning tractor (page 37). On March 18, 1918, this Company became the property of Deere and Co., Moline, Ill.

The Avery Co. (Peoria) introduced its 12-25 model during the year; the Diamond Iron Works, Minneapolis, Minn., its "America" (page 34), the Nichols Shepard Co., Battle Creek, Mich., the "Oil-gas" (page 36) of the Red River Line; the Nelson Blower and Furnace Co., Boston, Mass., its "Nelson Four Wheel Drive" (page 36), the Lambert Gas and Gasoline Engine Co., Anderson, Ind., the "Lambert Steel Hoof" (page 35); the Morton Plow Co. of Columbus, Ohio, The "Morton" (page 37; the Olmstead Gas Traction Co., Great Falls, Mont., the "Four Wheel Pull" (page 35); and the Pioneer Tractor Co., Winona, Minn., its 30-60 (page 36). This tractor had drive wheels 8 feet high. A later experimental model had a six-cylinder horizontal engine with some bore and stroke as the 30-60 but with drivers nine feet high and was rated as 45-90 H.P. R. L. Huntsberger, who worked for the Company 4 or 5 years, states that the Pioneer Co. was in operation from about 1911 to 1920. The Savoie Guay Co. of Quebec, Canada, produced the "one-man motor plow" (page 35) and the Wallis Tractor Co., Racine, Wis., was incorporated the same year.

The Backer Tractor of Royal Oak, Mich., placed on the market a tractor which was guided and controlled by reins. It was unique in that it was operated by electric motors mounted on the axles within the drive wheels. Electric current was supplied by a generator driven by the tractor engine.

Some 1912 Models

1912 Diamond Iron Works

"America," 40-60 H.P., 600 r.p.m., 4-cylinder (7-1/4 x 8) vertical engine mounted crosswise. Speeds 1.5 and 3.5 m.p.hr.

1912 Nevada Auto Plow

60 H.P. 4-cylinder vertical motor cast in pairs, mounted lengthwise. Bosch duel ignition; Holley carburetor with hot air tube connection, spring mounted horizontal tube radiator cooling; three wheeler, rear wheel steering and front wheel drivers; chain drive to each front wheel; forward speed 3 m.p.hr. self-steering unit mounted at front; plows, carried under frame, raised and lowered by power.

1912 Savoie

Three plows. 550 r.p.m., 3-cylinder vertical engine. Thermosyphon cooling with radiator and fan; battery and magneto, jump spark ignition; splash oiling system; cone clutch; chain final drive; 2 forward speeds, 1 to 1-1/2 m.p.h., and 2 to 2-1/2 m.p.h.

1912 Olmstead (Four wheel pull)

28 H.P., 400 r.p.m., 4-cylinder vertical Clifton engine. "Four Wheel Pull" with chain drive. Had 75-gallon cooling system, with tank placed in front out of heat zone of engine. Tubular radiator cooled by fan at its rear, Remy magneto ignition; Detroit force-feed oiling; Schebler carburetor.

1912 Hart-Parr 40

40 H.P., 400 r.p.m., 2-cylinder vertical engine. Oil cooled; forced-feed lubrication; K-W magneto; three wheels, with front wheel steering; shoe clutch, final drive, bull pinion and internal bull gear; 2 forward speeds, 2.2 to 4 m.p.hr.

1912 Lambert Steel Hoof

20 B.H.P., 450 to 900 r.p.m., 4-cylinder vertical engine. Battery and magneto ignition; cooling system, radiator, pump, and fan. Transmission selective - 3 forward speeds, one reverse. One front steering wheel; two rear-drive wheels equipped with solid rubber tires and adjustable steel "hoofs". By shifting at center, "hoofs" can be drawn in for road work or extended for field work. Independent brake for each rear wheel to facilitate turning.

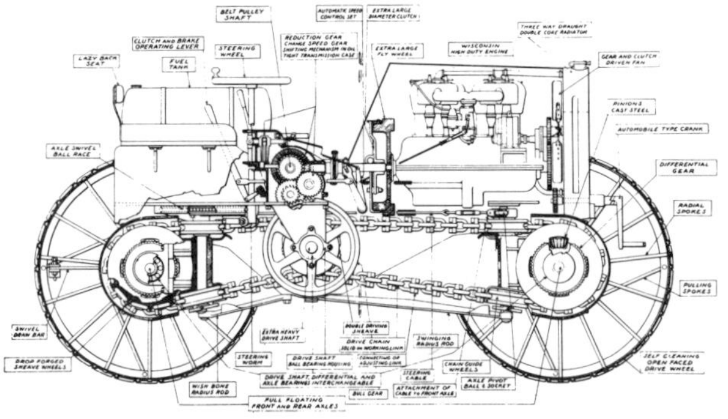

1912 Nelson (4-wheel drive)

Four-cylinder vertical Wisconsin engine; three models; 15-24 at 1,000 r.p.m., 20-28 H.P. at 900 r.p.m., and 35-50 H.P. at 800 r.p.m. Four-wheel drive and four-wheel steering. Power transmitted from clutch through 2-speed transmission to double driving sheave which drives both front and rear axles with hardened chains. Axle clearance, 15-1/4 inches; body clearance, 19 inches with 20-inch maximum body tilt.

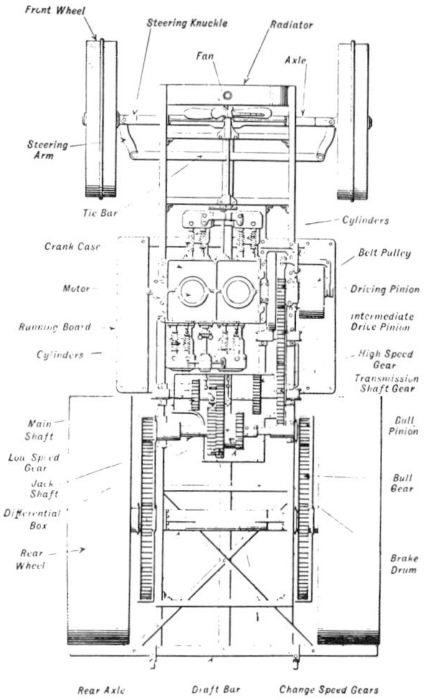

1912 Pioneer

30-60 H.P., 650 r.p.m., 4-cylinder double-opposed engine. Pump, fan, and radiator cooling; splash lubrication, oil being kept at constant level in crankcase; high-tension magneto ignition with impulse starter; clutch, expanding shoe in flywheel; straight open spur gearing from flywheel (belt pulley) to rear wheels, all lubricated by drip oilers. This model was later increased in size and rated 45-90 H.P.

1912 Nichols-Shepard

35-70 H.P., 375 r.p.m., 2-cylinder twin vertical engine mounted crosswise; speed governed by steam-engine-type governor; 360° cranks; cooling, pump, fan, and square radiator with horizontal tubes; force-feed and splash lubrication; high-tension magneto ignition; expanding shoe-type clutch; open spur gears from clutch to rear wheels; one forward speed, 2 m.p.hr. Also built in 25-50 H.P. size.

1912 Mogul 15/30

15-30 at 400 r.p.m., one cylinder, horizontal engine; make-and-break ignition; pump and overflow mixer. Madison-Kipp force-feed oiler cooling, centrifugal pump and cooling tower with induced draft from exhaust. Friction clutch and spur gear final drive; forward speed 2 m.p.hr. - friction reverse, floating rear axle.

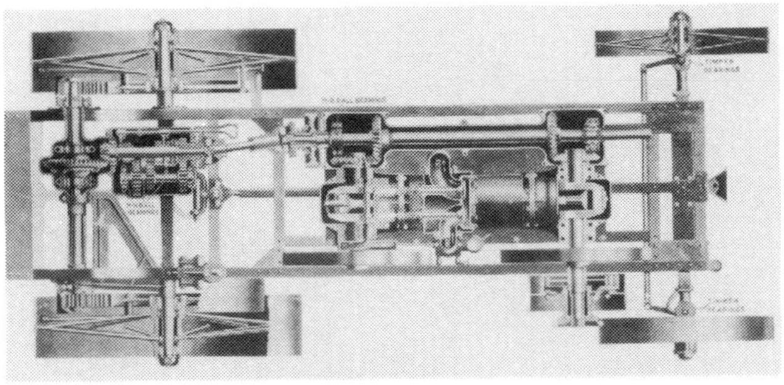

1912 Morton

Two sizes, Model F, 20 H.P., and Model H 26 H.P., both at 500 to 550 r.p.m. Single-cylinder horizontal engine with two pistons and crankshaft at each end. Ignition by HT magneto; lubricated by mechanical oiler; cooling, pump, fan, and radiator. Jack shaft on left side of engine took power from both crankshafts by bevel gear, thence through expanding clutch to two-speed transmission. Integral plow, controlled from seat. Furrow wheel adjustable to depth of furrow so tractor may be kept level.

1912 Waterloo Boy

Model "R", 12-24 H.P. at 750 r.p.m., 2-cylinder twin horizontal engine. Started on gasoline, ran on kerosene (gravity feed), with water injection on heavy load; H.T. magneto with impulse starter; band clutch operated by hand lever; belt pulley mounted on extension of crank shaft; final drive, bull gears and pinions. Chain and roller steering. One speed forward and one reverse.

1913

The year 1913 was a banner year for the introduction of new makes and models. The Dayton-Dick Co. of Quincy, Ill. produced the Leader tractor (page 38); the Avery Co. produced their 40-80 (page 38), and the Case Co. their 12-25 (page 39).

The Hoke Tractor Co. of South Bend, Ind., introduced the Hoke Light Farm Tractor (page 38), an approach to the general purpose tractor, the Denning Tractor Co. of Cedar Rapids, Iowa, the Denning Pug Tractor, and the Diamond Iron Works, the Diamond, Jr.; the Master Huffman by the Huffman Tractor Co.; the Farquhar 40 by the A. B. Farquhar Co., Ltd., York, Pa., and the Hume 50 H.P. by the Hume Mfg. Co., Hume, Ill.; the "Big Four 20" (page 38) by the Emerson-Brantingham Co.; the Gramont one-man auto plow (page 38) by the Gramont Traction Plow Co., Springfield, Ohio, the latter two being of the so-called motor-plow type; and the "TC fifteen" by Minneapolis Steel and Machine Co.

The John Lauson Mfg. Co., New Holstein, Wis., built its first gasoline and kerosene tractor. It had a 2-cylinder vertical engine, conventional steam-type steering, and open bull wheel and pinion drive.

A significant development of the year was the introduction of the Bull tractor (page 39) by the Bull Tractor Co. of Minneapolis. This tractor was the forerunner of the small units toward which practically all tractor manufacturers were turning in this period. A second small tractor which also appeared was the Wallis "Cub" (page 39). This was a general type of tractor, redesigned with refinements for straight-line assembly, and became more or less the standard type of farm tractor for the next several years. The model was the result of several years experience with the Wallis Bear, started in 1902, but was built to pull four to six plows. It was on this model that the first "patented boiler plate U-shaped combination frame and crankcase was first used."

The Waterloo Gasoline Engine Co. placed on the market the Waterloo "Catapillar," tractor in the nature of a half track (page 39). This appears to be the result of H.W. Leavitt's experience with track-type trailers when he produced the Leavitt tractor (page 89). According to the records uncovered by R. M. Merrill of the Deere Co., "Mr. Leavitt came to work for the Company to build a 'Catapillar' tractor in 1913" and his "endeavors were confined to building Catapillar tracks using the "LA" frame, shafts, gears and motor as much as possible." The Model LA, tractor, built about 1911, resulted from several years experimental work and embodied a two-cylinder opposed engine on a four-wheeled chassis.

The Union Iron Works of Minneapolis constructed a small model of a Gearless tractor (page 39) which appeared to have considerable merit. The first full-size tractor, however, did not work satisfactorily, and the company abandoned the idea of a gearless machine.

The Fairbanks-Morse Co. produced their 30-60 (page 40) and the Ward Tractor Co. of Lincoln, Nebraska, began production of a one-man plowing outfit (page 40). The plow was close-coupled to the tractor at the rear, and was controlled from the driver's seat. This outfit was probably one of the first approaches to the so-called "integral" or "attached" plow.

The International Harvester Co. produced the 12-25 Mogul. This tractor had a horizontal opposed engine, with single plate clutch, chain drive from clutch sprocket to countershaft, and chains from countershaft to rear wheels. The rear end was spring-mounted. Ignition was by a Wico ignitor (stationary magneto with vertically oscillating armature). See page 66, some early ignition systems.

The year marked the discontinuance of the Winnipeg, Canada, tractor trials, with the result that the United States became the center of power farming demonstrations. The 20th Century Farmer sponsored the first of its large-scale tractor demonstrations at Fremont, Nebraska.

Some 1913 Models

1913 Leader

12-18 H.P., 750 r.p.m., 2-cylinder opposed-type motor; high-tension magneto with impulse starter; Kingston carburetor; and lubrication by Detroit six-feed oiler. Chain drive through clutch to first countershaft. Selective-type sliding gear transmission; two forward speeds, 1.8 and 2.5 m.p.hr., and one reverse speed; automobile-type steering.

1913 Emerson-Brantingham (Big 4)

20-35 H.P., 700 r.p.m., 4-cylinder vertical engine; fuel, gasoline, or kerosene; K-W high-tension ignition; splash oiling; radiator cooling system (water); frame spring mounted; cone clutch; final drive, bull gear and pinion: 2 forward speeds, 1.88 and 2.52 m.p.hr.

1913 Gramont

35 H.P. and 50 H.P., 2 cylinder horizontal opposed engines; mechanical oiler; channel frame; three wheeler - two front wheel drivers, with rear or steering wheel offset to left to allow for suspension of the 3- or 5-furrow plow; spur-gear final drive; two forward speeds, 2 and 3 m.p.hr.

1913 Avery

40-80 H.P., 500 r.p.m., 4-cylinder horizontal double-opposed engine; thermosyphon cooling system with induced draft in cooling tower; double carburetor for gasoline and kerosene fuel; high-tension ignition; pump and splash oiling, friction clutch in belt pulley; final drive, bull gear and pinion two forward speeds, 1-3/4 and 2-2/3 m.p.hr., effected by sliding frame. For a couple years each tractor was provided with 3 separate pinions to slip on end of crankshaft to alter road speeds by use of sliding frame. Later models had two speeds incorporated in the transmission.

1913 Hoke

15-30 H.P., 700 r.p.m., 4-cylinder vertical motor; cooling system, pump fan and radiator; ignition, Eisemann high-tension; oiling system, constant level splash; clutch, external contracting band for reverse and internal expanding two-shoe for forward; front wheel drivers; final drive, roller chain from countershaft to large sprocket in drive wheels; transmission; one forward speed and reverse, varied by throttle from 2 to 4 m.p.hr; when plowing, one wheel ran in furrow and leveling device was used to keep frame level.

1913 Bull

5-12 H.P., 150 r.p.m., 2 cylinder horizontal opposed engine. Had one drive wheel, making differential unnecessary; idler wheel mounted at left on crank axle for leveling; cone clutch and standard magneto. Superseded in about 1915 with the more powerful "Big Bull" 20 H.P. tractor.

1913 New Gearless

7-12 H.P. engine. Three wheels - front steering, left rear idler, and right rear driver; no gears, chains, or sprockets. The rotating balance wheel, through a system of linkages, caused two dogs, one after the other, to grip and ring on the drive wheel, making it rotate. Weight 1,600 pounds.

1913 Waterloo "Catapillar"

12-30 H.P. - 4-cylinder vertical engine mounted crosswise. Cooling by tank with induced draft by exhaust; force feed and splash lubrication; jump, spark ignition.

1913 Wallis "Cub"

26-44.2 H. P., 650 r.p.m., 4-cylinder vertical engine. Mechanical oiler; K-W high-tension ignition: Bennett carburetor for gasoline, kerosene, or distillate. First of the frameless type in which the one-piece, boiler-plate steel crankcase and transmission housing provided a backbone for the machine. All moving parts were enclosed in this housing except final drive gears.

1913 Case 12/25

12-25 H.P. at 600 r.p.m., 2-cylinder opposed engine. Lubrication - Madison-Kipp force-feed. Ignition - high-tension magneto with impulse starter. Cooling pump, fan and radiator. Kerosene tank mounted over right rear wheel and water tank mounted over left rear wheel. Two-shoe friction clutch, sliding gear transmission and bull gear and pinion final drive - floating one piece rear axle. Forward speed 1-3/4 and 2-1/2 m.p.hr.

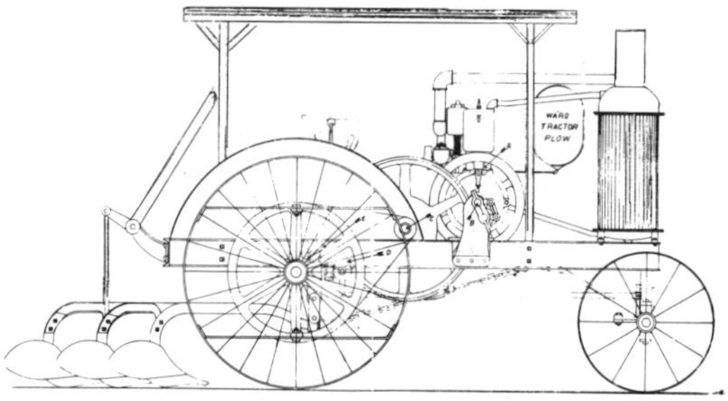

1913 Ward

15-20 H.P., 550 r.p.m., twin-cylinder vertical Cushman engine. Throttle governor; plate clutch in flywheel; clutch pinion transmitted power through train of gears to small sprockets driving main wheels by chain; two forward speeds and one reverse. This outfit had the plow hung under the tractor frame and was pulled by chain from front of tractor; plows were raised and lowered by lever in reach of operator.

1913 Fairbanks-Morse

30-60 H.P., 350 r.p.m., 2-cylinder twin-horizontal engine. Cooling, centrifugal pump, induced draft by exhaust in cooling tower; lubrication, force feed and splash; high-tension, dry-cell - magneto ignition; expanding shoe-type clutch; open-spur gearing direct to the rear wheels.

1914

The B. F. Avery & Sons Co., Louisville, Ky., built their first Louisville Motor Plow (page 42). After a momentary success, sales fell off and its manufacture was discontinued in 1917.

The International Harvester Co. developed two small 2-plow-tractors—the Mogul 8-16 with a single-cylinder, low-speed, horizontal engine and the Titan 10-20 with a twin-cylinder horizontal engine (page 41).

The Wolfe Tractor Co., LaPorte, Inc., (page 41) built a 2-plow so-called all-purpose (noncultivating) tractor, which was designed to plow small fields and fence corners. The company was purchased in 1915 by the Leo Rumely Tractor Co., LaPorte, Ind. and the tractor then renamed "The Hoosier."

Models were produced by the Common Sense Gas Tractor Co., Minneapolis, Minn. (page 41), the Steel King Plow Co. (page 42), and the Lawter Tractor Co., St. Mary's, Ohio, (page 42).

The Townsend Mfg. Co., Janesville, Wis. began producing a tractor powered by a gas engine. This was one of the last gas tractors to retain the outward appearance of a steam traction engine (page 43). The Company was later taken over by the LaCrosse Boiler Co., La Crosse, Wis.

The Emerson-Brantingham Co. was producing the "Reeves 40" (page 41), which had been offered to the trade a couple years earlier, and the Andrews Tractor Co., Minneapolis, Minn., started production on their 25-horsepower "2-cycle" model (page 41).

The Minneapolis Steel and Machine Co. offered their TC 16-30 model of the automotive type (all enclosed) to replace the TC 15. The Allis-Chalmers Co. of Milwaukee, Wis., built their 10-18 three-wheeled tractor (page 44), and Buffalo-Pitts produced their 40-65 using a Twin City engine (page 43). The Big Bull tractor was introduced to replace the little "Bull," which did not have sufficient power.

The Moline Plow Co. of Moline, Ill., purchased the Universal Tractor Co. of Columbus, Ohio and started production on the Moline-Universal. This was one of the earliest practical approaches to a general-purchase-type tractor (page 44).

The Huber Co. produced their 30-60 (page 43), the Samson Tractor Co., Stockton, Calif., built the "Sieve Grip" (page 42) and the Minneapolis Machinery Co. their 15-25 and 24-45 models. The Holmes Little Giant (page 41), a product of the Holmes Mfg. Co., Port Clinton, Ohio, appeared, the "Joliet" (page 44) was produced by the Joliet Oil Tractor Co., Joliet, Ill., and the McIntyre Mfg. Co. offered the "Farmer Boy" (page 43).

New companies that entered the tractor manufacturing field were: Strite Tractor Co., Minneapolis, Minn., Auto Tractor Co., Russell Tractor Co., Massillon, Ohio, G. I. Dell Tractor Co. of Harrisburg, Ark., and the Unity Steel Tractor Co. of Antigo, Wis.

The second National Power Farming demonstration was held at Fremont, Nebr.

Some 1914 Models

1914 Holmes

12 H.P., 750 r.p.m., 2-cylinder horizontal opposed engine. Enclosed cooling system, 4-gallon capacity, with circulating pump; lubrication by mechanical oiler; high-tension ignition from Wico ignitor; planetary gear transmission with change gear giving three speeds forward and two reverse; final drive, chain and sprocket; low speed 3/4 and working speed 1-3/4 m.p.h.

1914 Reeves

40-65 H.P., 500 r.p.m., 4-cylinder vertical motor mounted crosswise. Ignition, K-W high-tension; cooling, pump, fan, and tower; Kingston carburetor; clutch, expanding shoe-type; spur-gear transmission; final drive, bull gear and pinion to each rear wheel; forward speed, 2-1/4 m.p.h.

1914 Andrews

25 H.P., 4-cylinder, 2-cycle, horizontal, double-opposed crossmounted engine. Forced-air cooled; lubricated through gasoline feed; friction drive; chain and sprocket final drive to single rear wheel driver; two front wheels for steering; forward speed, 1 to 3 m.p.h.

1914 Mogul

8-16 H.P., 400 r.p.m., single-cylinder horizontal engine. Hopper cooler; make-and-break ignition with "Accurate" oscillating type magneto; planetary-gear transmission; single chain final drive to sprocket encircling differential in left rear wheel; belt pulley mounted on flywheel driven through friction clutch controlled by idling hand wheel; one forward speed 2 m.p.h., and one reverse speed.

1914 Wolfe

10-18 H.P., 4 cylinder vertical Waukesha engine. High-tension magneto ignition; enclosed cooling system with tubular radiator and centrifugal pump; lubrication, pump and splash constant level; cone clutch; enclosed gear transmission; final drive, chain to right rear wheel, left rear wheel idler; one forward speed, 1-1/2 to 3 m.p.h., adjustable by engine speed.

1914 Common Sense

20-40 H.P., 12 r.p.m., 8-cylinder V-type engine; K-W high-tension magneto ignition with impulse starter; choice of Schebler, Zenith, or Kingston carburetor; full-pressure oiling; expanding shoe-type clutch in flywheel; belt pulley operated at 1/3 engine speed; one rear drive wheel driven on each side by chain and sprocket.

1914 Lawter

45 H.P. engine. Three wheel-front two drivers, rear steering. Short turning was accomplished by a separate lever - controlling planetary transmission for each drive wheel. The underhung plows were controlled by foot lever by which the three plows were progressively power-lifted and then all dropped together.

1914 Titan

(1916 model shown) - 10-20 H.P., 500 r.p.m., twin-cylinder horizontal engine, (burned kerosene). Valve automatically supplied water to fuel mixture in proportion to fuel being used; K-W high-tension magneto ignition with impulse starter; lubrication by six feed mechanical oiler; thermosyphon impulse cooling; spur gear, selective type transmission; double chain drive to rear wheels turning on dead axle; automobile type steering with worm and gear assembly enclosed; two forward speeds, 1.85 and 2.5 m.p.h., and one reverse speed.

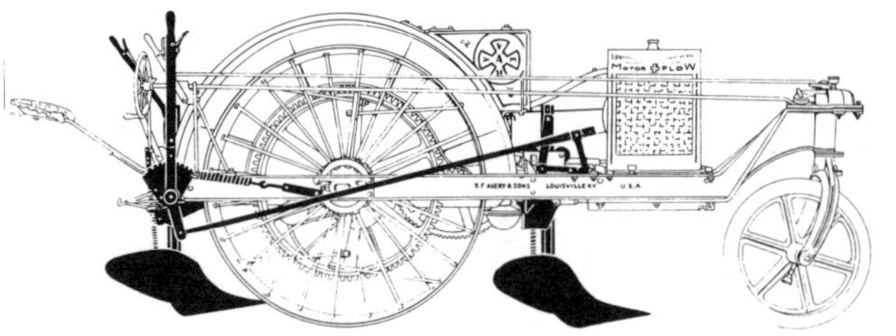

1914 Louisville Motor Plow

10-20 H.P., 675 r.p.m., 2-cylinder opposed engine. Mechanical oiler; pump and radiator cooling; high-tension magneto ignition with impulse starter. The plows (or discs) were mounted on vertical standards, which in turn, were attached to the frame of the machine and controlled by two independent levers in reach of the operator. Steering worm could be disengaged from the driver's seat, so that front wheel was free to follow the furrow - self steering. Leading wheel was V-shaped. The one drive wheel ran in the furrow. One forward speed, 2-1/2 m.p.h.

1914 Samson Sieve-Grip

12-25 H.P., 650 r.p.m., 4-cylinder vertical engine. Holley distillate carburetor; Bosch duplex high-tension magneto; pressure feed and splash lubrication; three wheels, two rear-wheel drivers; rims of all wheels open to permit passage through of sand and dirt, hence "sieve grip"; gooseneck frame; roller-pinion final drive; two forward speeds, road 3-1/2 m.p.h. and plowing 2-1/4 m.p.h.

1914 Steel King

16-18 H.P., 800 r.p.m., 2-cylinder horizontal opposed engine. Three wheels - front wheel steering directly ahead of right driver which was chain driven; left rear wheel was idler. The two furrow plow was semi-integral type - front end attached to toggle on engine frame and controlled by lever, rear supported on castor wheel and controlled by a lever.

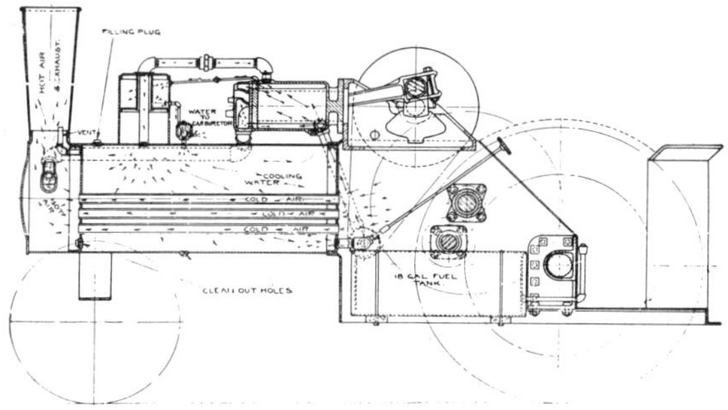

1914 Townsend

30-60 H.P., 2-cylinder twin engine. Boiler contained the cooling system which consisted of a tank with numerous tubes through which air passed, induced by the exhaust in the stack, thus cooling the water in the tank. Later five sizes were built ranging from 10-20 to 30-60 H.P.

1914 Farmer Boy

10-18 H.P., at 800 to 1,200 r.p.m., Waukesha 4-cylinder vertical motor. Pump, fan, and radiator cooling; Atwater-Kent high-tension ignition with battery; splash lubrication; multiple disc clutch; sliding gear transmission; three wheels, front steering, right rear driver, and left rear or land wheel adjustable so machine could be kept level regardless of depth of furrow, one forward speed, 2 to 2-1/2 m.p.h.

1914 Buffalo-Pitts

40-65 H.P., 500 r.p.m., 4-cylinder engine, mounted crosswise, cylinders cast separately. Cooling, fan, pumps, and tubular radiator; high-tension ignition; two 4-shoe-expanding clutches - one for forward and one for reverse; final drive, enclosed internal bull gears and pinions; two forward speeds, 1-1/4 and 2-1/2 m.p.h.

1914 Huber 30/60

30-60 H.P., 600 r.p.m., 4-cylinder vertical engine with cylinders arranged lengthwise; cooling capacity 95 gallons, cooled with horizontal tubular radiator and circulating pump and fan; lubrication, mechanical force feed oiler; ignition by high-tension magneto; band-type clutch; transmission, bevel gears from engine to cross-shaft; final drive, spur gears. Model 13-22 had 2-cylinder opposed engine, dry-cell high-tension ignition, expanding clutch, and chain and sprocket final drive.

1914 Allis-Chalmers

10-18 H.P. 720 r.p.m., 2-cylinder, horizontal opposed engine. High-tension magneto ignition; centrifugal type governor. Countershaft carried the differential and pinions fitted with rollers which meshed with the one-piece internal ring gear in the drive wheels. Three wheel, with steering wheel in line with right driver.

1914 Moline Universal

10-12 H. P., 800-1,200 r.p.m., 2-cylinder horizontal opposed engine. Dixie high-tension ignition; automatic force feed oiling; pump, fan, and radiator cooling; two front wheel drivers; steering through curved horizontal rack and pinion from steering wheel; cone clutch; final drive, bull pinion to internal bull gear in both drive wheels; forward speed, 1/2 to 3 m.p.h.

1914 Joliet

22-40 H.P., 600 r.p.m., 4-cylinder enclosed engine. Cooling system, radiator and cooling coils; lubrication by force feed oiler; ignition, high tension; sliding-gear transmission; frame had 3-point suspension and mounted, both front and rear, on semi-elliptical springs; three forward speeds, 2, 3-1/2, and 8 to 10 m.p.h.

1915

The Hart-Parr Co. put their "Little Devil" (page 46) on the market. The LaCrosse Co. began building the "Happy Farmer" (page 45) in Minneapolis, Minn., and then moved to LaCrosse, Wis. where the Company was liquidated in 1924. Allis-Chalmers started producing their 8-16; the Case Threshing Machine Co. a 10-20 kerosene tractor (page 45); and the International Harvester Co. their 4-cylinder (double twin) 15-30 Titan (page 45). In 1918 the name of this tractor was changed to International.

The Michigan Tractor Co. of Detroit, produced a three-wheel drive tractor (page 46) with "all the weight on the drivers", but few seem to have been manufactured. Apparently this company was reorganized in 1916 under the name of the A. T. Harrow Co. The Dakota (page 46) with angle iron drive wheel rims, the Detroit Line Drive Tractor (page 47) which could be hitched to pull any wheeled vehicle, and the Rumely "Ideal Pull" (page 46) were placed on the market. The assets of the Detroit Line Drive Tractor Co. were bought by the Line Drive Tractor Co. and then known by the latter name.

The Minneapolis Steel and Machine Co. introduced their TC 10-20 tricycle with one drive wheel; the Ford Tractor Co. of Minneapolis, Minn., produced the Model B Ford Tractor and the Ebert-Duryea Farm Tractor Co., Chicago, Ill., their gearless roller-rim-drive (page 45). This tractor was the invention of Chas. E. Duryea, pioneer automobile inventor, who designed the car which bore his name in the early days of the automobile industry. Henry C. Ebert, Vice President of this tractor Company, was a former Vice President of the Westinghouse Co. and spent several years with the tractor "reducing its construction to a practical basis." The Anderson three wheel tractor (page 46) appeared; and the Rock Island Plow Co. produced another model of their "Heider" friction drive (page 46).

Probably no single factor was more responsible for farmer acceptance of the tractor than the tractor demonstrations or shows and this year the interest seemed to be at fever heat. According to Arch Merrifield of the Farm Equipment Retailing, "Farmer interest in tractors was intensified in 1915 by three factors — (1) the increased demand for food production occasioned by World War I; (2) the advent of smaller and more versatile tractors and (3) the series of tractor demonstrations held in various mid-western states.

"To the average farmer the score sheets of the Winnipeg contests (held from 1908 to 1912) were difficult to translate into terms of work on their own farms. In 1911 one of the prominent farm papers (The 20th Century Farmer) sponsored a field demonstration which was held annually near Omaha and Fremont, Nebraska. By 1915 this show was so well-attended that similar demonstrations were promoted by farm and trade magazines, commercial clubs and other organizations, in South Dakota, Illinois, Missouri, Nebraska, Kansas, Indiana, Oklahoma, and Texas. Thirty tractor manufacturers had entries at the Enid, Oklahoma demonstration in 1915.

"On March 29, 1916, a committee of the National Tractor & Thresher Manufacturers Assn. met in Chicago to formulate rules and to make arrangements for holding a series or circuit of demonstrations in strategic agricultural centers. Starting in Dallas, Texas, on July

17, 1916, four-day demonstrations were held near each of the following towns; Hutchinson, Kans., July 24-28; St. Louis, Mo., July 31-Aug. 4; Cedar Rapids, Ia., Aug. 14-18; Bloomington, Ill., Aug. 21-25; Indianapolis, Aug. 28-Sept. 1; Madison, Wis., Sept. 4-8; closing with later demonstrations at Fargo, N.D. and Aberdeen, S.D. The tractors and personnel moved by train from one location to the next.

"The idea of a tractor circuit was abandoned in 1917 and single national demonstrations were sanctioned each year at selected points such as Salina and Wichita, Kan.; Fremont, Nebr., Atlanta, Ga., and Fargo, N.D., where the last of the great national demonstrations was held. A tragic outcome of this meeting was the death of five of the horses entered in competition with the tractors—martyrs to a lost cause.

"In addition to the field demonstrations, the tractor committee authorized huge winter tractor and implement shows in Kansas City and other points, the last being held in Minneapolis in 1922. The display of component parts was a feature of these shows. During these seven years of tractor demonstrations and shows farmers everywhere had an opportunity to view actual performance of every type of tractor under normal field operations. Each entry was allotted a portion of the field according to the number and size of plow bottoms pulled."

Some 1915 Models

1915 Ebert-Duryea

12-16 H.P. at 700 r.p.m. Enclosed roller chain drive to countershaft. Final drive - roller rim.

1915 La Crosse (Happy Farmer)

12-24 H.P., 750 r.p.m., 2-cylinder twin horizontal engine. Kingston kerosene carburetor; Atwater-Kent high-tension dual-magneto ignition; mechanical oiler; pump, fan, and radiator cooling; sliding-gear transmission; roller pinion, internal gear, final drive; one forward speed, 2-1/2 m.p.h.

1915 Case 10/20

10-20 H.P., 800 r.p.m., 4-cylinder vertical motor, mounted crosswise. High-tension magneto ignition with impulse starter; cooling, pump, fan, and radiator; pump and splash oiling; steering wheel on right in line with main driver keyed to axle; left rear wheel idle except in pulling when it could be locked to axle by jaw clutch to give added traction; cone clutch inside belt pulley; brake stopped belt pulley and when in gear stopped forward motion of tractor; bull gear and pinion final drive; one forward speed, 2 m.p.h.; reverse speed accomplished by jaw clutch and chain drive.

1915 I.H.C. Titan 15/30

(Replaced by International 15-30 in 1918) - 15-30 H.P., 575 r.p.m., 4-cylinder horizontal engine. Mechanical oiler to all pistons and bearings with centrifugal ring to connecting rod bearings; high-tension magneto ignition; mixer had four fuel needle valves and one water needle valve; contracting-band-type clutch. Spur gear transmission; double-chain drive to rear wheels; two forward speeds, 1.9 and 2.4 m.p.h.

1915 Heider

10-20 H.P., 800 r.p.m., 4 cylinder Waukesha engine. Cooling system, radiator and fan; splash oiling; Kingston carburetor; ignition, Kingston magneto and batteries; friction drive with seven speeds - road 4 m.p.h. and plowing 2-1/4 m.p.h.

1915 Hart Parr (Little Devil)

15-22 H.P., 600 r.p.m., 2 cycle, 2-cylinder twin, engine. Thermosyphon water cooling; automatic force feed lubrication; one rear drive wheel (center) and two front steering wheels; channel frame construction; spur-gear transmission; two forward speeds, 3-1/3 and 2-1/4 m.p.h.

1915 Anderson

25 H.P., 800 r.p.m., 4-cylinder vertical engine. High-tension magneto ignition; cooling by cross tubular radiator, pump, and fan; splash lubrication; spur-gear transmission; final drive, spur gear in single rear drive wheel; one forward speed.

1915 Michigan (3 drive)

2-plow and 4-plow sizes. 4-cylinder vertical engine, mounted lengthwise. Chain from transmission drove overhead shaft which in turn drove front steering wheel by chain; rear end of this shaft drove the rear wheels through a cross shaft, sprockets, and chains.

1915 Dakota

15-27 H.P., 4-cylinder vertical engine - cylinders cast in pairs. Ignition - high-tension magneto with impulse starter; carburetor, "Linga Vaporizer" equipped with air cleaner; clutch - Bierman expanding-shoe type; three wheels - front two steering, and 60-inch wide rear drive wheel with bent angle irons attached to the wheel rims to minimize slippage; drive to rear wheel by chain and sprocket, otherwise spur gear; belt pulley on engine shaft.

1915 Rumely Ideal Pull

22 H.P. at belt, four cylinder vertical engine mounted lengthwise; speed controlled by governor; water cooled with pump, fan and radiator; Bosch high tension ignition; power lift for undermounted plow; double acting clutch, giving both forward and reverse; internal bull gear drive; two wheels in front, one driver, other idler; rear wheel steering; forward speed 2-1/4 m.p.h.

1915 Detroit (Line drive)

15-36 H.P., 1,200 r.p.m., 4-cylinder vertical Waukesha engine. High-tension magneto ignition; cooling, pump, fan, and radiator; lubrication, force feed and splash, Kingston carburetor; Bennett Air Cleaner; cone clutch (asbestos lined) running in oil; sliding gear transmission; spur gear final drive; forward speeds, 1 to 4-1/2 m.p.h. Tractor control by means of lines.

1915 Ford (Minneapolis)

8-16 H.P., 2-cylinder horizontal opposed engine. Three wheels - two front drivers, one rear steering; ignition - low-tension magneto and coil for jump spark; flyball governor; spur-gear transmission; bull gear final drive; clutch, internal expanding; mechanical oiler; "U" drawbar with traveling clevis.

1916

The Wallis Tractor Co. produced the Wallis Cub Jr. or Model J, a 3-wheeler (page 48), the Huber Co. their Light-4 (page 48), and the Dauch Mfg. Co. their 10-20 Model J. Sandusky tractor (page 49). The Eagle Mfg. Co., Appleton, Wis, marketed their first 4-cylinder tractor (page 49). This Company, which was established in 1888, manufactured engines and silo fillers. It began experimenting with 2- and 4-cycle engines about 1907 and produced its first tractor of 40-60 H. P. in 1912, apparently a 2-cylinder model. After 1912 additional models were produced and in the early '20's a 6-cylinder tractor was developed of the road and cultivator type. The Minneapolis Steel and Machinery Co. marketed their Twin City 60-90 (page 47), the Nichols and Shephard their 25-50 model (page 49); the International their single-cylinder, 10-20 Mogul, a slightly larger edition of the 8-16 Mogul but with a 2-speed sliding gear transmission.

The Emerson-Brantingham Co. offered to the trade their three-wheel 12-20 H.P. Model L. tractor (page 48). With the two wheels in front and a single drive wheel in the rear, opportunity was afforded for the unit to be self-steering when plowing by running the right front wheel in the furrow.

The Altgelt gasoline traction engine was patented by H. E. Altgelt (page 48). This was a walking tractor. Each of the two ambulatory drive wheels had nine spokes, and a foot consisting of a spring steel plate hinged to the end of each spoke. The wheels were mounted on crank axles which permitted them to move up and down as they rolled from spoke to spoke without imparting up and down motion to the body of the tractor. This design through field tests indicated that a traction wheel did not necessarily require a rim, and in certain sticky soils the rim was objectionable. This probably led a number of manufacturers in that period into producing traction wheels with very narrow rims.

Dent Parrett of the Parrett Tractor Co., Chicago, placed his 12-25 H.P. (page 49) tractor on the market. This outfit had considerabiy more ground clearance than many tractors of the time and was of the then all-purpose type.

The Waterloo Gas Engine Co. marketed their Model "N" Waterloo Boy (page 50) rated at 12-25 H.P. at 750 r.p.m. It had two speeds: forward and one reverse.

The Russell & Co. started production on its 30-60 H.P. (page 48) model, later called the "Giant" and the Minneapolis Threshing Machine Co., Hopkins, Minn., added another model to their line, the Minneapolis 20 (page 50). The John Lauson Mfg. Co., New Holstein, Wis. was producing two models, the Lauson 15-25 and Lauson 20-35.

The first cooperative tractor catalog in the industry's history was published in the September 14, 1915 issue of "Farm Machinery and Equipment" now "Farm Equipment Retailing" and revised in the March 28, 1916 issue of this Journal. This showed a directory and specifications of tractor models produced by 62 different manufacturers.

Some 1916 Models

1916 Twin City

60-90 H.P. (rerated from earlier model of 60-110 H.P.) 500 r.p.m., 6-cylinder vertical motor. Tubular cooling tank; high-tension, ignition; mechanical oiler; Bennett carburetor for gasoline, kerosene, or distillate; expanding clutch; spur gear drive to live rear axle; one forward speed, 2 m.p.h. Rear wheels, 84 inches D.

1916 Altgelt "Walking tractor"

20 H.P. at 900 r.p.m., 4-cylinder vertical engine; pump, fan, and radiator cooled; high tension ignition; cone type clutch; transmission and final drive gears especially designed for the job. Tractor speed 2-1/2 m.p.hr. at normal engine speed but could be varied by means of variable speed governor.

1916 Emerson-Brantingham L 12-20

12-20 H.P. at 800 r.p.m., 4-cylinder vertical "Big Four" engine; K.W. high tension ignition; splash oiling; radiator cooling; one rear driver and two front steering wheels; cone clutch; forward speeds 1 2/3 and 2 1/3 m.p.h.

1916 Russell 30-60

30-60 H.P. at 400 r.p.m., 4-cylinder vertical motor; high tension and battery ignition; Madison-Kipp oiler; pump, fan, and radiator cooling; forward speed 1 3/4 to 4 1/2 m.p.h.

1916 Wallis "Cub Junior"

13-25 H.P., 650 r.p.m., 4-cylinder vertical engine, mounted lengthwise. Fuels, gasoline, kerosene, or distillate; carburetor, Bennett; ignition, K-W high tension. (Alexander model was introduced in 1913 as the first frameless type.) A one-piece boiler-plate-steel crankcase and transmission housing provided a backbone for the machine. Three wheeler-single front steering, rear drivers.

1916 Huber Light Four

12-25 H.P., 1,000 r.p.m., Waukesha 4-cylinder vertical motor. High tension magneto ignition, impulse starter; fan, pump, and radiator cooling; Kingston carburetor - gasoline, kerosene, or distillate; combination force feed and splash oiling system; two-shoe expanding clutch; spur gear drive to bull gears in rear wheels; two speeds forward, 2-1/2 and 4 m.p.h.

1916 Eagle

Twin-cylinder horizontal engine built in 12-22 and 16-30 H.P. sizes at 450 r.p.m. Pickering governor, Schebler carburetor, and Dixie H.T. magneto with impulse starter. Cooling, radiator, fan, and centrifugal pump. Fuel, gasoline or kerosene. Oiled by mechanical force-feed oiler. Expanding shoe clutch (wooden block) which pushed out against rim of belt pulley.

1916 Parrett

12-25 H.P. at 100 r.p.m., 4-cylinder vertical Buda eigine; Eisemann H. T. magneto ignition; water-type air washer, gravity feed fuel system; direct drive on both high and low gears accomplished by means of a double-faced intermediate gear and corresponding drive pinoins on crankshaft, through clutch running in oil; belt pulley on end of crank shaft.

1916 Sandusky

10-20 H.P., 1,000 r.p.m., vertical 4 cylinder motor. Centrifugal pump, fan, and radiator cooling; pressure feed and splash lubrication; Dixie and Sevison high-tension magneto ignition with impulse starter; Kingston dual gasoline-kerosene carburetor; sliding gear transmission; internal worm final drive; two forward speeds, 2 and 3 m.p.h.

1916 Nichols-Shepard

Two models: 25-50 at 350 r.p.m. and 35-70 H.P. at 300 r.p.m. Two cylinder twin horizontal engine, valves in cages and set at an angle; float feed carburetor and Pickering governor; high tension ignition; tubular radiator cooling; two shoe expanding clutch; spur gear transmission and spur gear differential.

1916 Waterloo Boy N 12-25

12-25 H.P. at 750 r.p.m. Two cylinder twin engine with opposed crankshaft. HT magneto with impulse starter. Pressure feed and splash lubrication; pump, fan, and radiator cooling; cone clutch; final drive, open spur gear and pinion - forward speed 2 1/4 and 3 m.p.h.

1916 Minneapolis 20

20 H.P. at 600 r.p.m., 4 cylinder horizontal motor; K.W. high tension ignition; Detroit force feed oiler; pump, fan, and radiator cooling; shoe clutch; 2 speeds forward 2 and 2 1/4 m.p.h.

1917

In 1917, the year the United States entered World War I, tractor production more than doubled. A total of 62,742 were manufactured as 85 new companies entered the field to bring the number of manufacturers to more than 200. Approximately 15,000 tractors were exported.

The Moline Universal introduced its Model D (page 53), about the first tractor to make use of a storage battery for ignition, starting, and lighting. The Ford Motor Co. of Detroit, Mich. produced for the trade, its first tractor—the Fordson (page 52). This was the first tractor to make use of cast iron unit frame construction and it was not long until practically all of the tractor manufacturers adopted this type of design. The previous model was constructed with the worm drive overhead but because of lubrication difficulties the worm was placed underneath. The heating caused by the overhead worm would cause the housing to become so hot that the driver's seat became unbearable. The reason the name Ford was not given to the new tractor is stated as follows by H. E. Everett of the Implement and Tractor:

> "Henry Ford was known to be developing a tractor, although no public announcement had been made. Many farmers were waiting for this tractor as it was expected to be a popular-priced unit which would fill a place in the tractor field comparable to that occupied by the Model T in the automobile field.
>
> "Taking advantage of this farmer psychology, a group of promotors in Minneapolis, Minnesota hired a young man by the name of Ford and organized the Ford Tractor Company. This group was legally entitled to use the Ford name. Only a few Ford tractors were ever manufactured.
>
> "You may recall that Henry Ford's first tractor was known as the Fordson and this was because name Ford had been pre-empted by the Minneapolis outfit. . ."

New models introduced during the year were the Avery 8-16 (page 51), the Rumely 12-24 (page 51, The Besser 15-30 (page 51), by the Besser Mfg. Co., Alpena, Mich., the Russell 12-24 (page 52), and the Lang by the Lang Tractor Co., Minneapolis (page 51). The last named was unique in that it made use of power steering and departed, in other respects, from the convention design of the period. Apparently very few were produced. Also the Allen Water Ballast (page 50), the Fogeol (page 52) by the Great Western Motor Co., San Jose, Calif., the Electric Wheel, Allwork, (page 52), and the Square Turn (page 51) by the Albaugh-Dover Co., Chicago, Ill., were newcomers. the last named was rerated the next year from 15-30 H.P. to 18-35 and the Company's name changed to the Square Turn Tractor Co. of the same city.

The demand for tractors was so great because of wartime agricultural needs that several companies made adaptations for the Ford and other cars so they could serve as tractors. An example of this activity was the Feeney Tractor Addition built for a Ford chassis by the Farmers Tractor Co. of Chicago (page 53).

General Motors Corp. entered the tractor field by buying the Samson Sieve Grip Tractor Co. of Stockton, Calif.

Some 1917 Models

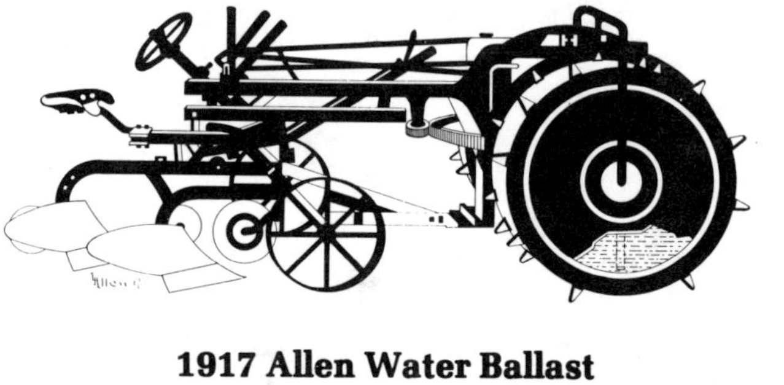

1917 Allen Water Ballast

Two-plow. Front two wheels drivers with provision for filling them with water for additional traction. Rear supported by wheels of the machine being pulled. Steering by curved rack and pinion.

1917 Besser

15-30 H.P., 900 to 1,000 r.p.m., 4 cylinder vertical motor mounted crosswise. Cooling, pump, fan, and radiator; lubrication with mechanical oiler; high-tension magneto ignition with impulse starter; Kingston carburetor provided with air filter; dry plate (3) clutch; sliding gear transmission; final drive, spur gear to enclosed bull gear; two forward speeds, 2-1/2 (plowing) to 4 m.p.h.

1917 Lang (power steering)

15-30 H.P., 900 r.p.m., 4-cylinder vertical Waukesha engine with Dixie HT magneto; lubrication pump and splash water cooling. Three wheels with two front for driving and steering; cone clutch; sliding gear transmission; final drive roller chains and sprockets enclosed and running in oil. Forward speed 2-1/2 to 5 m.p.hr. Power steering.

1917 Avery 8/16

8-16 H.P., 600 r.p.m., 2-cylinder horizontal opposed engine. Kingston carburetor for gasoline and kerosene; high-tension ignition; pump and splash lubrication; thermosyphon cooling; expanding shoe clutch; final drive to rear wheels, bull gear and pinion; two forward speeds, 1-3/4 and 3 m.p.h.

1917 Square Turn

15-30 H.P., 950 r.p.m., 4-cylinder vertical Waukesha engine. Ignition Dixie high tension magneto with impulse starter; Bennett dual gasoline and kerosene carburetor; cooling, centrifugal pump, fan, and radiator; lubrication automatic splash. Power from engine transmitted from two fiber driving cones keyed to crankshaft. For forward motion the large right steel cone is forced forward against the front fiber cone and the large left cone backward against the rear cone, simultaneously, by means of levers actuating eccentric stub shafts to which the drive pinions are keyed. For backward travel the direction of thrust of the large cones is reversed. Steering is also accomplished with these cones. (pages 76-77) Final drive, internal bull gears and spur pinions; speed 2 to 3 m.p.hr.

1917 Rumely

8-16 H.P., 600 r.p.m. Four-cylinder vertical engine; Kingston dual carburetor for gasoline or kerosene; Kingston high tension magneto with manually controlled impulse starter; Manzel force feed and splash lubrication; three wheel-front right wheel driver, left front wheel idler, and steering by rear wheel.

1917 Electric Wheel (Allwork)

14-29 H.P., 800 r.p.m., 4-cylinder vertical engine mounted crosswise. Lubrication by combination splash and force feed; fan and radiator cooling; ignition, high tension magneto with double impulse starter; sliding gear transmission; final drive, bull gear and pinion to each rear wheel; two forward speeds, 1-1/4 and 2.4 m.p.hr.

1917 Russel

12-24 H.P., 900 r.p.m. Waukesha 4-cylinder vertical motor. Splash system lubrication; Bosch high-tension magneto ignition; fan, pump, and radiator cooling; sliding gear transmission; final drive, spur gears to rear wheels; forward speed, 1-1/2 to 3-3/4 m.p.h.

1917 Fordson

20 H.P., 1,000 r.p.m., 4-cylinder vertical engine mounted lengthwise. Ignition by low-tension flywheel magneto and high tension coils; clutch, multiple disc in oil; worm-gear final drive; two-piece cast-iron frame bolted together at middle; splash lubrication - oil from fly-wheel splash caught in funnel and conducted to individual trays in which connecting-rod cap projections dipped.

1917 Fageol

8-12 H.P., 1200 r.p.m., Lycoming Model CT 4-cylinder vertical motor; Dixie HT magneto ignition. The Fageol "walked" on its "legs" or grousers. It was claimed that because of the wedging action of the ground between two adjacent grousers, the drive wheels would not sink deeply even in loose sand. In the drum of each driver was an internal expanding clutch which coupled the solid live axle to the drive wheel; no differential on the axle.

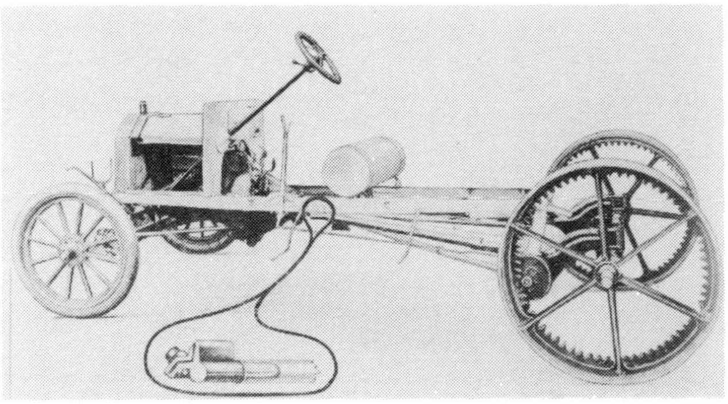

1917 Ford Chassis with Feeny Tractor Addition

This addition consisted in replacing the rear wheels of the car with an extension to the rear axle housing carrying an axle on which the traction wheels, fitted with internal gears, were mounted. The final drive was from spur pinions mounted on the ends of the regular rear axle and meshing with the traction wheel gears.

1917 Moline

9-18 H. P., 1,800 r.p.m., 4-cylinder vertical engine. Lubrication, force feed to crankshaft bearing through hollow crankshaft; high-tension ignition with storage battery - also electric starting and lighting system; thermosyphon cooling with fan and enclosed radiator; front two wheels drivers and steering, rear supported on two wheels or on pulled machinery; spur and bevel-gear transmission; final drive, bull pinion and internal gear; forward speed, 3/4 to 3-1/2 m.p.h. Electric governor.

1918

In 1918, the United States in its second year of war, was faced with an acute shortage of labor and work animals. The farm tractor played an important part in meeting the situation and that year 132,000 tractors were produced.

Ford announced that his Fordson tractors would be sold only to state and national governments. By May several states had made arrangements for allotments of approximately 7,000 Fordsons. The nation-wide distribution of Fordsons through Ford's car dealer agencies greatly aided in relieving the labor and work animal shortage. By August 1920 Ford claimed that 100,000 Fordsons had been sold.

The year marked the discontinuance of the Bull tractor—the tractor, which became the sensation of the industry when first produced. The Bull, in 1914, held first place in tractor production. It declined to second place in 1915, to fourth in 1916, and to seventh in 1917. The Company discontinued manufacture in 1918, reorganized, but did not resume tractor production. The next year it combined with the Madison Motors Corp., and in 1920 was sold by receivers to the American Motor Parts Co.

An important development of the year was the introduction by the International Harvester of a practicable power take-off mechanism on its tractors. This permitted operating and controlling mounted and drawn equipment by the tractors own engine, through a special shaft and under control of the operator. It was not many years until all of the leading manufacturers had their tractors so equipped.

The Dart Truck and Tractor Corp. of Waterloo, Iowa, entered the tractor field with the Dart "Blue J" (page 54). Case produced its 9-18, first with channel frame and round spoked drive wheels—then changing to one piece cast iron frame and flat spoked wheels, the first tractor in the case line with unit-frame construction (page 56). The 15-25 Kinnard-Haines (page 58) appeared.

The International Harvester produced the International 8-16 (page 54), the Kardell Tractor and Truck Co., St. Louis, Mo., its Kardell "Utility" and Kardell "4 in 1" (page 55), the Massey-Harris (page 54), and the Lauson Co. each a single model. The Farquhar 18-35 (page 57) was in production by the A. B. Farquhar Co., Ltd., York, Pa.

The McIntyre Mfg. Co. of Columbus, Ohio, built the "Morton Motor Plow," (page 58) using a conventional 4-cylinder vertical engine. Prior to 1912 the tractor was powered with a single cylinder, 2-pistion engine and was built by the Morton Motor Plow Co., Columbus, Ohio. Aultman-Taylor offered to the trade 18-36, 25-50, and 30-60 H.P. model tractors.

Other tractors that appeared during the year were the Uncle Sam 20-30 H.P., manufactured by the U.S. Tractor and Machinery Co., Menasha, Wis., the Minneapolis (page 56), the Illinois (page 57), by the Illinois Silo and Tractor Co., Bloomington, Ill., the Boring, by the Boring Tractor Co., Rockford, Ill., (page 55), and the Bethlehem (page 57), by the Bethlehem Motors Corp., Allentown, Pa.

General Motors, which had entered the tractor field the year before by purchasing the Samson Sieve Grip Tractor Co., bought the Janesville Machine Co., and shortly after began constructing a new plant there. This Company liquidated its tractor business 2 years later.

The Gray Tractor Co., Inc., Minneapolis, Minn., started production on their 18-36 H.P. wide Drum Drive tractor (page 57). It will be noted that some of the features that this Company developed earlier for orchard use were maintained in this model and that a self-steering device as an aid in plowing was introduced.

The first model of the tractor known as the Gray was built in 1908 by W. Chandler Knapp, a prominent fruit grower near Rochester, N.Y. It was a 2-cylinder type designed mainly for orchard cultivation. Another machine somewhat heavier was built in 1909 but its drive wheels were too narrow and traction difficulties arose. "In 1910 the drive wheel was composed of two ordinary drive wheels mounted together." The two cylinders did not give the needed power so that the 1912-13 model was equipped with a more powerful 4-cylinder motor and the double drive wheel "was changed to a wide Drive Drum with tight heads and no spokes."

At about this time the Gray management took over and moved the executive and manufacturing departments to Minneapolis where in 1914 they produced the first unit to be known as the Gray Tractor.

In 1916 the Gray Tractor Co. of Minneapolis, Minn. placed two models of their wide Drum Drive tractor on the market, the Model 13, 15-25 H.P. and the Model A, 20-35 H.P. In 1925 the 22-40 Canadian special was produced, a larger powered unit than previous models but of the same general design.

The Keck-Gonnerman Co. of Mt. Vernon, Ind., entered the tractor field by producing their 12-24 H.P. model (page 56). It will be noted that this Company had been producing a line of steam traction engines and threshers.

The Hart-Parr Co. started production on their New Hart-Parr 15-30 (page 56) which was of considerably different styling than previous models.

The Acme Harvesting Machine Co., Peoria, Ill., introduced their Acme 12-24 H.P. to the trade. This tractor was furnished either with rear drive wheels or rear driving tracks (page 55) which could be readily interchanged.

The Fairbanks Morse Co. offered their Fair-Mor 10-20 H.P. model (page 55) together with the Fair-Mor 12-25. This latter followed the styling of the conventional steam traction engine and resembled very closely the design of the Townsend (page 43).

Another model — Flour City Jr. 14-24 H.P. (page 55) was produced by the Kinnard & Sons Mfg. Co., Minneapolis, Minn.

The Whitney 9-18 was placed on the market by the Ohio Mfg. Co., Upper Sandusky, Ohio. This was more or less of the conventional 4 wheel design — rear driving, front steering and powered by a two cylinder opposed horizontal engine. In 1921 this company seems to have combined with the Post Tractor Co., Cleveland, Ohio, which produced the Post 12-20 beginning about 1918, to form the Post-Whitney Co., Cleveland, Ohio and which were marketing both tractors.

Some 1918 Models

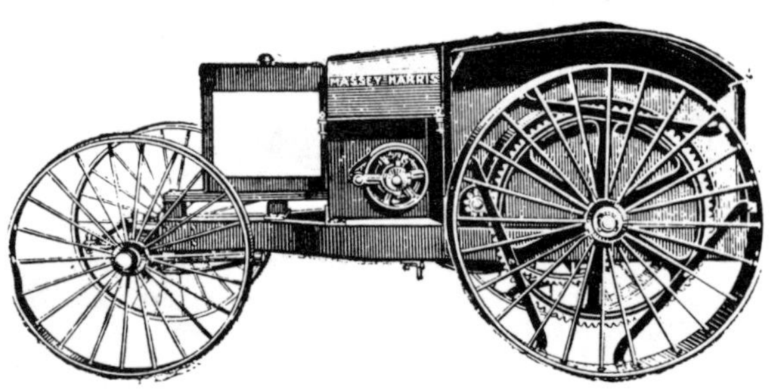

1918 Massey-Harris 12/25

12-25 H.P., 1,000 r.p.m., 4-cylinder vertical Buda engine. Lubrication, force feed through drilled crankshaft; cooling, radiator pump, and fan; ignition, high-tension magneto with impulse starter; air cleaner; 3-plate disc clutch; sliding gear transmission; final drive, spur gear; 2 forward speeds, 1.63 and 2.68 m.p.h.

1918 Dart "Blue J"

40 B.H.P., 950 r.p.m., 4-cylinder vertical engine, cast in one piece. Ignition, high-tension magneto with manual advance; combination pump and splash oiling system; centrifugal pump, fan, and radiator cooling system; drive from motor through disc clutch running in oil, thence through transmission, jack shaft, and worm which operated the large enclosed bull gear; three forward speeds, 1-1/4, 2-1/2 (direct drive), and 5 m.p.h.

1918 International 8/16

8-16 H.P., 1,000 r.p.m., 4-cylinder vertical engine with removable cylinder sleeves. Lubrication, force feed and splash; cooling, thermosyphon, fan, and radiator; ignition, high-tension magneto with impulse starter; multiple disc dry clutch; sliding gear transmission; final drive, roller chains to rear wheels; 3 forward speeds, 1-3/4, 2-1/4, and 4 m.p.h.

Flour City 14-24

14-24 H.P., 800 r.p.m., 4-cylinder vertical engine, cylinders cast in pairs. High tension ignition with K.W. magneto and impulse starter; radiator and circulating pump cooling system; splash lubrication for engine; sliding gear transmission; final drive bull pinion to bull gear in drive wheels; two speeds forward 2-1/4 and 3-1/4 m.p.h.

Acme 12-24

12-24 H.P., 900 r.p.m., 4-cylinder motor cast en bloc and mounted crosswise. Crawlers in rear are interchangeable with wheels. K.W. high tension magneto ignition; radiator, pump, and fan cooling system; force feed and splash lubrication; sliding gear transmission; open spur gear final drive; forward speeds 2-1/4 to 3-1/8 m.p.h.

1918 Boring

25 H.P., 1,000 r.p.m., Waukesha 4-cylinder vertical motor. Lubrication, force feed through drilled crankshaft; cooling, radiator pump, and fan; ignition, high-tension magneto with impulse starter; dry-type air cleaner; twin-disc clutch; sliding gear transmission; traction, two-front drive and one rear castor wheel; forward speed, 1 to 5 m.p.h.

1918 Kardell "4-in-1"

18-35 H.P., 900 r.p.m., 4-cylinder vertical Waukesha motor. Lubrication, force feed and splash; Dixie high-tension ignition with impulse starter; three wheeler, two front drivers and rear steering; spur-gear transmission, selective type; final drive, chain and sprocket; forward speeds, 2 to 8 m.p.h. As a *motor plow* a three-bottom plow was suspended under the frames; as a *motor truck*, it could be converted into a 3-ton truck by use of extension frame; as a *tractor* it could be hitched to various pulled machines; and for *farm power* it was fitted with a belt pulley to drive stationary machines.

Fair-Mor 10-20

10-20 H.P., 600 r.p.m., 2-cylinder twin horizontal engine. High tension ignition with Dixie magneto; radiator and circulating pump cooling system; force feed oiling system; mixer type carburetion; sliding gear transmission; final drive, enclosed bull gears in rear wheels - forward speeds 1-1/2 to 2-1/4 m.p.h.

Keck-Gonnerman 12-24

12-24 H.P., 650 r.p.m., 2-cylinder twin horizontal motor cast en bloc; Bosch high tension magneto ignition with impulse starter, barrel type radiator, pump, and double fan cooling; pump and splash lubrication; Schebler carburetor fitted with air cleaner; contracting band type clutch; sliding gear transmission; spur gear final drive; forward speed 2 to 4 m.p.h.

1918 Case 9/18

9-18 H.P., 4-cylinder vertical engine, mounted crosswise. First Case tractor with one-piece cast-iron frame; wheels pressed-steel construction; high tension magneto with impulse starter; gear driven fan; expanding shoe clutch inside flywheel; hand-operated brake lever actuated break shoe on pulley — stopped the pulley when doing belt work and braked the tractor when in gear.

New Hart-Parr 15-30

15-30 H.P., 750 r.p.m., 2-cylinder twin horizontal engine. High tension magneto with impulse starter ignition; radiator, pump, and fan cooling system (water). Madison-Kipp force feed lubrication; sliding gear transmission; final drive, cast internal gear; forward speeds 2 to 3 m.p.h.

1918 Minneapolis

17-30 H.P., 775 r.p.m., 4 cylinder vertical motor cast in one piece, cross mounted; lubrication, force feed; ignition, high tension magneto; cooling, pump, fan and radiator; spur gear transmission from plate clutch to live two-piece rear axle.

Gray 18-36

18-36 H.P., 900 r.p.m., Waukesha 4-cylinder vertical engine with cylinders cast in pairs and mounted crosswise; K. W. high tension magneto ignition; radiator and pump cooling system; constant level splash and circulation oiling system; sliding gear transmission; enclosed chain final drive; forward speed 2 to 2-1/2 m.p.h. Self-steering device mounted on right front axle. Electric starting and lighting equipment optional.

1918 Farquhar

18-35 H.P., 550 r.p.m., 4-cylinder vertical, cylinders cast in pairs and engine mounted longitudinally; lubrication, Detroit force feed oiler, and splash; cooling, tubular radiator and circulating pump; ignition K. W. high tension magneto.

1918 Illinois

18-30 H.P., 600 to 900 r.p.m., 4-cylinder vertical engine. Lubrication, force feed through drilled crankshaft; cooling, radiator, pump and fan; ignition, high tension magneto with impulse starter; crossover gear transmission (two gears mounted on outside of transmission housing which can be interchanged for lower speeds on heavy pulling); final drive, internal planetary gears to rear axle and through shock absorbing spiders to wheel rims; forward speed, 1-3/4 to 6 m.p.h.

1918 Bethlehem

18-36 H.P., 1,000 r.p.m., 4-cylinder vertical motor mounted lengthwise. Cooling by pump, fan, and radiator; lubrication, Beaver splash system; Bosch magneto high-tension ignition; selective gear transmission; spur-gear final drive; forward speed, 3.3 m.p.h.

1918 Kinnard-Haines

15-25 H.P., 650 r.p.m., 4-cylinder vertical engine. Lubrication, splash system with pump feed to crankcase; cooling, pump, fan, and radiator; ignition, high-tension magneto; sliding gear transmission; spur gear final drive; forward speeds, 2-1/2 to 4 m.p.h.

1918 Morton Motor Plow

(First one built about 1908). 20 and 26 H.P., 500 to 550 r.p.m., 4-cylinder engine. Specifications same in both except cylinder diameter of latter was greater. High-tension magneto ignition; cooling through fan-cooled radiator; mechanical oiler; belt pulley could be attached to crankshaft or side shaft; sliding gear transmission; final drive, bull pinion to internal bull gear; furrow drive wheel adjustable to furrow depth to keep machine level; three speeds forward, 2-3/4, 3-1/2, and 6 m.p.h.

1919

The Case Threshing Machine Co. announced their 15-27 model (later rated 18-32) with cast iron frame, 3-bearing crankshaft (previous models had 2), and force feed lubrication, and the J. I. Case Plow Works their Model K (15-30 H.P.) — the previous Wallis Cub Jr. or Model J three-wheeler which had been converted into a four wheeler.

The Samson Co. brought out the Samson "Iron Horse" (page 63), the Chase Co. of Syracuse, N.Y., the Chase (page 63), the Appleton Co., Batavia, Ill., the Appleton (page 62), and the Minneapolis Steel and Machinery Co. placed their Twin City 12-20 (page 63) (later rated 17-28) on the market. The Russell 20-40 (page 62) appeared as did also the Post Model C, 12-20 (page 63) by the Post Tractor Co., Cleveland, Ohio, which departed radically from conventional design.

The LaCrosse Model M (page 63) was shown for the first time at the Wichita (Kans.) Power Farming Show. Production was started on the Victor (page 62) by the Victor Tractor Co. another radical departure from conventional design, and on the Nilson (page 61) by the Nilson Tractor Co., both of Minneapolis, Minn., the latter unique in that it had 3 rear wheel drivers. The Fitch Four Drive Tractor (page 62) built by the Four Drive Tractor Co., Big Rapids, Mich. was placed on the market. The Aultman Taylor Co. discontinued its 18-36 and 25-50 models and started production on 15-30 (page 61) four cylinder vertical and 22-45 horizontal engine models and continued in production the30-60 model. The Frick Co. of Waynesboro, Pa. produced the Frick (page 62) and the Advance-Rumely Thresher Co. produced another model, the 16-30 Oil Pull. The Lauson Co. discontinued the manufacture of the Lauson 15-25 at years end.

The majority of tractors which appeared up to 1916 or '17 were designed mainly for drawbar work and possibly to a lesser extent for belt pulley. They were heavy and not adapted to cultivation. The need for tractors which would perform this operation, especially in corn, was rapidly becoming apparent and the so-called motor-cultivator began to appear. After a few years of development the first Corn Belt cultivator demonstration was staged at Blue Mound, Ill., in 1919. Participants included Avery, Emerson-Brantingham, Illinois, Moline Plow Co., Toro Motor Co., Allis-Chalmers, and International Harvester Co. This latter was the result of some experiments by Carl Mott, designing engineer for this project at that time. See page 64—Some motor cultivators (wheel type). These light machines, which were initiated about 1916, were soon to be replaced by the all-purpose tractor.

With close of World War I in 1919, the steel quotas which had restricted automobile manufacture were removed, and many automotive manufacturers promptly deserted the tractor field and returned to the manufacture of cars and trucks.

At Kansas City the fourth annual national tractor show was held at which 85 models were shown. A survey of these models indicates the trend among tractor manufacturers of that time. Sixty-four models had channel frames, and 19 had unit type frames, which were soon to be the accepted type of frame construction; 49 had vertical motors and 24 horizontal; 64 models had engines with 4 cylinders, 6 had 2 cylinders, 2 had 1 cylinder and 1 had 6 cylinders.

In 1919 tractors pulling plows were tested at several locations in Ohio by Ohio State University. These tests resulted in some interesting comparisons (page 59).

In the same year the State of Nebraska passed "a bill to provide official tests for gas, gasoline, kerosene, distillate, or other liquid fuel traction engines in the State of Nebraska and to compel the maintenance of adequate service stations for same."

Previous to drawing up this bill there had been considerable agitation, beginning about 1915, by the farm equipment industry and others that there be a standardized rating for tractors and that the Federal Government should establish a National Testing Station. It was "touch and go" as to whether the Bureau of Standards in the Department of Commerce or the office of Farm Management under the Department of Agriculture was to do the job, providing funds became available. Then a bill at the request of the American

Official Report of Results of Tractor Tests made by the Agricultural Engineering Department, Ohio State University

Tractor	Rating	Cly.	Motor Size	R. P. M.	Type	Position of Wheels	Wt.	Plow'g Speed	Plows	Previous Crop	Footing	Normal Test: Lbs. Pull	Normal Test: Rate of Travel Ft. Min.	Normal Test: Rate of Travel Mi. Hr.	Normal Test: H. P.	Normal Test: Depth Plowed	Maximum Test: Lbs. Pull	Maximum Test: Rate of Travel Ft. Min.	Maximum Test: Rate of Travel Mi. Hr.	Maximum Test: H.P.	Maximum Test: Depth Plo'd.
COLUMBUS TESTS. Clay Loam. Soil, Hard and Dry Heavy Sub-Soil																					
Cletrac	12-20	4	3¾x5½	1250	Crawler	On land	3150	3½	2-14"	Wheat	Good	1448	243.6	2.76	10.69	7.69	1892	233.3	2.65	13.38	9.40
J. T.	16-30	4	5 x6½	1000	Crawler	On land	6900	2½	3-14"	Wheat	Good	2038	210.0	2.39	12.95	7.62	3450	110.0	1.25	11.50	9.47
Avery	12-25	2	6½x7	570	4-wheel	On land	7500	1¾	3-14"	Wheat	Good	2235	153.2	1.74	10.38	7.65	2710	156.2	1.37	11.29	8.62
Case	15-27	4	4½x6	950	4-wheel	On land	5600	2¼	3-14"	Wheat	Good	1947	220.0	2.50	12.98	8.06	2720	187.5	2.13	15.45	8.60
Moline-Univ	9-18	4	3½x5	1600	2-wheel	In furrow	3380	3½	2-14"	Wheat	Good	2275	272.0	3.09	18.72	8.20	2705	203.3	2.31	16.67	9.22
Fordson	10-20	4	4 x5	1000	4-wheel	In furrow	2700	3	2-14"	Wheat	Good	1304	272.5	3.09	10.78	7.68	1687	173.8	1.97	8.88	8.25
La Crosse	12-24	2	6 x7	750	4-wheel	In furrow	4200	2½	3-14"	Wheat	Good	1748	205.5	2.33	10.88	7.00	1838	190.7	2.16	10.62	7.65
La Crosse	12-24	2	6 x7	750	3-wheel	In furrow	4000	2½	3-14"	Wheat	Good	1926	198.5	2.26	11.58	8.06	2000	192.0	2.19	11.65	8.31
Case	10-18	4	3⅞x5	1050	4-wheel	On land	3600	2¼	2-14"	Wheat	Good	1418	214.0	2.43	9.22	7.30	1475	182.6	2.07	8.16	7.75
Illinois	18-36	4	5 x6½	800	4-wheel	In furrow	5200	2½	4-14"	Wheat	Good	2540	191.5	2.17	14.87	7.94	3125	175.0	1.98	16.59	8.12
Wellington	10-20	4	4 x6	1000	4-wheel	In furrow	4750	3	2-14"	Wheat	Good	1418	247.0	2.81	10.61	7.75	1584	254.2	2.89	12.18	8.60
MIDDLETOWN TESTS. Sandy Loam. Wet Surface. Hard, Dry Sub-Soil. Raining During Tests of Reliable and Wallis.																					
Huber	12-25	4	4½x5¾	1000	4-wheel	In furrow	5000	2½	3-14"	Wheat	Fair	1693	250.0	2.84	12.86	7.83	2158	205.0	2.33	13.40	8.94
Parrett	12-25	4	4¼x5½	1000	4-wheel	In furrow	5200	2⅜	3-14"	Clover	Fair	2020	200.0	2.27	12.25	6.75	2074	204.0	2.32	12.82	7.66
Reliable	10-20	2	6 x7	600	4-wheel	In furrow	3800	2	2-14"	Wheat	Poor	893	168.3	1.91	4.55	6.25	1000	172.0	1.96	5.20	6.86
Waterloo-Boy	12-25	2	6½x7	750	4-wheel	In furrow	5900	2½	3-14"	Clover	Fair	1905	210.0	2.39	12.13	7.00	2480	147.8	1.68	11.10	8.69
Wallis	15-25	4	4¼x5¾	850	4-wheel	In furrow	3250	3	3-14"	Wheat	Poor	1900	264.0	3.00	15.20	7.18	2300	216.0	2.45	15.05	8.58
Titan	10-20	2	6½x8	500	4-wheel	On land	5525	2½	3-14"	Wheat	Poor	1690	146.0	1.66	7.47	7.06	1800	140.0	1.59	7.63	7.22
International	8-16	4	4 x5	1000	4-wheel	On land	3300	2½	2-14"	Wheat	Poor	1440	186.0	2.11	8.12	7.46	1600	175.0	1.99	8.48	7.68
Ault-Taylor	15-30	4	4¾x6¾	800	4-wheel	In furrow	7800	2	4-14"	Wheat	Poor	2160	184.3	2.09	12.08	6.52	2800	176.6	2.05	14.98	7.19
Monarch	18-30	4	4¾x6	800	Crawler	On land	7400	2¼	3-14"	Wheat	Fair	2010	248.0	2.82	15.12	7.25	2600	202.0	2.29	15.91	8.42
E-B	12-20	4	4¾x5	900	4-wheel	On land	4355	2⅓	3-14"	Wheat	Poor	1820	215.0	2.44	11.86	7.66	2130	170.0	1.93	10.96	8.53
Shelby	9-18	4	3¾x5¼	1100	4-wheel	In furrow	3600	1¾	2-14"	Wheat	Poor	1090	194.3	2.21	6.42	6.83	1240	181.8	2.06	6.83	7.50
Hart-Parr	-30	2	6½x7	750	4-wheel	In furrow	5100	3	3-14"	Wheat	Poor	1520	330.0	3.75	14.46	6.71	1510	322.0	3.66	14.77	8.08
Heider	12-20	4	4½x6¾	750	4-wheel	In furrow	6000	2¼	3-14"	Wheat	Poor	1922	168.2	1.91	9.81	7.56	1923	164.0	1.86	9.56	7.75
Heider	9-16	4	4¼x5¾	800	4-wheel	In furrow	4000	2¼	2-14"	Wheat	Poor	1290	180.0	2.04	7.03	7.53	1312	182.5	2.07	7.26	8.25
Bates Steel Mule	15-22	4	4 x6	900	Crawler	On land	3500	2½	3-14"	Wheat	Poor	2212	209.0	2.37	14.02	7.75	2455	203.2	2.31	15.10	8.91
Frick	15-28	4	4½x6	900	4-wheel	In furrow	6000	2¼	3-14"	Wheat	Poor	1640	250.0	2.84	12.42	7.17	2028	236.1	2.68	14.52	8.00
Whitney	9-18	2	5½x6½	750	4-wheel	In furrow	3000	2½	2-14"	Wheat	Poor	1295	204.0	2.31	8.02	6.87	1357	191.5	2.17	7.88	7.16
FOSTORIA TESTS. Heavy Dry Loam, Hidden Rocks.																					
Whitney	9-18	2	5½x6½	750	4-wheel	In furrow	3000	2½	2-14"	Timothy	Ex'lent	1420	230.0	2.61	9.00	6.60	1520	231.5	2.63	10.66	6.62
Frick	15-28	4	4½x6	900	4-wheel	In furrow	6000	2¼	3-14"	Timothy	Ex'lent	2470	195.0	2.21	14.58	8.40	2560	192.2	2.18	14.92	8.67
Bates Steel Mule	15-22	4	4 x6	900	Crawler	On land	3500	2½	3-14"	Timothy	Ex'lent	2600	182.5	2.07	14.39	7.80	2710	185.7	2.11	15.25	7.95
Shelby	9-18	4	3¾x5¼	1100	4-wheel	In furrow	3600	1¾	2-14"	Wheat	Fair	1500	134.5	1.52	6.12	7.35	2100	125.2	1.42	7.97	7.95
Heider	12-20	4	4½x6¾	750	4-wheel	In furrow	6000	2¼	3-14"	Wheat	Fair	1972	180.0	2.04	10.76	6.37	2040	172.5	1.96	10.66	7.02
Hart-Parr	-30	2	6½x7	750	4-wheel	In furrow	5100	3	3-14"	Timothy	Ex'lent	2760	288.0	3.27	24.08	9.92	3220	271.5	3.08	26.50	10.25
E-B	12-20	4	4¾x5	900	4-wheel	On land	4355	2⅓	3-14"	Timothy	Ex'lent	2600	167.5	1.90	13.20	7.95	3000	147.5	1.67	13.42	9.71
Monarch	18-30	4	4¾x6	800	Crawler	On land	7400	2¼	3-14"	Wheat	Good	2778	175.0	1.99	14.73	8.57	3300	147.2	1.67	14.72	8.87
Ault-Taylor	15-30	4	4¾x6¾	800	4-wheel	In furrow	7800	2	4-14"	Timothy	Ex'lent	2800	185.0	2.12	15.69	7.85	2930	180.5	2.05	16.02	8.05
International	8-16	4	4 x5	1000	4-wheel	On land	3300	2½	2-14"	Wheat	Good	1642	187.5	2.15	9.31	7.41	1800	138.2	1.57	7.54	7.60
Titan	10-20	2	6½x8	500	4-wheel	On land	5525	2½	2-14"	Wheat	Good	1845	171.5	1.95	9.60	7.02	1990	200.7	2.28	12.10	7.67
Wallis	15-25	4	4¼x5¾	850	4-wheel	In furrow	3250	3	3-14"	Wheat	Good	2260	268.0	3.05	13.35	8.67	2287	285.8	3.25	19.82	9.42
Waterloo-Boy	12-25	2	6½x7	750	4-wheel	In furrow	5900	2½	3-14"	Wheat	Good	2480	210.0	2.39	15.78	7.12	2710	176.0	2.00	14.43	7.95
Reliable	10-20	2	6 x7	600	4-wheel	In furrow	3800	2	2-14"	Wheat	Good	1578	165.1	1.88	7.90	8.72	1575	168.0	1.91	8.02	8.72
AKRON TESTS. Sandy Loam, Dry and Hard, Hidden Rocks.																					
Parrett	12-25	4	4¼x5½	1000	4-wheel	In furrow	5200	2⅜	3-14"	Timothy	Good	1802	240.0	2.73	13.11	8.02	1815	237.0	2.69	13.04	8.35
Huber	12-25	4	4½x5¾	1000	4-wheel	In furrow	5000	2½	3-14"	Wheat	Poor	1962	193.0	2.19	11.47	7.62	1818	226.0	2.56	12.45	8.25
Wellington	10-20	4	4 x6	1000	4-wheel	In furrow	4750	3	2-14"	Wheat	Poor	1666	196.0	2.22	9.89	8.26	1777	180.0	2.04	9.70	8.78
Illinois	18-36	4	5 x6½	800	4-wheel	In furrow	5200	2½	4-14"	Wheat	Poor	2730	191.0	2.17	13.72	7.62	2480	196.0	2.23	14.73	8.02
Case	10-18	4	3⅞x5	1050	4-wheel	On land	3600	2¼	2-14"	Wheat	Poor	1170	240.0	2.73	8.50	7.81	1290	238.2	2.70	9.33	8.04
La Crosse	12-24	2	6 x7	750	3-wheel	In furrow	4000	2½	3-14"	Wheat	Fair	2020	187.5	2.13	11.48	8.42	2160	196.5	2.23	12.85	8.55
La Crosse	12-24	2	6 x7	750	4-wheel	In furrow	4200	2½	3-14"	Wheat	Fair	2020	182.5	2.07	11.17	8.45	2080	182.5	2.07	11.51	8.70
Fordson	10-20	4	4 x5	1000	4-wheel	In furrow	2700	3	2-14"	Wheat	Fair	1320	222.5	2.53	8.90	7.75	1540	165.0	1.87	7.70	8.76
Moline-Univ	9-18	4	3½x5	1600	2-wheel	In furrow	3380	3½	2-14"	Wheat	Poor	1643	270.0	3.07	13.45	8.32	1708	253.0	2.87	13.09	9.02
Case	15-27	4	4½x6	950	4-wheel	On land	5600	2¼	3-14"	Wheat	Poor	1848	205.5	2.35	11.60	8.40	1995	197.0	2.24	11.70	9.08
Avery	12-25	2	6½x7	570	4-wheel	On land	7500	1¾	3-14"	Timothy	Fair	1403	224.0	2.55	9.52	8.02	1950	153.0	1.74	9.04	8.45
Heider	9-16	4	4¼x5¾	800	4-wheel	In furrow	4000	2¼	2-14"	Wheat	Poor	1266	178.7	2.03	6.86	6.92	1370	156.9	1.78	6.52	7.85
J. T.	16-30	4	5 x6½	1000	Crawler	On land	6900	2¼	3-14"	Timothy	Fair	2070	189.0	2.15	11.85	8.87	2285	187.5	2.13	12.99	10.07
Cletrac	12-20	4	3¾x5½	1250	Crawler	On land	3150	3½	2-14"	Wheat	Fair	1095	262.0	2.98	8.69	7.90	1688	258.0	2.93	13.19	8.85

Society of Agricultural Engineers was introduced favoring the "creation of a separate commission under the Department of Agriculture, part of whose job would be the testing of tractors, to be known as the Bureau of Agricultural Engineering." The ASAE in 1917 had already adopted a code of tractor testing that had been set up previously by the Society of Automotive Engineers. (Farm Implement News, September 25, 1952) Several requests for funds for initiating the tests were made but without success.

In commenting on this matter March 19, 1952, Robert A. Jones, Executive Secretary of the Farm Equipment Institute indicated, there was a statement by T. H. MacDonald, prepared for a meeting of the Tractor & Thresher Department, in December 1919 wherein he said, "A little bit in the historical line, as early as 1915, in December, the Bureau of Public Roads & Rural Engineering, of the Department of Agriculture, called a conference of representatives of manufacturers and of the agricultural engineers and one or two government departments . . . At that time the question of a Federal rating of tractors was taken up, and a brief outline was made, and fairly definite plans for the work formulated."

Mr. Jones continued "Mr. MacDonald's statement went on to declare that appropriations had been requested regularly, but failed to carry."

The following excerpt from an article entitled History of the Nebraska Law by Prof. C. W. Smith throws further light on the legislation:

> "The Nebraska Tractor Law was introduced in the House of Representatives of the Nebraska State Legislature by Representative Wilmot F. Crozier of Polk County. Mr. Crozier was born on his father's (John W. Crozier's) Polk County homestead in 1883. He completed 8 grades in the country school, District 68; then graduated from the Osceola High School in 1903. He then taught country school two years. He continued his education by first going to Taylor University from which he graduated in 1909 with an A. B. degree. After passing a U.S. Civil Service examination he was sent to the Philippine Islands for four years as Division Supt. of Schools for the Province of Palawan. Some interesting notes relative to this experience are (a) that his salary was $100.00 per month, (b) he hired native teachers for the equivalent of $4.00 per month, (c) he met General John J. Pershing on several occasions who was in charge of U.S. Armed Forces in the Philippines at the time, (d) met his future wife, Maud Thurber, who was teaching at the time in the Philippines. They were married in Los Angeles in 1916.
>
> "Mr. Crozier returned to the Farm in Polk County. He was elected three times to the State Legislature, each time for a biennium, 1919-20, 1931-32, and 1933-34. He was defeated for reelection for the biennium 1935-36.
>
> "He could fore-see mechanization of agriculture and was one of the pioneer tractor farmers. He says the first tractor he bought was called a "Ford" but had no connection with Henry Ford. He purchased this tractor in 1916. It gave him so much trouble that he demanded a replacement. The replacement in 1917 was no better than the original. He then purchased (1917) a second-hand Bull tractor. This was unsatisfactory. In 1918 he purchased a so-called three-plow Rumely Oil-Pull tractor. He pulled five bottoms satisfactorily with this tractor which gave him absolutely no trouble. This caused Mr. Crozier to ask himself the question, "If one manufacturer can build a dependable tractor why can not all tractors be dependable?" As he "puts it" he returned to teaching in the early 20's to support the farm. At that time he sold the Rumely tractor and 25 years later that tractor was operating satisfactorily. During this period he was Supt. of Schools at White Salmon, Washington, took one year of graduate work at Stanford University; then taught Math 2 years in the Delano High School, California. Following this period of 7 years in the educational field he again returned to the farm in Polk County."

Here he was when Prof. Smith contacted him again in October 1953. Professor Smith continues — "This brief history of Mr. Crozier is given to correct any misconception that the Nebraska Law was introduced by a hay-seed who by chance found himself elected to the State Legislature. To the contrary Mr. Crozier is a good conversationalist; he appreciates and uses good English and has complete confidence in his own ability to formulate the wording of a law to accomplish what he has in mind. A false rumor was circulated that Mr. Crozier was elected to the Legislature on a Tractor-Testing Platform. This was false because Mr. Crozier says he never thought of a tractor testing bill until after he was elected to the Legislature: that to his memory the incident that led to his introducing the bill was his reading an editorial in a farm paper. This editorial in commenting on the undependable character of many tractors said that unless industry corrected that condition it would be necessary to have some regulation of tractor sales.

"It was necessary in those days of bi-cameral Legislature to have a bill presented in both chambers. Our present (1953) Lieutenant Governor Charles Warner was a member of the State Senate at that time and Mr. Crozier is very useful to explain that Charles Warner, in sponsoring the bill in the Senate should have equal credit with himself for without passing both houses it could not become law.

"Mr. Crozier was asked by the Implement and Tractor Trade Journal of Chicago to prepare an article for the Journal explaining why he introduced the Tractor Testing Bill in the Nebraska State Legislature. He compiled with this request and his article was published in the Journal in September, 1919. . . ."

Oscar W. Sjogren, who in those days was closely associated with L. W. Chase, Head of the Department of Agricultural Engineering, recollects the start of the tests as follows: "The man who sparked the movement was a farmer in Polk County, Nebraska, by the name of Wilmot F. Crozier. He had some unhappy experiences with tractor operation. . . .

"He introduced a bill the provisions which were that no tractor could be legally sold in the State of Nebraska without a permit, that such a permit would be issued by the State Railway Commission

Nebraska test car and tractor (Twin City 12-20) under test in 1919, delayed by snow

Nebraska test car and tractor (Case VAC) under test in 1949

upon certification of a Board of Engineers that such tractor had been tested by the University of Nebraska Agricultural Engineering Department and that the results of such test had been compared to the published claims of the manufacturer covering that tractor. If the claims were not substantiated by the test, then no permit could be issued. Provision was made for sale under temporary permit good until such time as the tractor could be tested and reported on. No tractor could be sold without a permit, under threat of penalty. This bill passed and became known as the 'Nebraska Tractor Test Law.'

"The program of organizing and conducting the testing procedure became the responsibility of the Agricultural Engineering Department of which L. W. Chase was then the Head. He was then on a leave of absence being a Major in the U.S. Army stationed in Washington, D.C. He returned soon after the bill was passed and with his usual energy and organizing ability set about to get the testing program under way. He induced Claude Shedd to come over from Ames, Iowa and take over as chief engineer of the project. This task was not an easy one. He had to direct the setting up of the test procedure and the design of the equipment necessary for conducting the drawbar tests. This required about a year's time so it was in the spring of 1920 when all was ready and the first complete test was run. . ." This was with the Waterloo Boy, "the Model N" 12-25.

An attempt had been made in the fall of 1919 to test a Twin City 12-20 but was interrupted by snow (page 61), postponed to March 21, 1920, and completed on April 4. When finally completed this was listed as test No. 19. Sixty-five tests of the 69 started were completed this year. The Nebraska tests for tractors have since become significant not only in Nebraska and the nation as a whole, but in a score of foreign countries. These tests have provided standards by which tractors are rated, and were responsible for improvements in tractors generally. They also eliminated many tractors of inferior design. (See page 61 for the 1949 Nebraska test car and tractor.)

(See page 2, Part II, 1920 Nebraska Test Report)

Some 1919 Models

1919 Aultman-Taylor

22-45 H.P., 600 r.p.m., 4-cylinder horizontal double opposed engine. Lubrication by mechanical oiler; cooling with own horizontal tubular radiator, pump, and fan. Magneto high-tension ignition; expanding shoe clutch; sliding gear transmission; spur-gear final drive; 2 forward speeds, 2.13 and 2.93 m.p.h.

1919 Nilson Sr.

22-45 H.P., 500 r.p.m., 4-cylinder vertical engines. Lubrication, constant level splash; high-tension magneto ignition with impulse starter; electric starter and lighting equipment available; cooling-pump, fan and radiator. Sliding-gear transmission; final drive, double-roller chain; five wheels with three rear drive wheels; 2 speeds forward, 2.5 and 5.5 m.p.h.

1919 Appleton 12/20

12-20 H.P., 1,050 r.p.m., Buda 4-cylinder vertical motor. Centrifugal pump, fan, and radiator cooling, motometer; lubrication, force feed and splash; ignition, Bosch high-tension; Borg and Beck disc clutch; sliding-gear transmission; final drive to each rear wheel through pinion and internal-type bull gear; 2 forward speeds, 2 and 3-1/2 m.p.h.

1919 Frick

12-25 H.P., 900 r.p.m., 4-cylinder vertical engine mounted crosswise. Lubrication, automatic splash; ignition, high-tension magneto with impulse starter; sliding-gear transmission; final drive, bull gear and pinion to each rear wheel; 2 forward speeds, 2.3 and 3.8 m.p.h.

1919 Fitch Four Drive

20-35 H.P., 900 r.p.m., climax 4-cylinder vertical motor. Lubrication, force feed through drilled crankshaft; high-tension magneto ignition with impulse starter; water-type air cleaner; four driving wheels; Cotta transmission; final drive, enclosed gears through two-piece live axles, front and rear; 3 forward speeds, 1.5 to 4 m.p.h. Also produced in 12-25, 12-24, 12-20, and 15-28 horsepower sizes beginning about 1916.

1919 Russel 20/40

20-40 H.P., 825 r.p.m., 4-cylinder vertical motor. High-tension magneto ignition with impulse starter; pump, radiator, and fan cooling system; force feed and splash lubrication; throttle governor; sliding-gear transmission; spur gear final drive; 2 forward speeds, 2.4 and 3.7 m.p.h.

1919 Victor

34 B.H.P., 850 r.p.m., 4-cylinder vertical engine mounted inside drive wheels. Front two wheels drivers, and small steering wheel at right rear; sliding-gear transmission; final drive, roller pinions to internal gear in each drive wheel; 2 forward speeds, 2-1/3 and 3-1/2 m.p.h.

1919 La Crosse 12/24

12-24 H.P., 750 r.p.m., 2-cylinder horizontal engine; force-feed lubrication; dry-cell battery ignition; sliding-gear transmission; final drive, roller pinion and internal gears; forward speed, 2.5 m.p.h.

1919 Post

12-20 H.P., 800 r.p.m., 4-cylinder vertical Waukesha motor. High-tension magneto with impulse starter; thermosyphon cooling; chain and bevel gear drive; traction—two centrally placed drive and steering wheels, one front and one rear; one "outrigger" support wheel on each side; 2 forward speeds, 1-3/4 and 5 m.p.h.

1919 Chase 12/25

12-25 H.P., 1,000 r.p.m., Buda 4-cylinder vertical engine. Lubrication, force feed through drilled crankshaft; cooling, pump, fan, and radiator; ignition, high-tension magneto with impulse starter; water-type air cleaner; Bierman clutch; transmission; first reduction, enclosed chain, second and final reduction, spur gears; final drive, spur gear; 2 forward speeds 1-3/4 and 2-1/2 m.p.h.

1919 Twin City 12/20

12-20 H.P., 1,000 r.p.m., 4-cylinder vertical engine mounted longitudinally. Ignition, high-tension magneto with impulse starter; cast-iron frame, unit construction; sliding-gear transmission; final drive, enclosed spur gear; belt pulley side mounted, 650 r.p.m.; forward speed, 2.2 to 2.9 m.p.h.

1919 Samson Iron Horse

4-cylinder vertical motor. High-tension ignition; transmission belt drive. Forward and reverse speeds with independent control for each side of tractor. No differential needed. All four wheels were drivers. Chain and sprocket final drive. Control was by levers acting as belt tighteners, to which could be attached reins permitting it to be operated from a distance, as from corn planter or binder.

Some Motor Cultivators (Wheel Type)

1916 International

15 H.P. 2-5/8 x 3 in. - 4 cylinder vertical Continental automobile type engine; high tension ignition; pump, fan and radiator cooling; final drive to rear wheels from engine by vertical shaft and bevel gears. One speed forward and one reverse controlled by jaw-clutch operated by separate levers - one forward, one for reverse. Power take-off was built into later models.

1918 Allis-Chalmers

6-12 H.P. 800 r.p.m. 4 cylinder vertical LeRoi motor, cast en bloc and mounted lengthwise. Plunger pump and gravity lubrication; thermosyphon cooling. Eisemann or Dixie high tension ignition. Sliding gear transmission enclosed, and final drive, exposed bull pinion and internal gears. Speed forward 1-1/2 to 2-4/5 m.p.hr.

1918 Bailor

Cushman motor, 2-cylinder 4 x 4 vertical engine. Gravity and splash lubrication; pump, fan, and radiator cooling system; sliding gear transmission; open chain final drive; speed forward 1-1/4 to 3 m.p.hr.

1918 Indiana

5-1 H.P., 4-cylinder vertical engine, cast en bloc; lubrication force-feed and splash; high tension ignition; cooling system radiator and fan; selective transmission; open chain final drive. Forward speed 1-1/2 to 4 m.p.h. Belt pulley.

1919 Parrett

6-12 H.P., 1,000 r.p.m. 4-cylinder vertical LeRoi motor, mounted lengthwise. Triple friction drive. Front wheel castor; steering controlled by hand wheel. Two gear driven drive wheels, outside the two rows to be cultivated. Forward speed 1-1/2 to 3.5 m.p.hr.

1919 Avery

5-10 H.P. 1200 r.p.m. Own L head, four or six cylinder motor cast en bloc. Splash lubrication; thermo-syphon cooling; K.W. high tension ignition; enclosed transmission; exposed double spur gear final drive. Speed forward 1-5/8 to 4-2/3 m.p.hr.

1919 Emerson-Brantingham

16 H.P. at 1000 r.p.m., four cylinder vertical LeRoi motor mounted lengthwise. Triple fraction drive. Front wheel castor, steering controlled by hand wheel. Two gear driven drive wheels, outside the two rows to be cultivated. Forward speed 1-1/2 to 3.5 m.p.hr.

1919 Traylor

6-12 H.P. 4 cylinder vertical LeRoi motor, 3-1/8 x 4-1/2. One plow or one row cultivator. Circulating splash lubrication; thermo-syphon cooling; final drive open using roller pinion; one speed enclosed transmission.

VI. Accessories and Component Parts

The early tractors were used principally for plowing the Buffalo range prairies extending from Oklahoma to Edmonton, Canada, which were being converted to growing wheat. Although some use was being made of them, in season, for driving threshers, it was not until about the middle of the 1909-20 period that their use for the purpose became general. Belt pulleys were available ordinarily only as special features on most models, and it was several years before the pulley became standard equipment.

The steam tractor was kept for threshing work long after it ceased to be a factor in plowing but was gradually replaced by gas tractors. With the prairies largely broken up demand for large tractors fell off and with the outbreak of the First World War in August 1914 labor to operate them began to get scarce. The next few years saw small one-man tractors in demand.

While there were many structural defects in engines and power transmission parts early engine failure were, in a large measure, due to lack of know-how, and poor oil offered to the trade, plus poor lubrication systems.

World War I, however, created its demand for tractive power and accelerated the development of tractors so that many of these difficulties were reasonably overcome.

The enclosed gearing in transmission cases, the use of heat treated steels, and the introduction and use of anti-friction bearings along about 1915 to 1920 produced a new era in the tractor design and construction which followed.

In 1918 the United States Department of Agriculture sent questionnaires to over 2,000 tractor operators requesting a report on that part of the tractor which gave the most trouble. Over 900 replies were classified as follows: magneto 299, spark plugs 110, gears 108, carburetors 104, bearings 80, cylinder and pistons 61, clutch 59, valves and springs 43, lubrication 29, starting 28. These figures indicate that the major troubles were with the ignition system, the combined ignition troubles constituting nearly half of the total classified troubles.

Ignition (pages 66-67)

Faulty ignition, resulting in failure of engines to start, was probably the first major difficulty to overcome in improving the gasoline tractor. Weak batteries, poor contact points, poorly insulated electrodes, weak magnets, damp spark coils, improper buzzer adjustments on vibrating coils, poorly insulated low and high tension wires and many other defects contributed to the troubles of the gas tractor operator's lot. A long tiring struggle, particularly on cold mornings, was often required to get the engine going, and once it was started it was frequently kept running through the "noon hour" for fear it could not be restarted.

The development of the high-tension magneto, generally fitted with impulse coupling, and the use of spark plugs solved the major part of ignition troubles. Some of the early magnetos of this type were the K.W., Dixie, Eisemann, Splitdorf, American Bosch, Simms, and the Michigan, and appeared on tractors about 1915. Prior to this time, dry cell batteries and spark coils connected to mechanically operated "ignitor plugs" or ignitors were used for the starting spark, after which a generator or magneto (often friction or belt driven) was switched into the circuit by use of a two-way switch. In low tension or "make-and-break" systems, oscillating magnetos, such as the Webster, Accurate and Wico, were frequently used.

The early insulation of the "fixed" electrode in the ignitor was a weak spot in the low-tension ignition

Some Early Ignition Systems

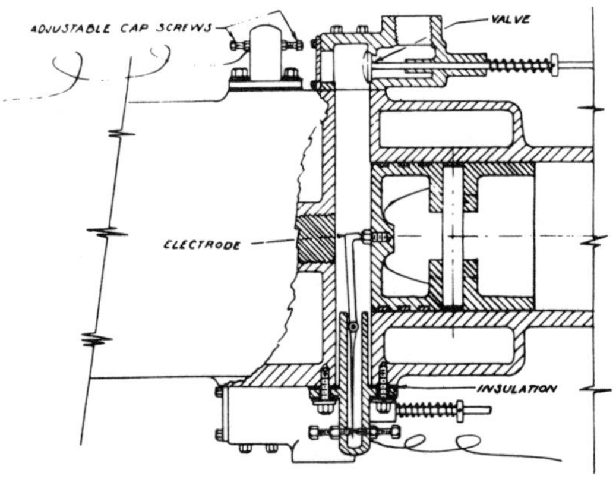

Ignition system used on Paterson's Balanced engine, 1892

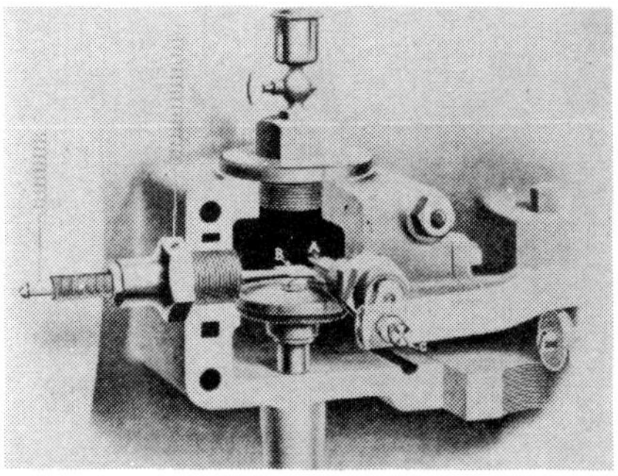

Wipe sparking device, about 1900

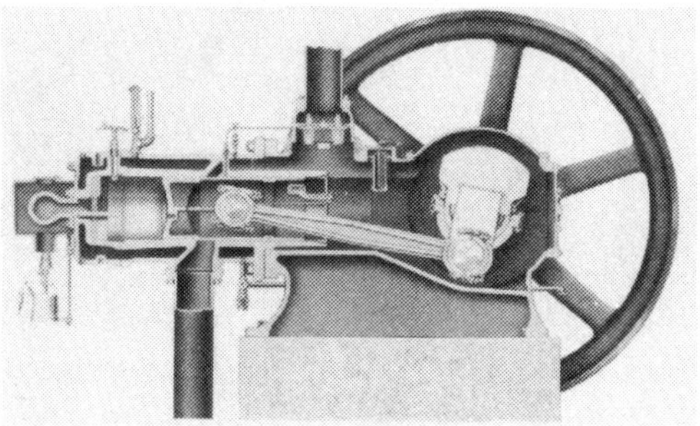

"Hot-bulb" ignition, about 1900

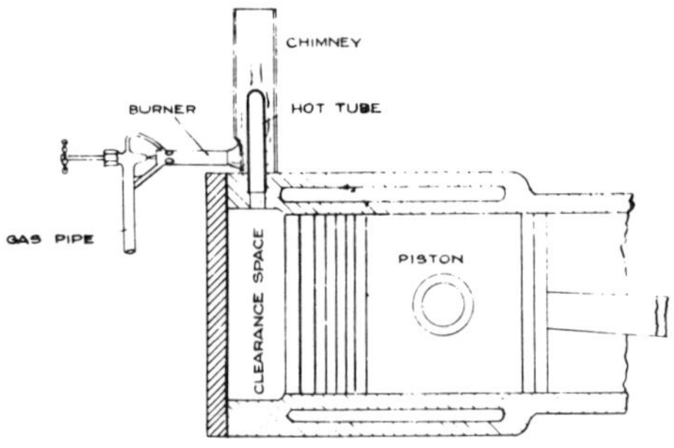

"Hot-tube" ignition, about 1900

View showing mounting of Auto Sparker

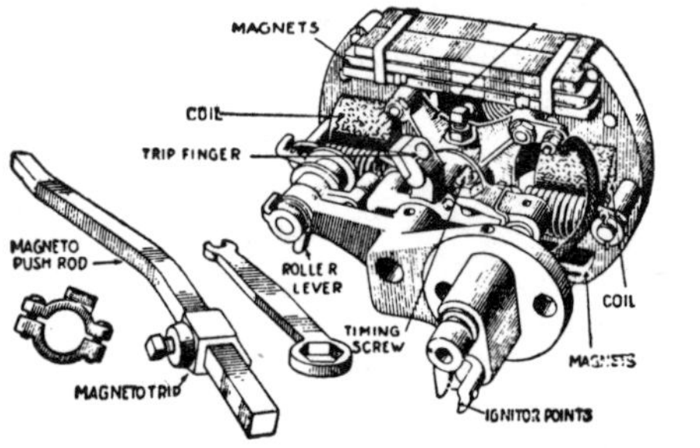

Oscillating low-tension magneto and integral ignitor, about 1908

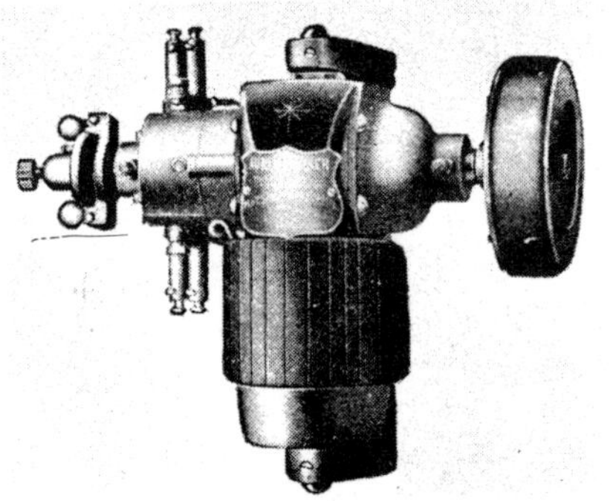

"Auto Sparker", 1910

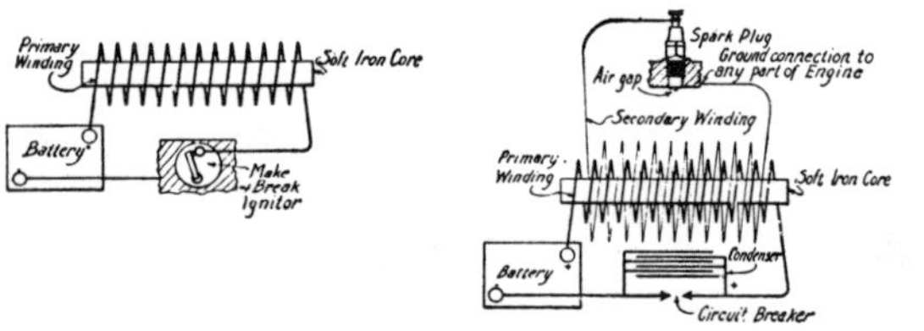

Principles involved in two electric ignition systems: "make and break" (low voltage) for slow-speed engines, and "jump-spark" (high voltage) for high-speed engines

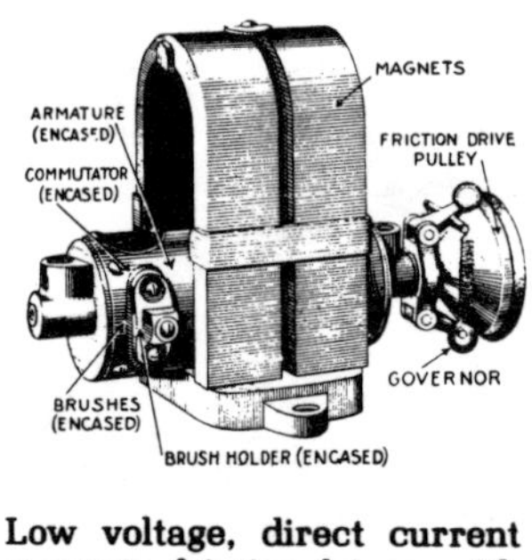

Low voltage, direct current magneto, friction driven, with governor, about 1910

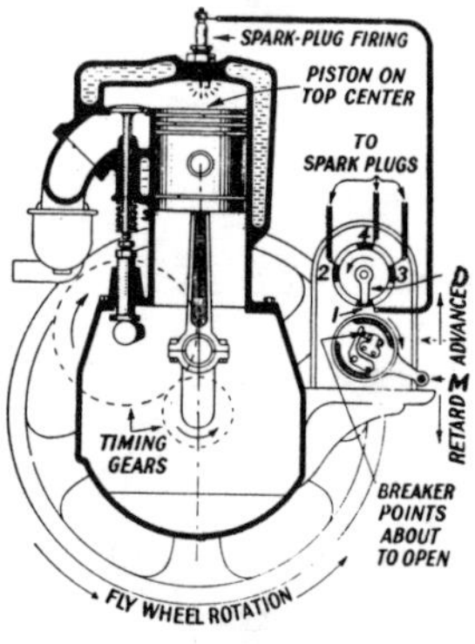

Ignition timing detail (D) distributing arm (M) breaker case arm

Some Early Ignition Systems

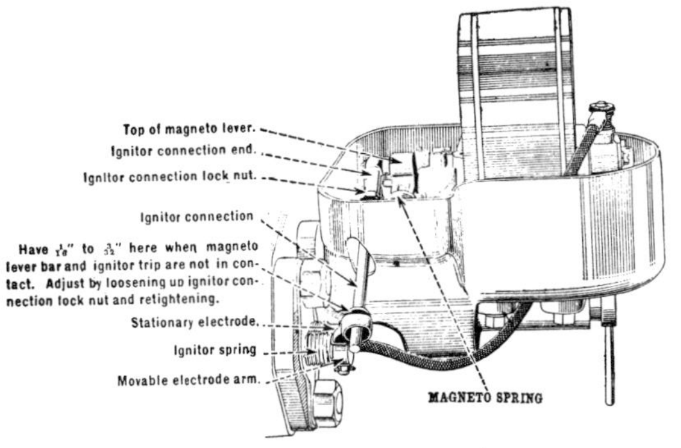

Low tension oscillating magneto and ignition mechanism

Oscillating magneto mounted on tractor engine

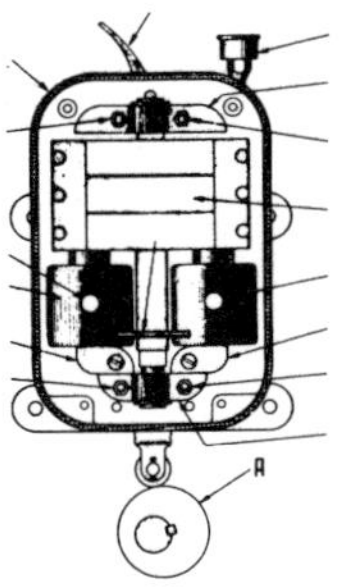

Wico igniter - cover removed. 1916. (actuated by eccentric "A" of camshaft)

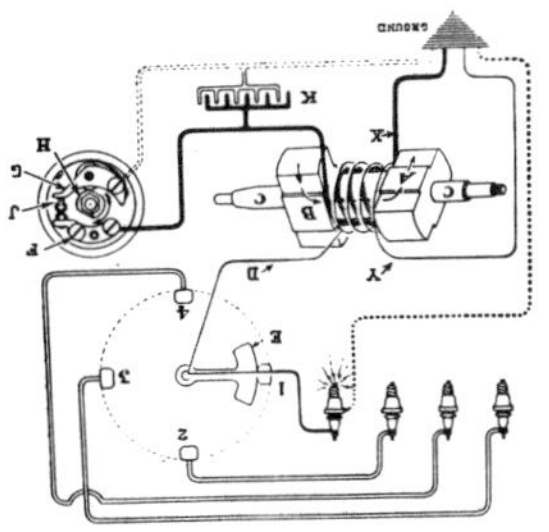

Induction type magneto and wiring for ignition system of a four-cylinder engine, 1916.
(A) and (B) rotor; (E) rotating contact arm of distributor; (F) breaker fixed arm; (G) breaker movable are; (H) breaker cam; (J) breaker points; (K) condenser; (X) primary ground; (Y) secondary ground

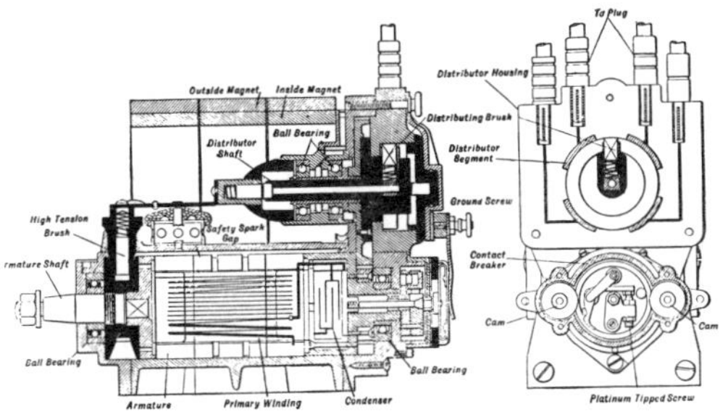

Section view of armature or shuttle type magneto for four cylinder engines

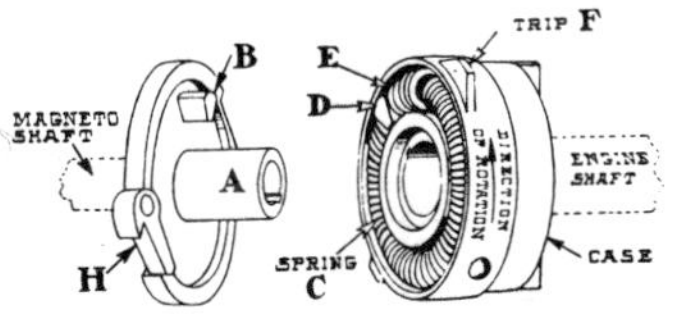

Diagram showing the principle of the K-W Impulse Starter, 1916. Spring (E) is a so-called cushion spring to absorb the shock of the operation, prevent breakage of the pins (B) and (D), and it also keeps the magneto in proper time

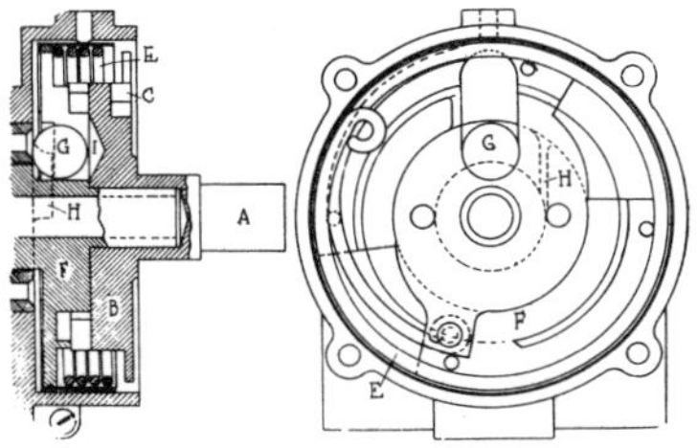

U & H Impulse Starter, about 1916. Spring (E) is wound by drive shaft (A) until steel ball (G) drops into conical depression (I). This releases spring which unwinds and gives armature speed

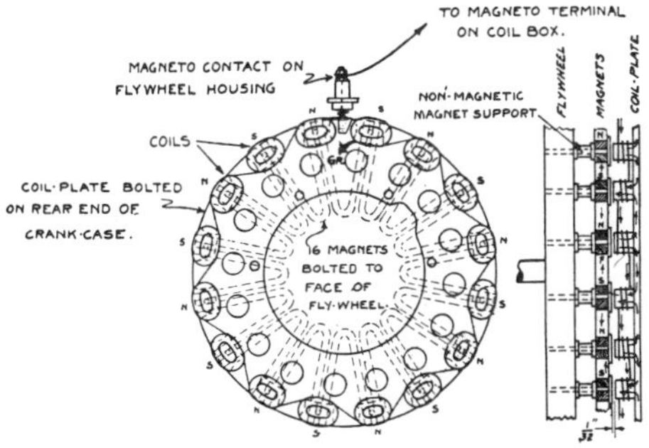

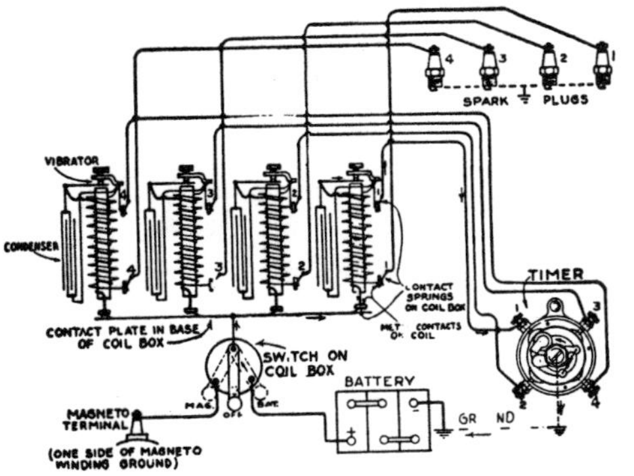

Diagram of four-cylinder ignition system using vibrating coil with fly wheel magneto, about 1916

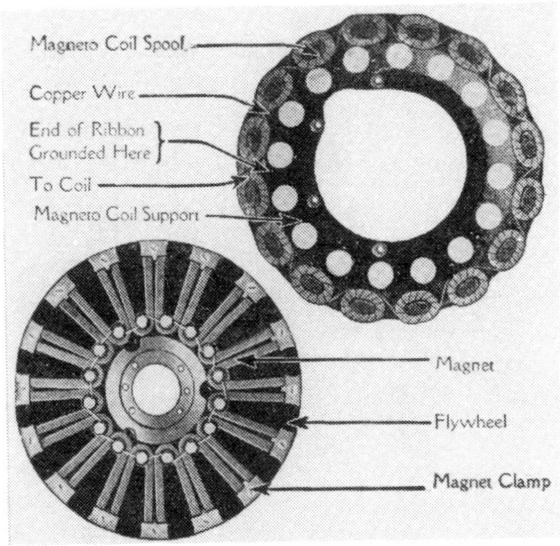

Fly wheel magneto, about 1916

systems. Sheet mica was wrapped around the electrode and soon resulted in a "short", which would prevent producing the necessary hot spark at the contact points. Later this condition was improved by using tightly held mica washers the outside of which was ground to a smooth finish and prevented the entrance of moisture and dirt.

Lubrication (page 69)

Another very important factor in the unsatisfactory performance of early tractors was the lack of proper lubrication.

The inadequate early lubrication systems and the frequent use of poor quality engine oils, which were unloaded on unsuspecting farmers by "slickers" in the early days, caused cylinders, pistons, rings, bearings, and gears to score and wear rapidly. "Seized pistons" and "hot boxes" were common. Before pressure oiling was used, frequent adjustment of connecting rods and main bearings was necessary. In fact, on some of the larger tractors these bearings required daily, or at least weekly inspection and adjustment, and plate-covered openings in crankcases were provided for the purpose. At first, lubrication of the cylinder and main bearings, and frequently bull gears, was accomplished with a drip or sight-feed oiler. This method was erratic in operation as the oil in the container when warm fed out rapidly and slowly when cold, or failed entirely.

Power-driven plunger pumps with metering control of oil (mechanical oilers) were soon developed to replace the erratic drip oiler. Leaders in this field of development included Madison-Kipp Co. of Madison, Wis., and the Detroit Lubricator Co. and McCord Co., both of Detroit, Mich.

From that time on, the lubrication of pistons (and of important bearings) was fairly satisfactorily accomplished, although the system was extravagant in the use of oil. The system failed, however, when some of the small, brass oil pipes became crushed or broken thereby interrupting the oil supply to a particular part or parts.

During the effort to improve the engine lubrication, the faulty lubrication methods used on gears and transmission shaft bearings—hand-operated grease cups and hard greases applied to gear teeth by wooden paddles—were replaced by enclosed gear housings filled with heavy, semi-fluid greases, and by tightly sealed transmission cases filled with oil.

Governing and Carburetion (page 70-71)

While many of the early tractors were powered with single-cylinder engines, others were powered with multiple cylinder units, and their speed was largely controlled by the "hit-or-miss" type flyball governor. When the engine was above speed, the flyball would cause a "catch" to hold the exhaust valve open, and the inlet closed so as not to receive a charge. When the engine got below speed, the governor would cause the "catch" to release the exhaust valve which would close and simultaneously permit the inlet valve to open. This would allow a charge to be taken in and the explosion would follow. This procedure would continue until the engine got above speed again, when the exhaust valve would again be held open and the intake closed. With this type of governor the speed of tractors in revolutions per minute often fluctuated considerably and their use in driving belt machinery was not entirely satisfactory.

Between 1910 and 1912, governors were developed which controlled engine speed by throttling the air and fuel supply—throttle type governor. They produced power impulses from each cylinder regularly, once every two revolutions of the flywheel on the 4 stroke cycle (4-cycle) basis—the intensity of each explosion depending upon the load and the corresponding throttle opening. Because of their more uniform speed control this type of governor, fully displaced the "hit-and-miss" type by about 1915.

Dust, along with faulty lubrication in early tractors, also caused excessive wear to engine parts, particularly to cylinders and pistons. Cylinders and pistons frequently became so badly worn, from dust drawn in on the suction strokes, as to be practically useless after a few weeks' operation, especially when operating under very dusty conditions. An air cleaner was necessary which would not interfere with the proper fuel-air mixture regardless of the variation in load placed upon the engine. Failure of the air cleaner to permit proper fuel-air mixtures under the varying engine loads resulted in faulty engine operation. Another requirement of an air cleaner was the ability to operate without frequent cleaning. Cleaners that required daily or more frequent cleaning, it was found, would be neglected. This neglect would result in loss of power, excessive fuel consumption, faulty ignition from carbon deposits due to over-rich mixtures, and in stifling exhaust fumes.

"Periscope" intake pipes, 3 to 9 feet long, were first used in an attempt to draw clean air into the engine. It was hoped that by extending these pipes above the dust zone, only clean air would be drawn in. The pipes, however, were difficult to support and the joints between sections could not be kept tight.

Frames covered with bolting cloth and connected with the air intake passage were next used as air cleaners. In dusty areas the cloth filled to "air choking" density in a very short time. Under extremely dusty conditions cleaning every hour was imperative. This type was short-lived.

Large containers, filled with water in warm weather and kerosene in cold weather, were developed to replace the cloth-covered frames as air cleaners. A floating baffle forced the air through the liquid. The cleaners were effective as long as the liquid level was maintained and the liquid was kept clean enough to remove the dust effectively from the air. Backfires, however, often set the kerosene-filled cleaners on fire.

About 1920 a self-contained oil-type air cleaner was developed. The air was cleaned by passing it through a mist of light-grade oil and a wire-mesh filter. This cleaner has been continually improved and is now the type generally in use. It requires cleaning at relatively infrequent intervals, and is easy to clean and replace.

Cooling (page 72)

The excessive use of water in the cooling system of many early tractors, and the frequent unavailability of clean water or water reasonably free of soluble salts, especially in regions of limited water supply, focused attention on the need of improving the cooling systems. This led to the replacement of the hopper and the open cooling systems by closed systems. The cooling air in the open system and some early closed systems was cir-

Some Lubrications Systems

Plain Compression Grease Cup | Flushing Oiler | Screw Compression Grease Cup | Self-Feeding Grease Cup

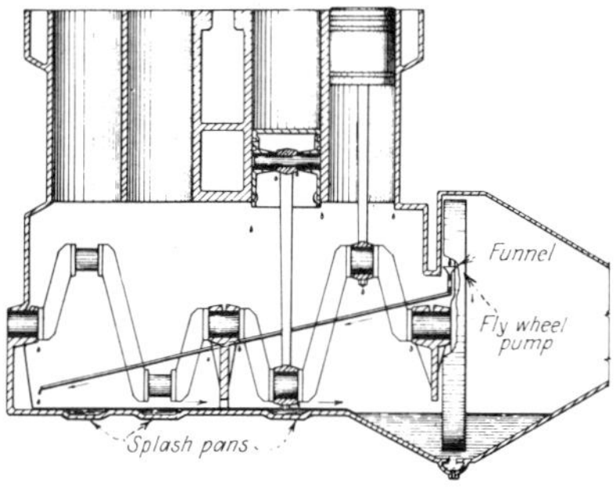

Splash lubricating system. The flywheel is depended upon to circulate the oil

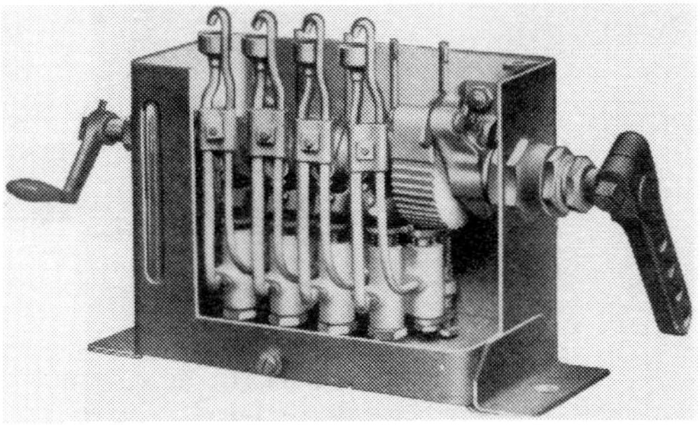

Madison-Kipp force-feed oiler, showing interior

Combination force-feed-splash system

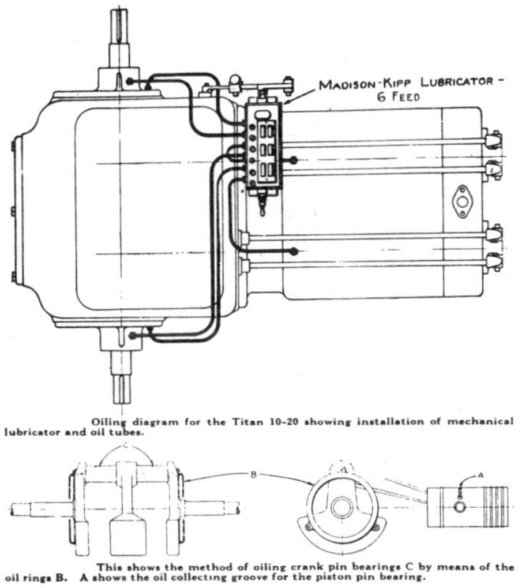

Mechanical oiler with oil tubes leading to centrifugal rings for oiling crank pins

Force-feed system, drilled crankshaft

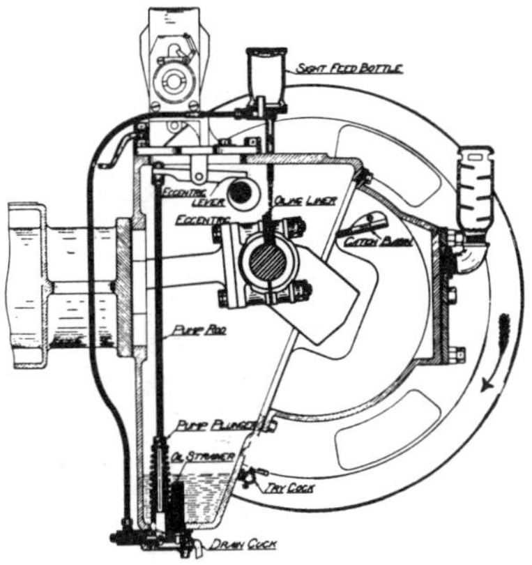

Combination force-feed splash lubricating system. Horizontal engine

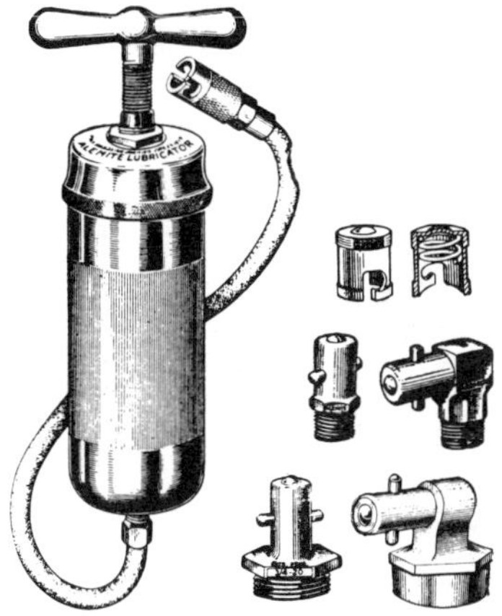

High pressure "Alemite" lubrication gun and fittings

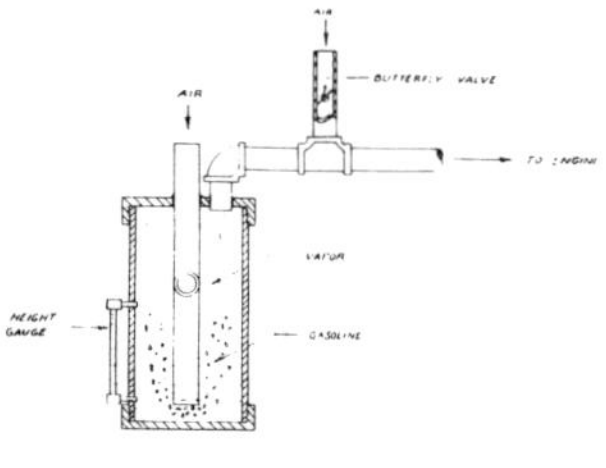

Carburetion installation on first Case tractor (Patterson)

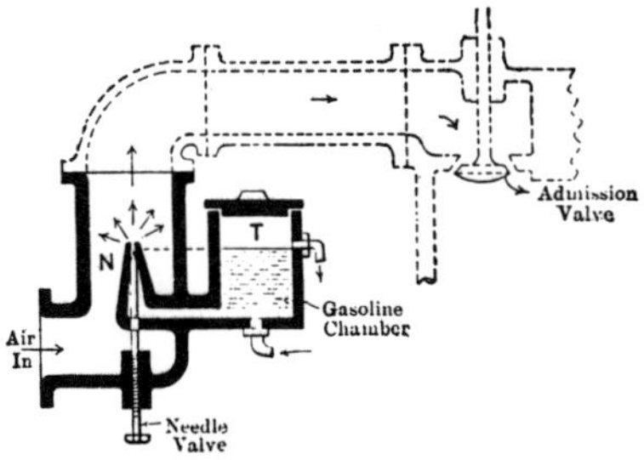

Simple jet carburetor furnished principle upon which carburetion for gasoline-burning equipment was based

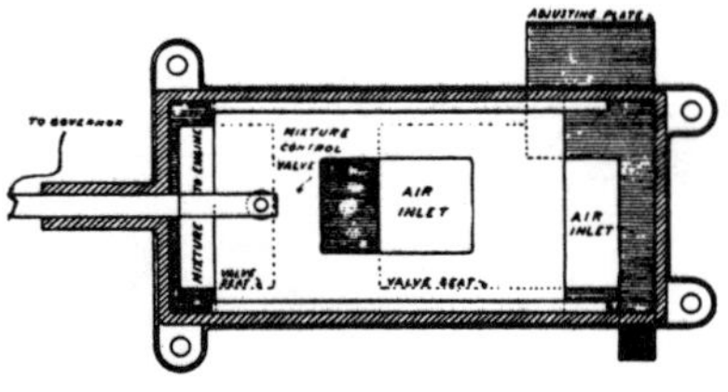

—Position of sliding valve at light load.

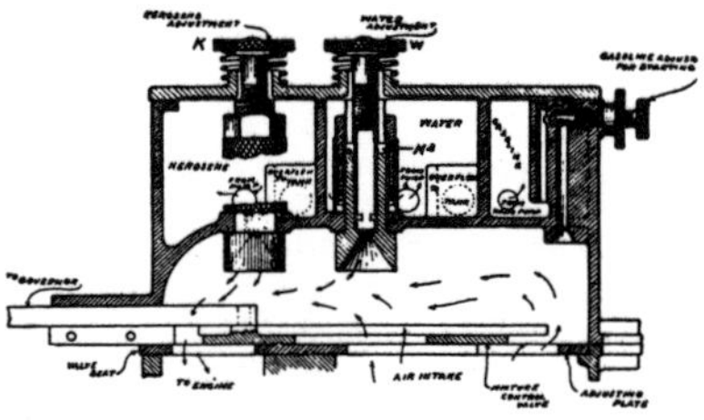

—Longitudinal section of carbureter.

Showing position of sliding valve at light load, and longitudinal section of Secor-Higgins carburetor used on Rumely "Oil-Pull" tractors

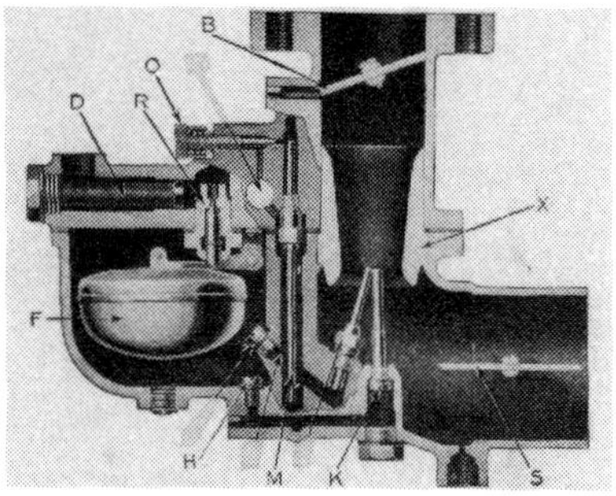

Zenith carburetor

B.	Throttle butterfly	H.	Main jet regulator
O.	Idling needle valve	M.	Idling jet
R.	Float valve	K.	Main jet
D.	Fuel screen	S.	Choke
F.	Float	X.	Venturi

Fuel mixer (pump and overflow type) designed for gasoline and lower grade oils. Used on Mogul 10-20 tractor.

A.	Air damper (choke)	D.	Water valve
B.	Heating jacket	E.	Kerosene valve
C.	Gasoline valve used for starting	F.	Butterfly valve operated by governor

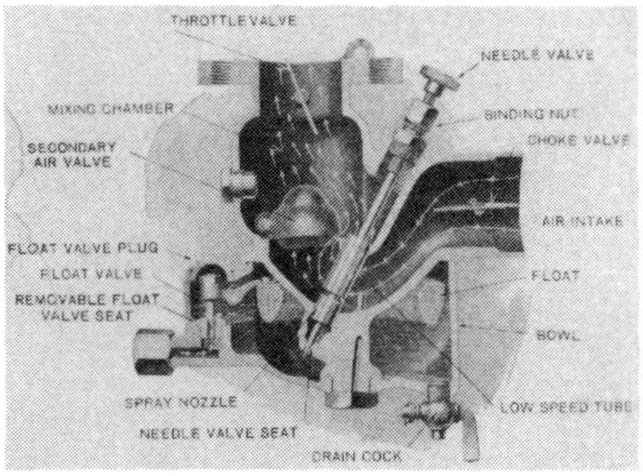

Kingston carburetor

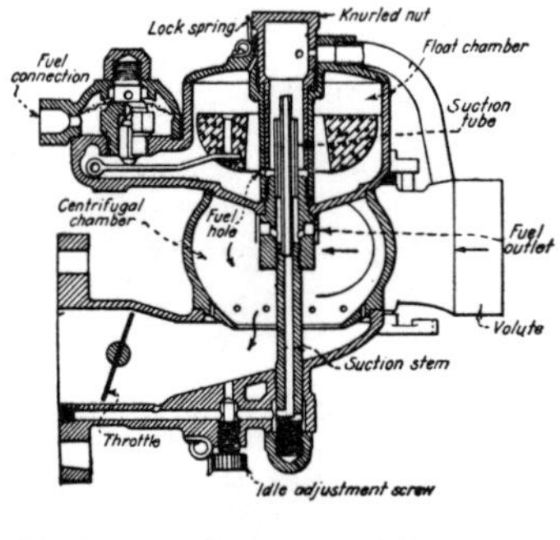

Ensign carburetor with vortex chamber

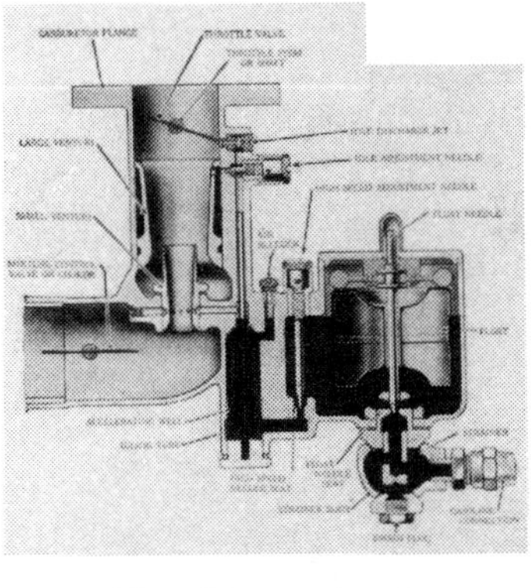

Stromberg carburetor

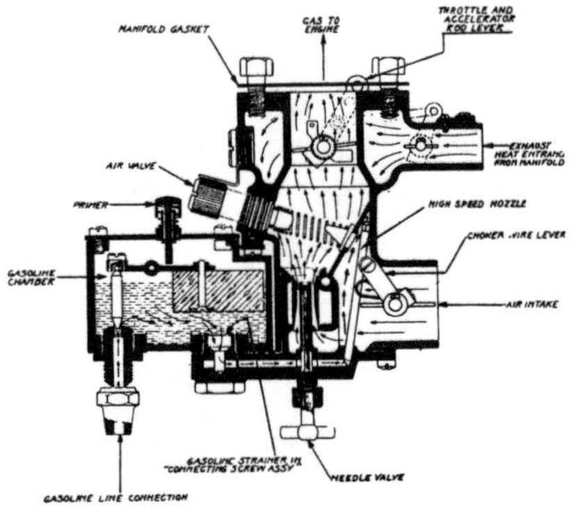

Marvel carburetor

Some Early Carbureting, Governing and Air Cleaning Systems

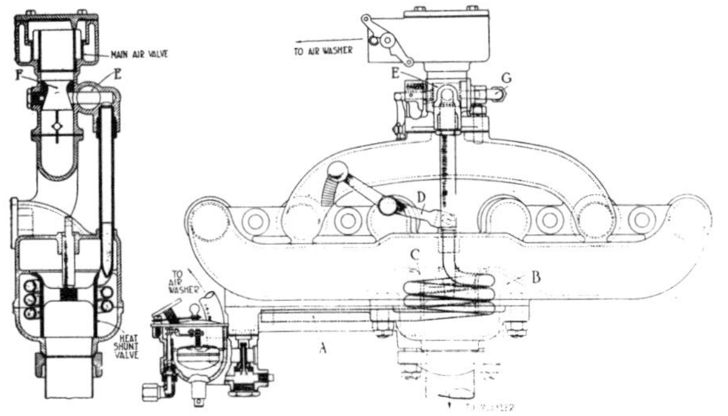

Holley vaporizing system for kerosene, showing method of conducting primary mixture through vaporizing coil and of controlling amount of heat used to vaporize primary mixture

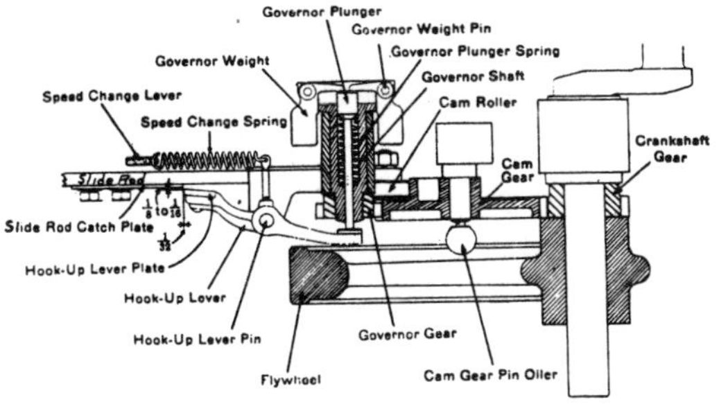

Hit-and-miss governor

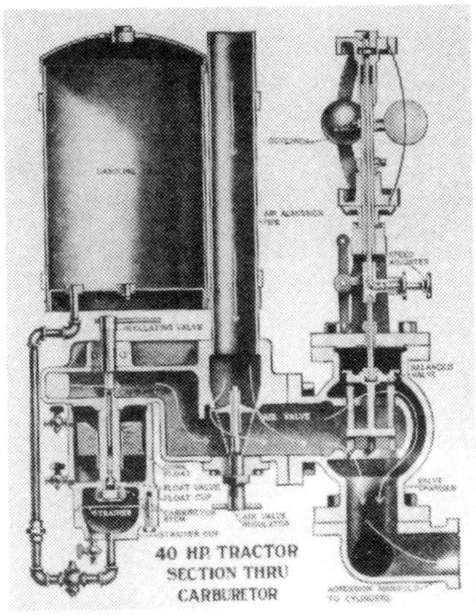

Throttling governor (Pickering)

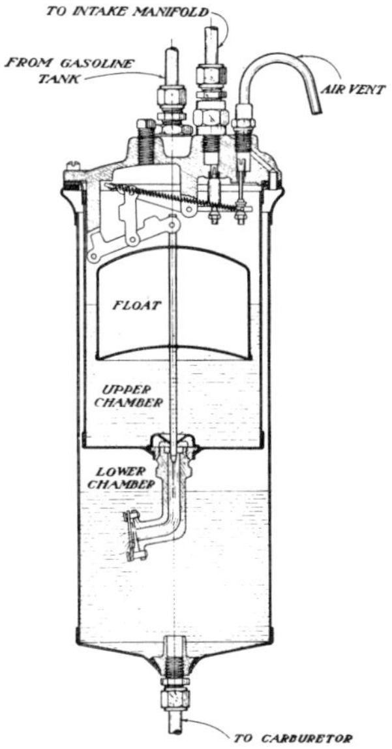

Vacuum tank used in feeding fuel to the carburetor where the fuel tank was located below the level of the carburetor

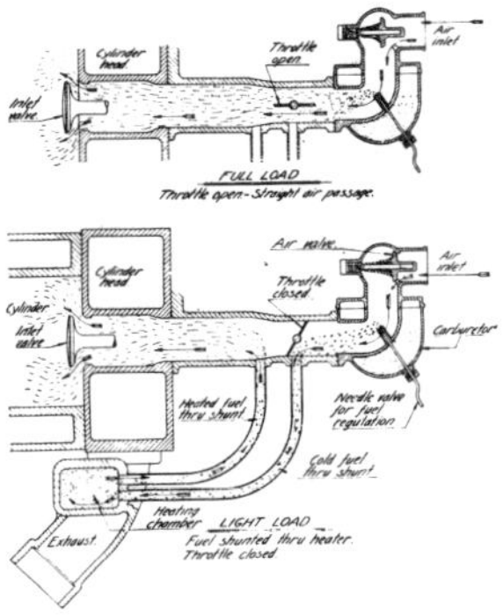

Shunt used on Hart-Parr tractor for feeding fuel to cylinder
(upper) with engine under full load
(lower) with engine under light load

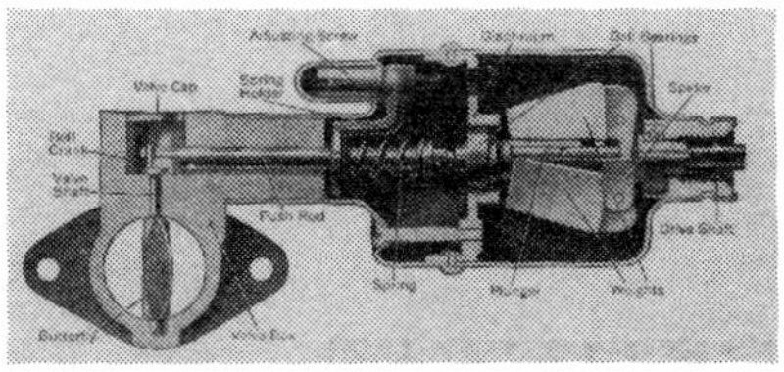

Pierce throttle-type governor

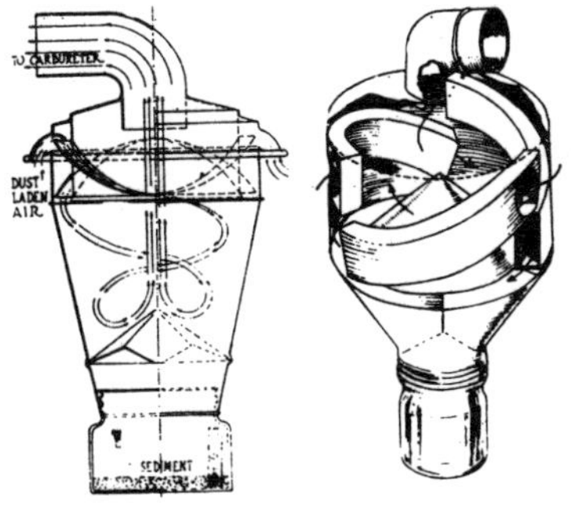

Two types of centrifugal air cleaners. Dust particles thrown out by centrifugal force collect in container at bottom

Air cleaner - water type. Dust-laden air enters at the top, passes down through inner tube, and is released below water surface

"Oil-mist and filter" cleaner. As the air rises, after being sucked downward into the cleaner, it carries with it a spray of oil from an oil chamber to a compartment filled with oily wool

Some Cooling Systems

Combination hopper and tank cooling

Hopper cooled

Residential radiator cooling, with air chamber

Open tank pump circulating cooling

Closed system with tubular radiator, fan and pump

Thermosiphon tank cooling

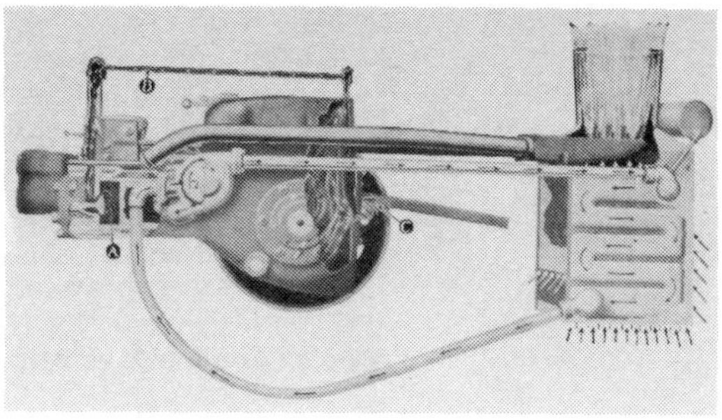

Cooling by induced draft from exhaust

Thermosiphon cooling - exhaust induces air draft through radiator

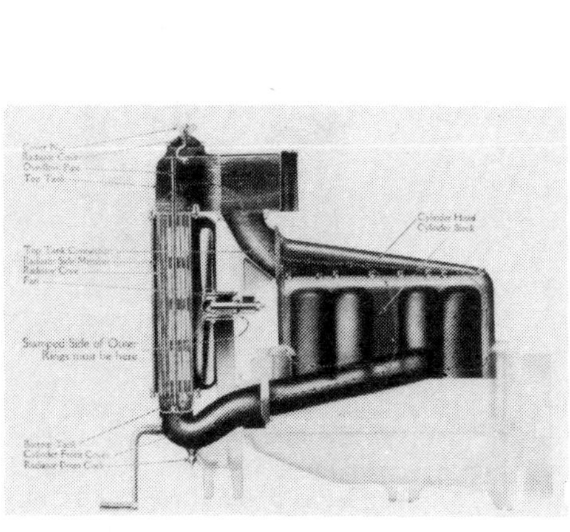

Thermosiphon

Forced circulating, cooling automatically controlled by thermostat

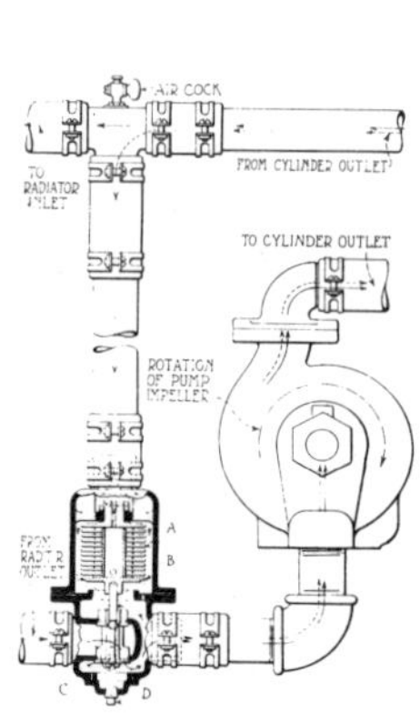

Thermostat

culated by induced draft from the exhaust or by fan, and the water was circulated by a plunger or rotary pump. In some of the open tank systems the water was caused to run down cooling screens into the tank. In some closed systems the water was circulated by plunger, or rotary pumps, and in others by thermosyphon or steam impulse.

Thermostats were introduced to maintain proper operating temperatures in closed systems, and the centrifugal pump, fan, and radiator soon became standard equipment. In some cases, adjustable curtains were used over the front of the radiator in order to raise more rapidly the temperature of the cooling water, particularly in cold weather. To overcome the danger from freezing some manufacturers made use of enclosed oil-cooling systems.

Clutches (page 74)

Efforts to improve bearings, clutches, and transmission essentials were made simultaneously with those directed toward improving the engine.

Bearings made by "poured" babbitt were replaced by bronze bearings which were machined and fitted into bored retainers, thus making replacement a simple matter. Later many of these bronze bearings were replaced by roller or ball bearings which were effective in reducing wear on the shafts. They also required much less space and so permitted tractor dimensions to be reduced.

The jaw clutch probably was first used to couple the engine to the transmission when the portable steam engine was made into a self-propelled tractor. In many instances, however, it was soon replaced or used in conjunction with a friction clutch. Some of the later types of double cylinder and cross compound types of steam tractors still made use of the jaw clutch only. Many types of friction clutches were developed for tractor use.

Transmissions (pages 76-77)

Transmissions in most of the early tractors were made of rough cast iron gears, which were exposed to dust and dirt and, wore rapidly. At first they provided only one forward speed of about 2 miles per hour and one reverse, but were soon adapted to provide two forward speeds. Replacing a heavy worn-out cast iron bull gear, differential or broken countershaft in the field was a man-size job and often taxed the ingenuity of "experts" to devise ways and means for handling these heavy parts. These cast iron gears, however, were soon replaced with lighter weight machine-cut, heat-treated steel designs. With the development of alloy steels and methods of hardening the wearing surfaces of the gears and shafts, it became possible to reduce the size of shafts, gears and gear teeth to such an extent that the over-all dimensions and weights of tractors were very much reduced. The reduction of dead weight increased the draw-bar capacity from a given engine horsepower development. Manufacturing costs were also reduced as better methods of production eliminated tedious hand-fitting methods in assembly, such as hand-filing high spots on pistons, scraping in bearings, and pouring babbit counter-shaft bearings.

Early transmissions included the friction drive, gear drive and friction reverse, sliding engine type to effect gear change, and the planetary gear drive. These were followed generally by the selective type sliding gear transmissions.

The final drives were of numerous types, including single and double chain, single and double spur gear, internal gear, worm and worm gear, and bevel gear. Types of rear axles were the one-piece axle either dead or floating, and live axle of one or two pieces.

Wheels and Tracks (page 78)

A wide range in types of wheels and wheel and track arrangements were used in the manufacture of early tractors, as well as types of lugs or grouters for improving the traction of these wheels. Lug types included the spade, angle iron, wedge, quick detachable, and retractable (page 81).

Thomson, when he built his Rubber Tired Steamer in 1871 (page 4), made the first recorded attempt to use rubber on agricultural tractor wheels. It was not until 1918 that the second attempt was made—this time by the International Harvester Co. on its 8-16 four cylinder tractor and L. B. Sperry,[8] tractor engineer who served the company many years, stated that, "the front wheels were fitted with solid-section rubber tires; the rear wheels with solid-section blocks moulded into metal detachable lugs and the blocks were made by the Firestone Tire & Rubber Co. The traction afforded by the block rubber treads when off the road, where the tractor was expected to be used, was far from satisfactory and the rubber adhesion to its metal frame was not at all satisfactory, so this construction did not get beyond the experimental stage."

Interest in rubber-tired wheels developed as an outgrowth of the rising sentiment against the use of tractors with lugs on the nation's rapidly expanding paved highways. Sentiment against the use of wheels equipped with lugs, or wheels with rivet heads protruding above the surface of the steel tires, resulted in laws prohibiting their use on paved rods. To move their tractors across paved roads, owners were then compelled to "plank" the roadway.

Tractor designers attempted to meet the pavement restrictions with the use of quickly detachable lugs or "over tires" but they both proved impractical, because of the time required to remove and replace them, and there were no means for carrying them when not in use. These "over tires" were continuous steel bands which were placed over the bolted or riveted lugs.

Although solid rubber tires were being used on industrial tractors in the early 1920's, they were not practical for agricultural tractors because they provided little or no traction on grassy or muddy fields, or on wet or icy pavements. The use of tire chains, to increase traction, posed the difficult problem of putting them on and taking them off.

It was perhaps inevitable, then, that tractor owners and the tractor industry looked to the rough-treaded, pneumatic, automobile-type tire for the solution to their traction and transport problems. Truck tire casings were mounted on steel rims, and even airplane landing-gear tires were tried. Finally, in 1932, specially designed pneumatic tractor tires were built by the Firestone Tire and Rubber Co. for Allis-Chalmers who introduced them on their tractor models that year.

---

[8]Mr. Sperry also reviewed Section I of the manuscript, and supplied a number of pictures and pertinent information concerning early tractor accessories, component parts and pneumatic tires. (See Section II-1928)

Some Friction Clutches

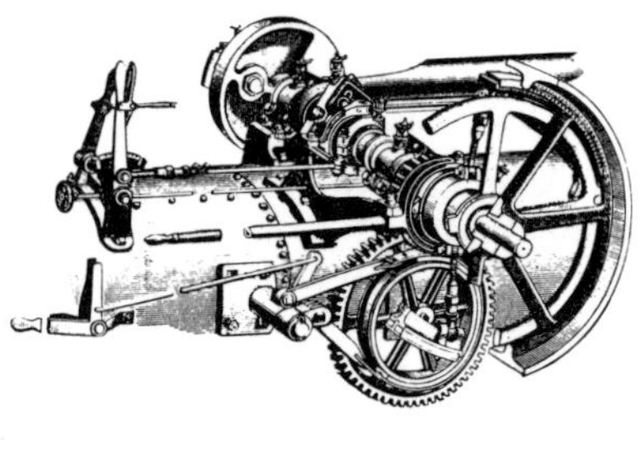

Early steam traction engine friction clutch

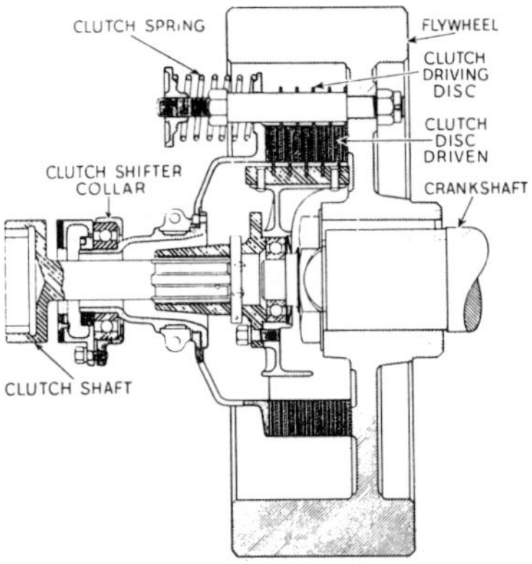

Multiple dry disk clutch, showing alternate driving (friction material faced) and driven saw-steel disks mounted within flywheel

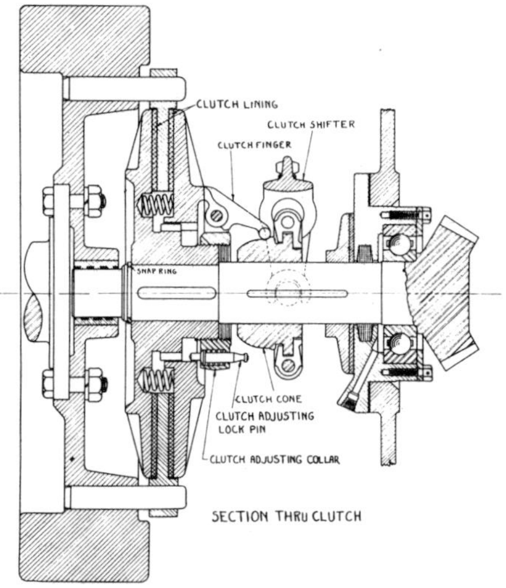

Twin-disk clutch. Engagement obtained by sliding cone-shaped collar. Light springs between driven disks insure their prompt release on disengagement

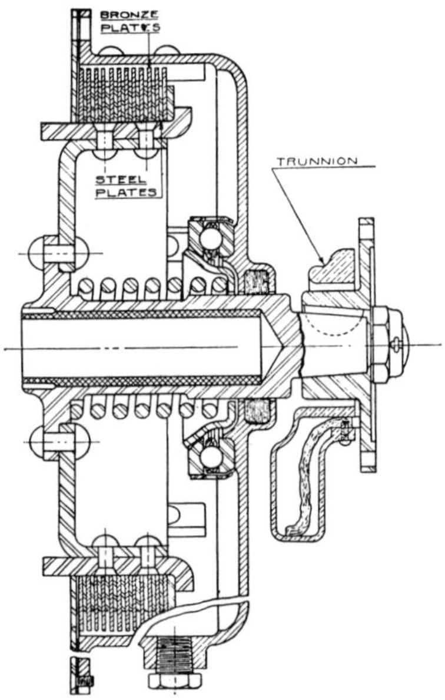

Multiple-disk clutch running in oil

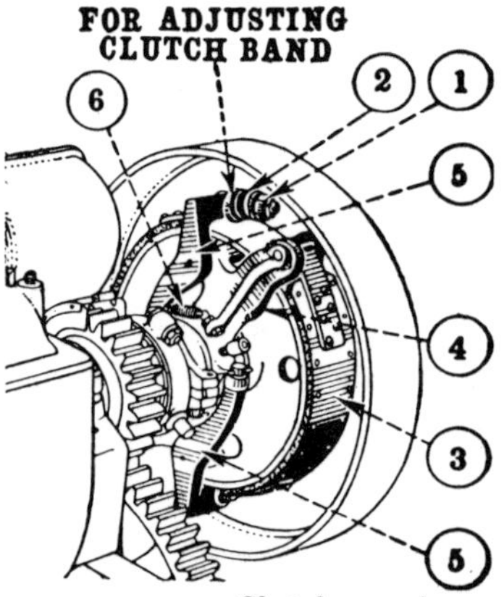

— Clutch used on Titan 10-20, Mogul 10-20 and International 15-30 tractors.

1. Clutch band adjusting nut.
2. Clutch band spring.
3. Clutch band.
4. Clutch spider.
5. Clutch lever.
6. Clutch lever releasing spring.

Contracting band clutch

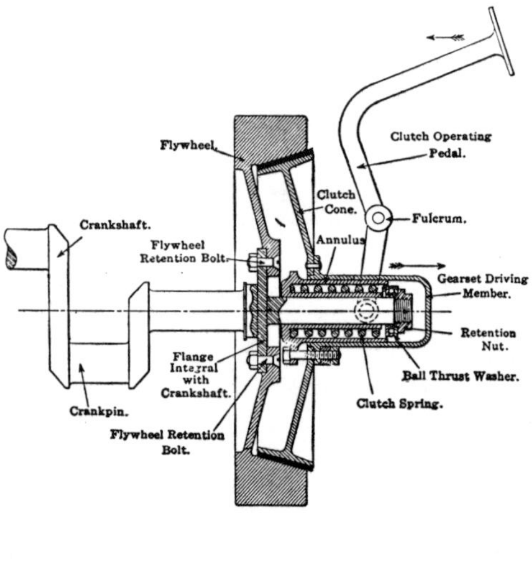

Cone clutch assembly showing clutch rim (flywheel) and cone-shaped friction-faced member held under spring pressure when engaged

Expanding shoe clutch with three counterweighted shoes, shifting fork and clutch brake

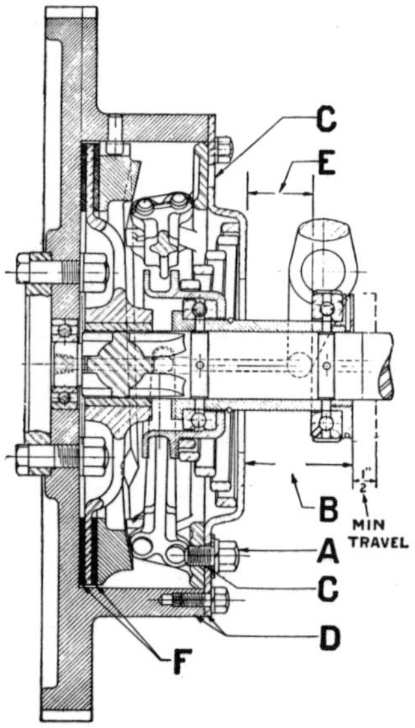

Plate clutch (Borg and Beck) A. Slot bolts for adjustment. B. Adjustment distance guide. C. Adjustment slot. D. Reference marks on clutch cover and flywheel rim. E. Space for block to lock out spring. F. Asbestos rings.

Diagrams of power transmission mechanism in different makes of farm tractors

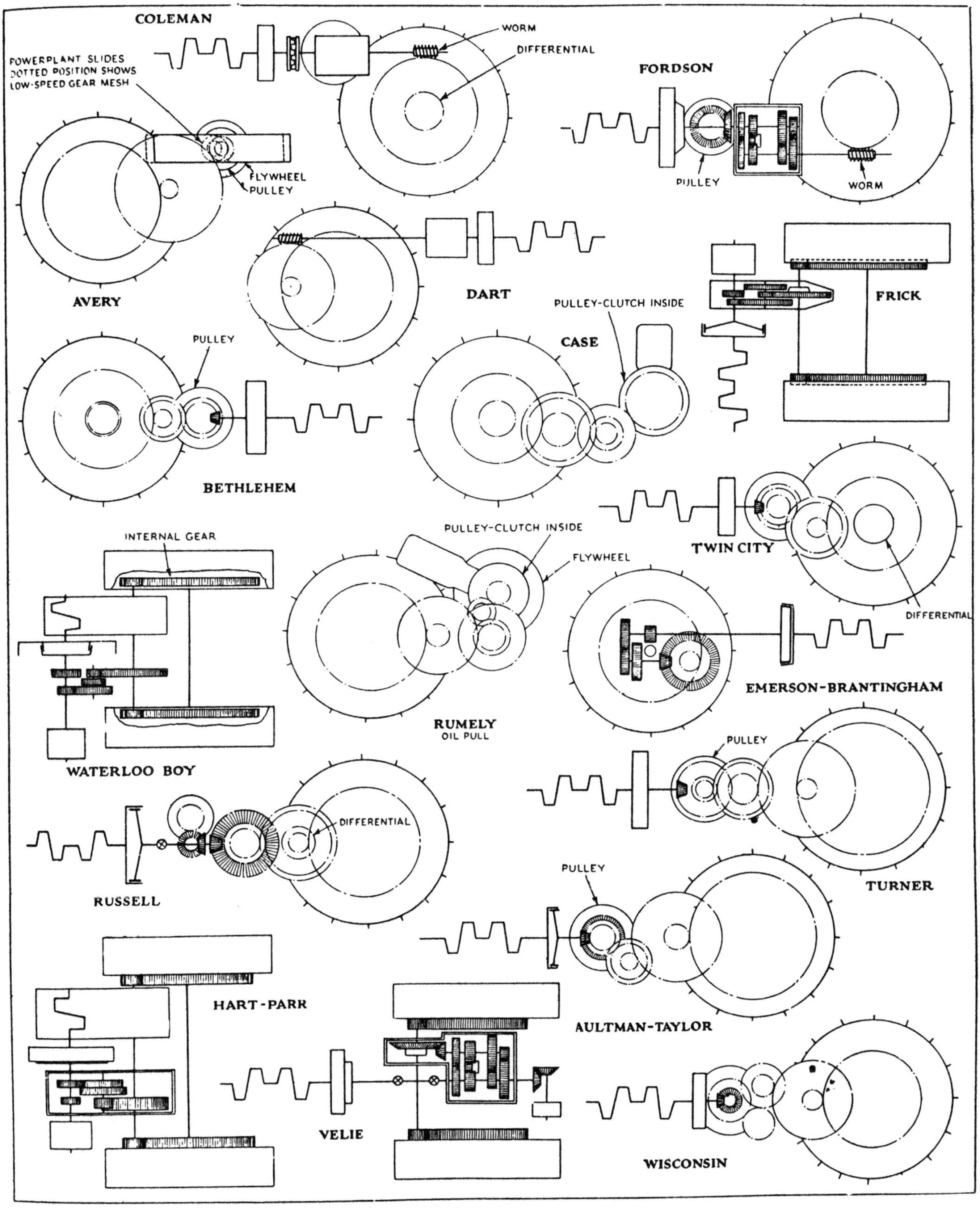

Some Early Transmissions and Final Drives

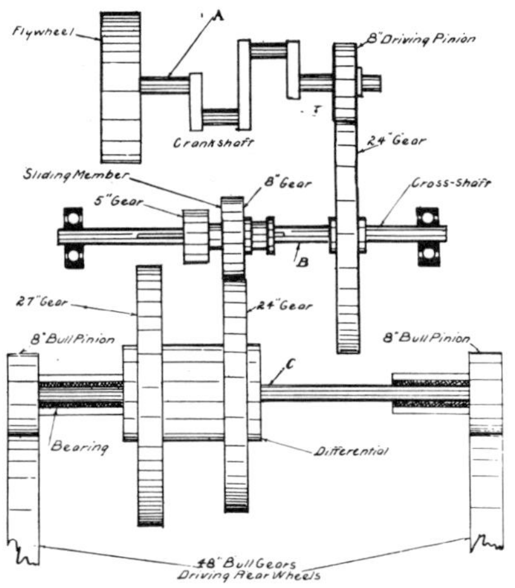

Diagrammatic illustration of a method of transmitting power from engine of tractor to drive wheels. Rear wheels on which bull gears are mounted, revolve on one-piece dead or floating axle

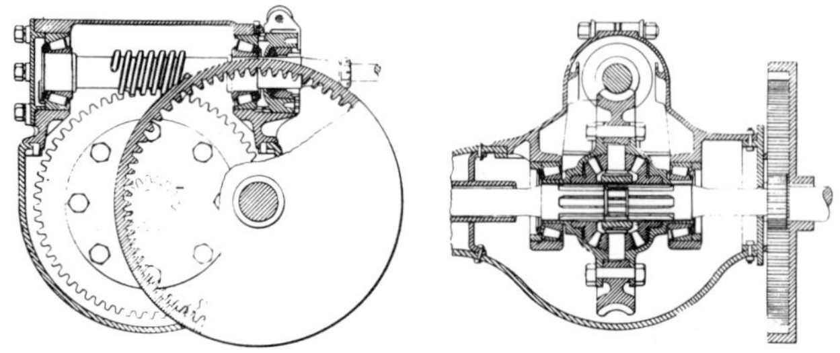

Side and end sections of combination worm and internal gear double reduction tractor axle

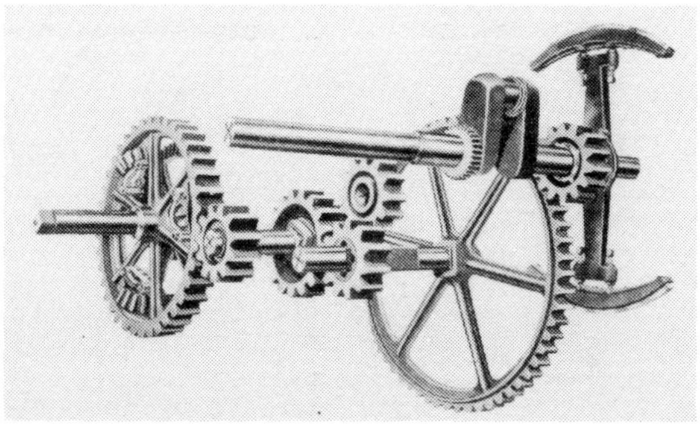

Transmission of single-cylinder tractor showing train of gears from engine to bull gear final drive

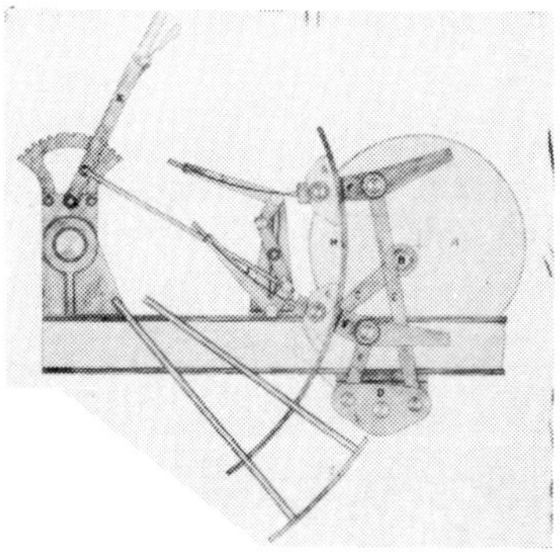

"Gearless drive." Offset flywheel crankpin (B) imparts rocking motion to rocker arm (D) causing grip dogs (F) to move up and down alternately. With control level (K) in forward position, control springs (J) hold grip dogs (F) so lower edge of jaw engages drive band (H), causing jaw to grip band, and pull it down causing drive wheel to revolve forward.

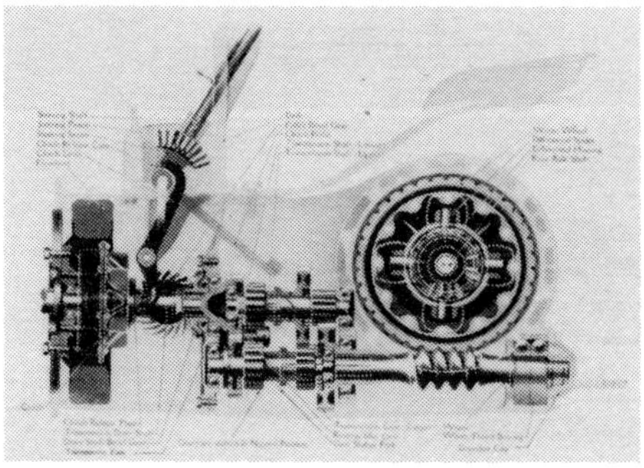

Sliding gear transmission and worm, and worm-wheel final drive. Rear axle two-piece live type

Friction clutch and transmission

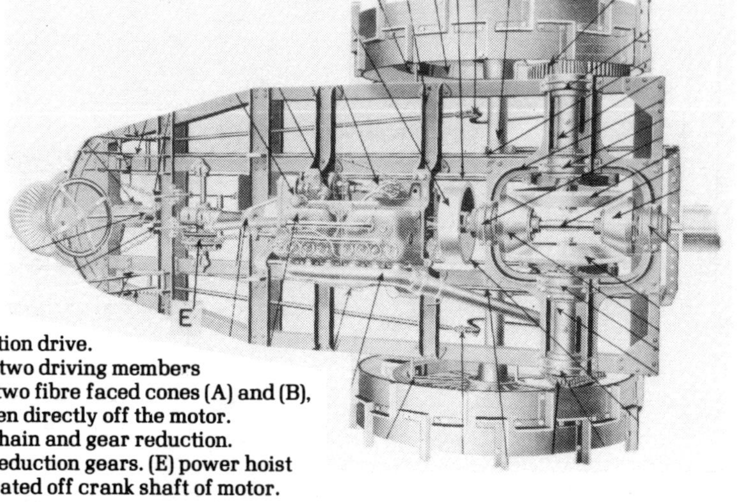

Friction drive.
The two driving members are two fibre faced cones (A) and (B), driven directly off the motor.
(C) chain and gear reduction.
(D) reduction gears. (E) power hoist operated off crank shaft of motor.

Some Early Transmissions and Final Drives

Inner half of planetary transmission with crank shaft pinion removed

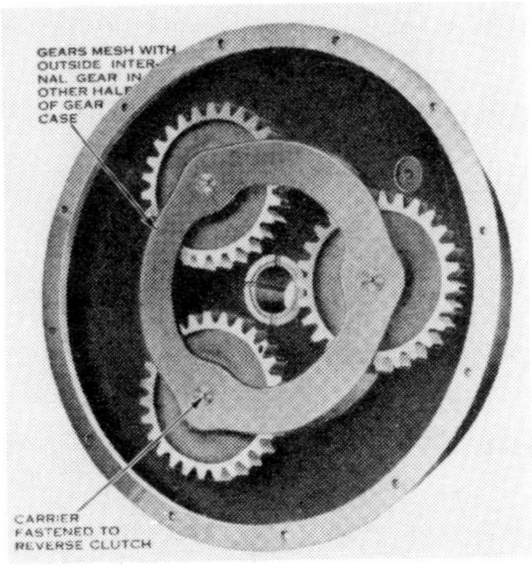

Outer half of planetary transmission with shaft and pinion removed

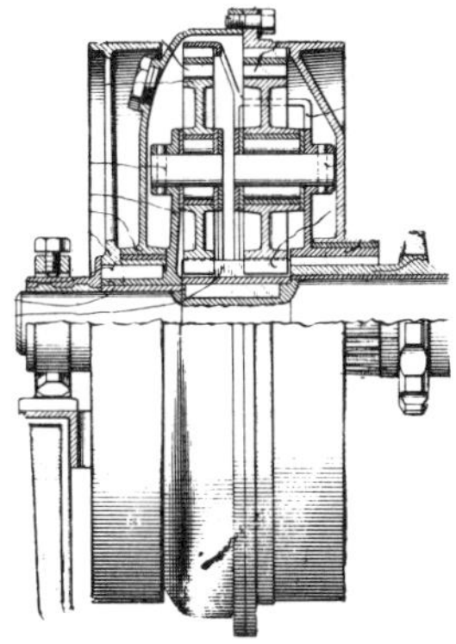

Planetary transmission assembled: forward motion produced with brake band held on (B), reverse with brake band held on (A)

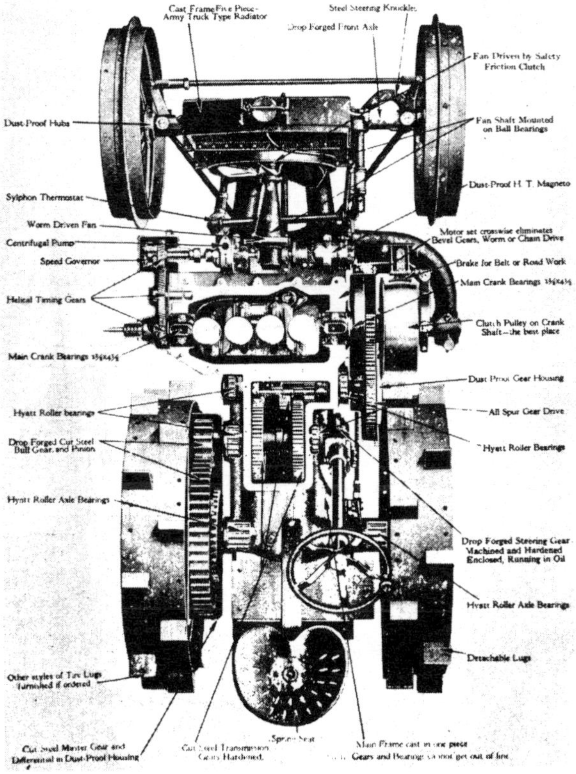

Showing cross-mounted tractor engine and transmission - one-piece live axle (rear)

Transmission showing direct drive with two-piece live axle (rear)

Hand lever slides the frame

Low speed pinion in mesh

Reverse gear engaged

High speed pinion in mesh

Pinions released for belt driving

Sliding frame transmission

Single chain final drive, one-piece live axle

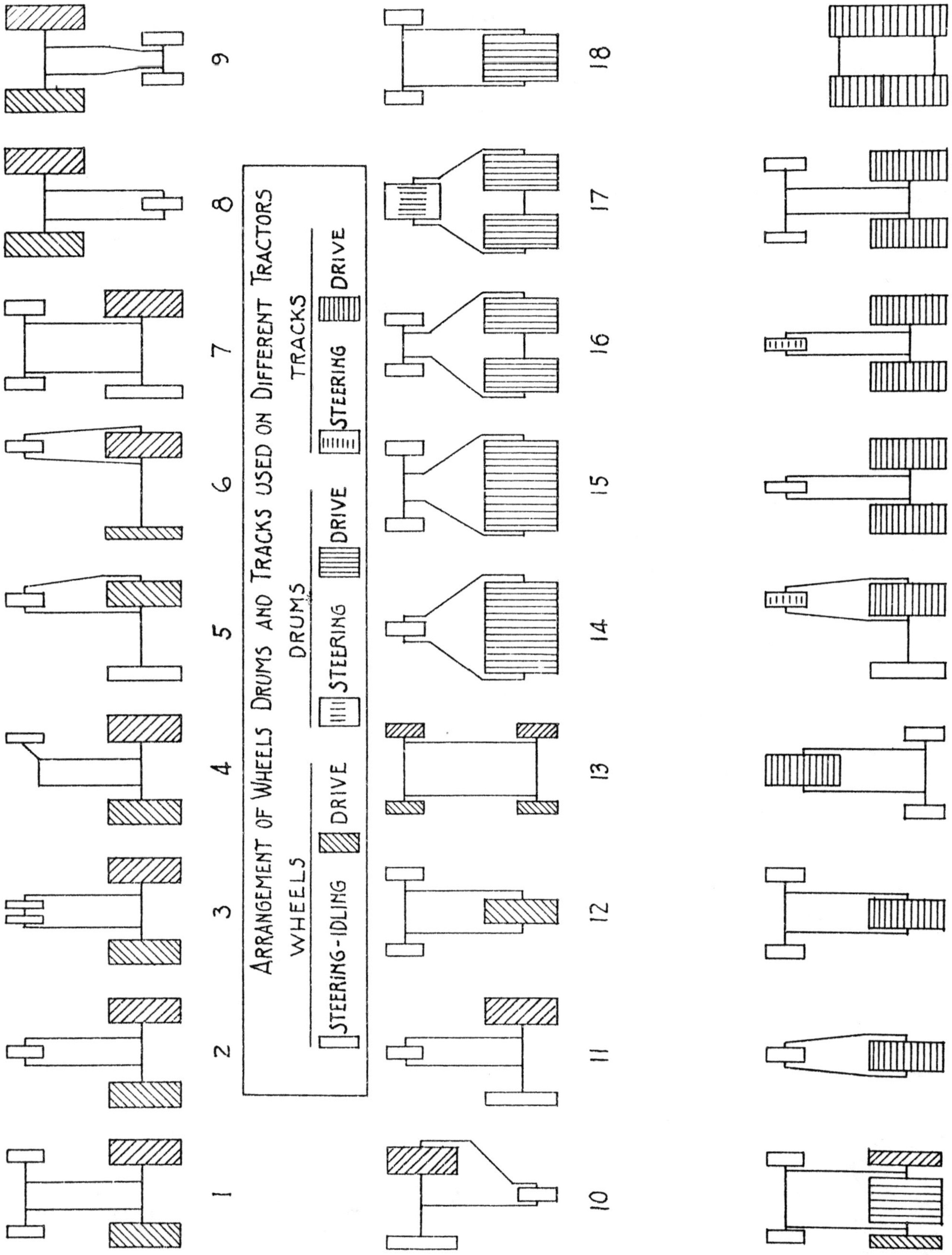
ARRANGEMENT OF WHEELS DRUMS AND TRACKS USED ON DIFFERENT TRACTORS
WHEELS
STEERING-IDLING
DRIVE
DRUMS
STEERING
DRIVE
TRACKS
STEERING
DRIVE
1
2
3
4
5
6
7
8
9
10
11
12
13
14
15
16
17
18
19
20
21
22
23
24
25
26
27

Allis-Chalmers, then, inaugurated the trend which has placed the tractor on air (See Part II.)— more than 70 years after rubber tires were first tried.

Some Miscellaneous Units (pages 80-82)

Several arrangements of pistons and cylinders were found in the early tractors, including single and two cylinder twin and opposed engines and crank shafts, as well as four cylinder vertical engines with their cylinders cast separately, in pairs, and en bloc — with and without detachable heads.

To minimize the physical efforts of the tractor operator, a variety of self-steering attachments were provided which ran in the open furrow left from the previous round.

Types of frame, and axle mountings — particularly the front, also varied considerably. Frames included built-up riveted channels and boiler plates, straight and formed channel irons, and unit frame construction — U-shape boiler plate steel and cast iron.

VII. Track-Type Tractors

Note — The discussion of the accessories for wheel tractors in the previous section is, in general, applicable also to track-type tractors.

The idea of an automotive machine laying its own track on which to travel is recorded as early as 1770 when R. L. Edgeworth of England was granted a patent on such an invention. The machine carried rails which were laid down in front of the machine, being picked up and relaid in a continuous operation.

Cayley, also of England, in 1825 produced a steam tractor which was supported by a type of belt built up of wood and iron. The belt ran over two notched pulleys, front and rear, with intermediate wheels to distribute the weight of the tractor along the chain. The engine was steered with a single wheel in front. In order to overcome the resistance to the necessary sidewise slipping or skidding of these long belts when turning, each section was provided on its outside surface with rollers which could roll sidewise. They also acted as grouters or cleats to dig into the soft earth and provide traction.

The British army during the Crimean War in 1854 used track-laying steam-traction engines patented in 1846 by James Boydell (page 83). This machine is considered by some to be the first real track-laying machine, and was considered the only practical one up to 1860. It was claimed that:

> "Boydell's Traction Engines and Endless Railway will draw heavier loads over good roads than any other traction engine, and on soft and bad roads will draw heavy loads where no other traction engine can move itself."

Andrew Dunlap of Glasgow, Scotland, took out several patents between 1861 and 1874 on track laying devices. A patent on a chain-drive, track-laying mechanism was granted Guillaume Fender of Buenos Aires in 1882.

Track-type tractor developments in the United States as recorded in the Patent Office started in the early 1850's. On June 11, 1850, Gideon Morgan of Calhoun, Tenn., received a patent for a wheel substitute. The description of his patent reads as follows:

> "My invention consists in the mode of employing endless chains around proper guides for supporting and carrying a car or locomotive and in a manner of keeping the chains tight. . ."

The language indicates that his patent was for an improvement in track-type tractor design; hence, the beginning of the crawler-type tractor development in the United States must have occurred prior to 1850.

In 1858 and 1859 a number of patents were issued for this type of construction to inventors living in widely separated locations: John F. Elliott of New Haven, Conn., James Ingersoll of Grafton, O., and Warren P. Miller of Marysville, Calif.

Warren P. Miller's patent was dated May 3, 1858, the preamble reading as follows:

> "The nature of my invention consists in providing the machine with portable tracks of a certain formation which, going around and over the leading and driving wheels hereinafter mentioned, are laid down on the ground as the machine moves forward so as to form safe, solid, firm tracks for the carrying wheels or trucks which support the car so as to enable it to run over soft or sandy roads or places with the same facility as solid roads, without sinking in the earth and at the same time furnishing a hard, smooth and comparatively straight surface for the carrying wheels to roll on."

Miller's description is very complete and one that well describes present-day track-type tractors. His first "Traction Locomotive Steam Car," built in the West by Benham and Booth Foundaries, was paraded through the streets of Marysville, Calif., during the State Fair in the fall of 1858.

In 1867, R. J. Nunn of Savannah, Ga. patented an "Improvement in Land Conveyance." This consisted essentially of two or more bands running over a series of grooved rollers mounted in a frame and driven through a larger roller powered by a steam engine. In 1867, Thomas S. Minnis of Meadville, Pa. patented a locomotive for plowing and in 1870 patented a Traction Wheel (page 83 — A Minnis Steam Tractor) mounted on three tracks — each rear track driven by steam engine, attached at rear, through pinion and drive gear. According to Hall Higgins,[9] "Iowa's first 'dirt farming tractor' was this Minnis Crawler from Pennsylvania that came out to the raw prairie within sight of the new Iowa State Agricultural College as the first students started attending classes within sight of its smoke."

In 1873 Robert C. Parvin of Illinois, built a steam tractor propelled by an endless chain of steel plates to which "feet," shod with 2-inch plank, were attached. It pulled six plows (page 83).

Inventors in the '80's and '90's who worked toward improving existing ideas of track-type tractors were: Edwards of Chicago, Benson of Baltimore, and Linn of Cleveland. Linn developed the ball tread idea.

G. F. Page of Baltimore, in 1884 patented a drive of two chains with only four wheels, two of which were drive wheels. The main interest in this machine was that it was steered by twisting the front wheels and thus led the chains from their otherwise straight course.

One of the most complete track devices to date was that invented by F. W. Baxter, an American, who secured a patent in 1888 (page 84).

---

[9] *Hal Higgins, a free lance writer and an authority on power farming, contributed a number of pictures and offered many valuable suggestions.*

Some Early Miscellaneous Units

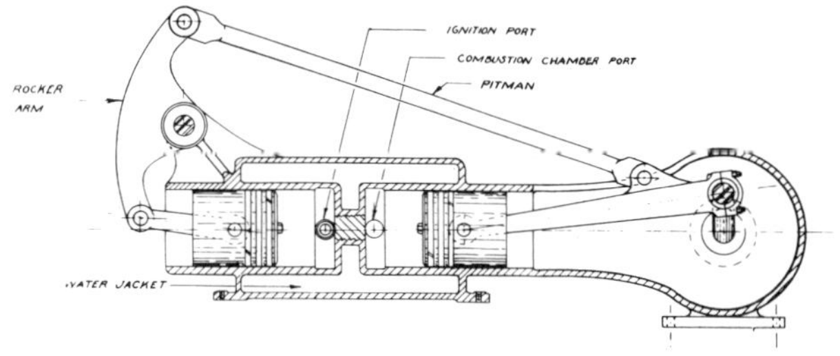

Sectional view of Paterson's Balanced engine, 1892

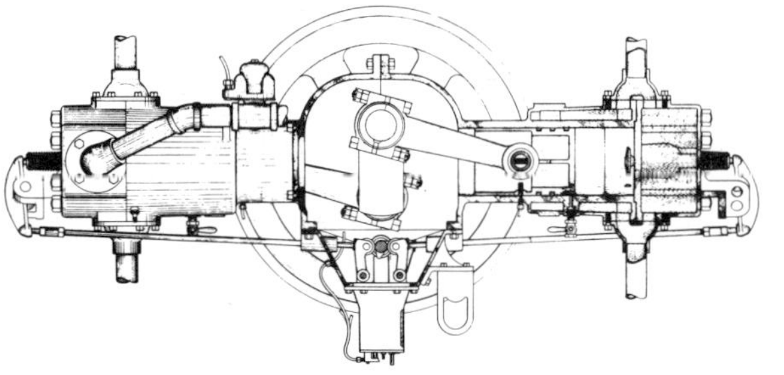

Two-cylinder horizontal opposed motor. Explosions occur every revolution. Universal

Crank for single cylinder engine

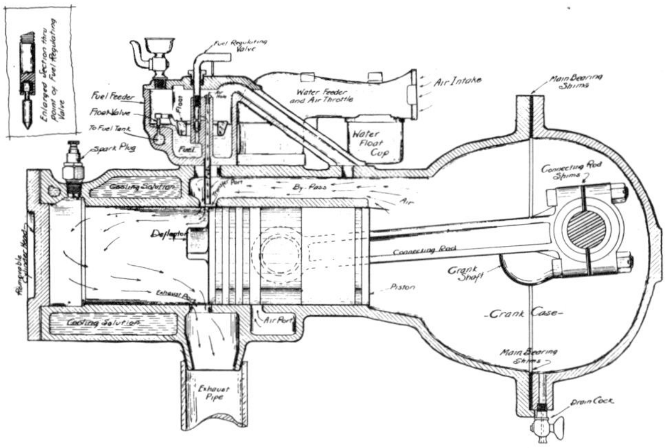

Two-stroke cycle engine of Hart-Parr "Little Devil" 22 horsepower oil tractor

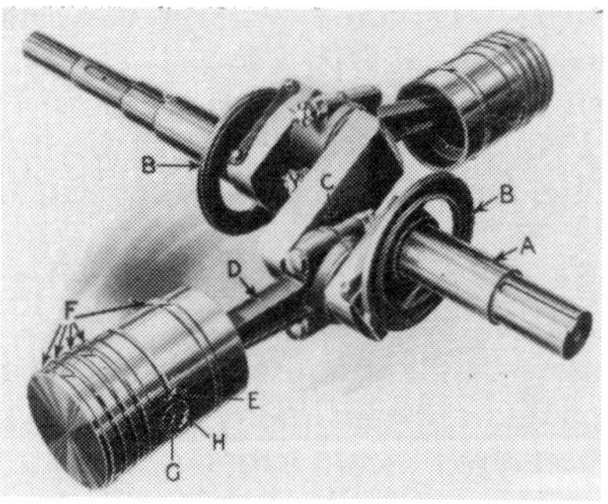

Opposed crankshaft of twin-cylinder tractor engine. A. Main bearing journal. B. Centrifugal oil ring. C. Connecting rod bearing. D. Connecting rod. E. Oil passage. F. Piston rings. G. Wrist pin. H. Wrist pin bearing.

Heavy-duty ball-bearing crankshaft, McCormick-Deering

Four-cylinder motor with cylinders cast in pairs. Strait

Single-throw crankshaft of twin-cylinder tractor engine

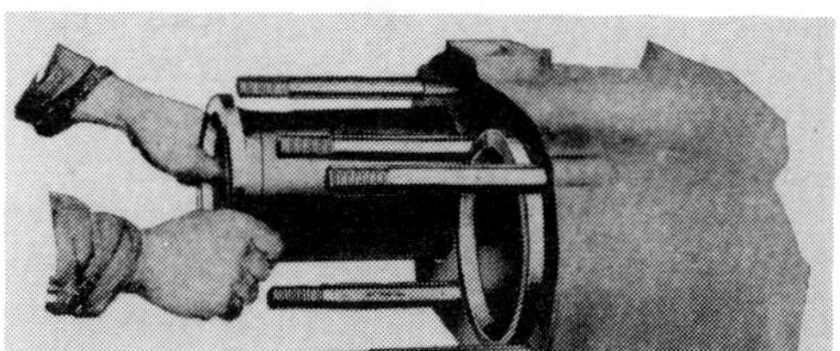

Renewable cylinder walls. The inner cylinder wall is cast separately from the main cylinder

Valve-in-head engine with cylinders cast separately. Caterpillar

Some Early Miscellaneous Units

Tractor lugs patterned after the foot of a mastadon, an extinct quadruped resembling the elephant

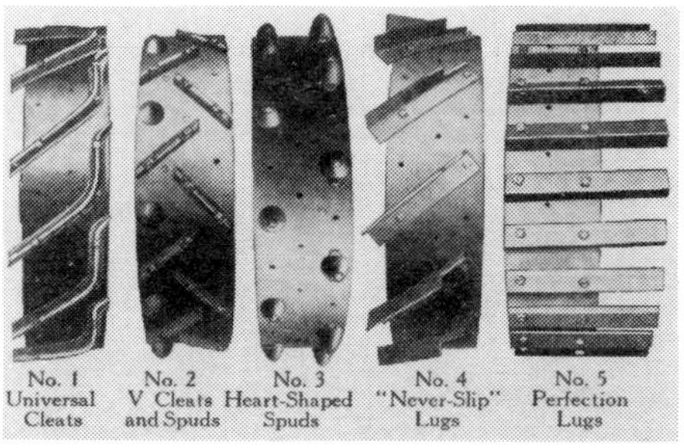

Wheel lugs. No. 5 for sand and light loose ground; No. 4 for soft ground; remaining 3 for varying conditions on heavier soils. Avery tractor

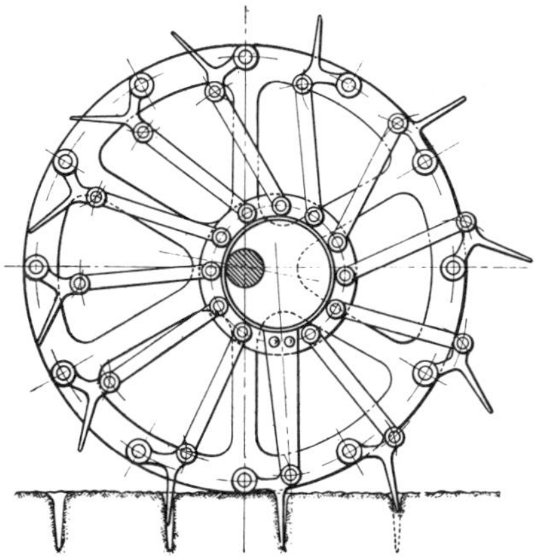

Tractor wheel with spades eccentrically mounted which enter and withdraw from the ground perpendicularly without packing the soil

The Craig tractor laid its own track by means of pads helċ against the rims of the wheels by springs

Drive wheel used on Samson "Sieve-grip" tractor

Tractor drive wheel with overtire for use on pavement

Three types of self steering attchments for tractors

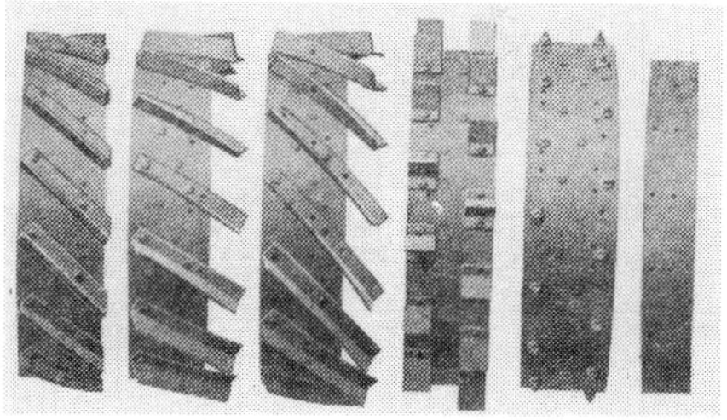

Legs and estension rims used on Hart-Parr tractors

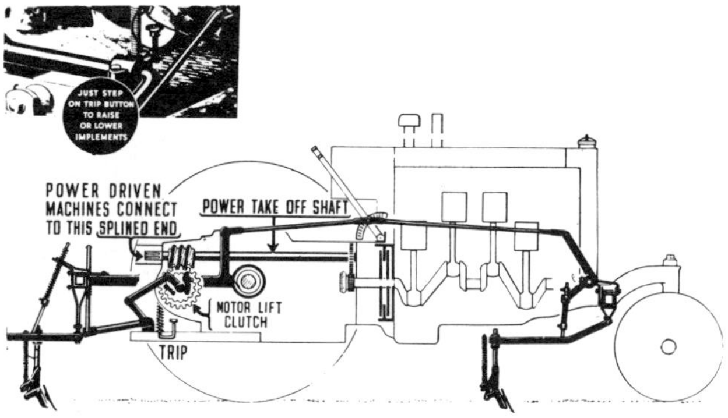

Diagrammatic view of Case Motor lift by which implements are raised and lowered by automatic control

Some Early Miscellaneous Units

"L" type head of four-cylinder, automobile type tractor engine. Parrett

U-shape boiler plate frame of Wallis tractor, 1912

Frame of two castings bolted together in the middle - a close approach to the one piece cast frame. Fordson tractor

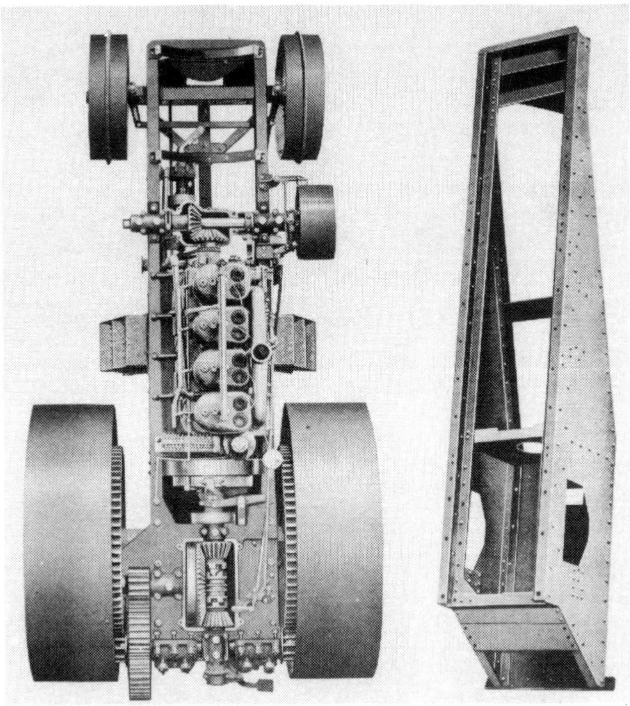

Top view and main frame of Twin City 20-40 tractor

Frame and transmission of Case 20-40 tractor

Transmission case acts as third frame member, Emerson-Brantingham

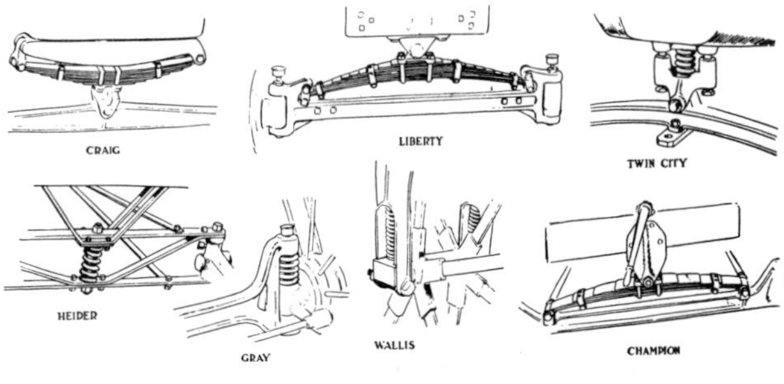

Tractor front end mountings (About 1920)

Early Track-Laying Engines

Track-laying steam traction engine of Boydell, 1846. Used by British Army during the Crimean War, 1854.

Minnis crawler, 1867 (Courtesy Hal Higgins)

Parvin's Steam Plow, 1873

Some Early Models of Track-Type Tractors (Steam)

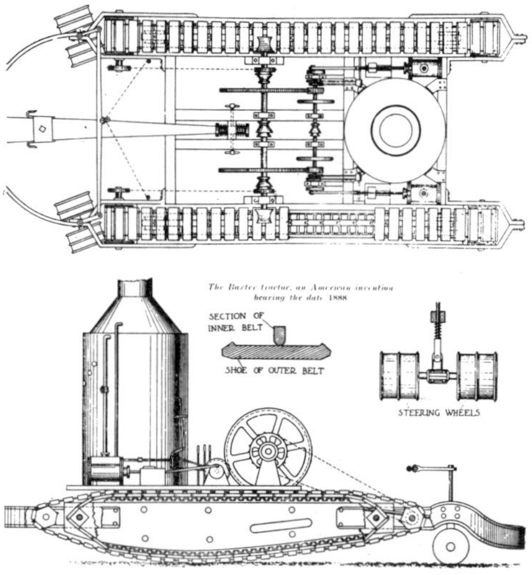

Baxter (1888) - The track consists of two sets of endless belts on each side of machine. The outer (ground) track is driven over hexagonal pulleys and the inner track over square pulleys from a forward countershaft, each track operating independently of the other.

Stratton's Traction Engine (1893) - 2 cylinder. Cold water tank in front of usual boiler and smoke box. Front end supported by wheels on pivoted axle, rear, by tracks. Axles of track carrying wheels and gears provided to permit oscillation. In addition to drive the tracks the horizontal engine actuates shaft which may be used to drive threshers and mowers- a so-called "power take-off."

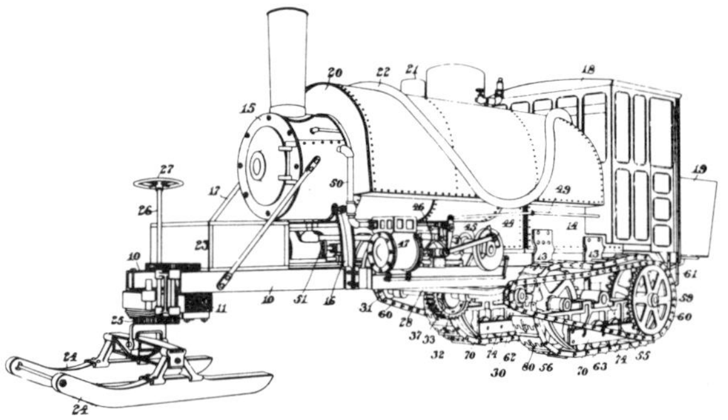

(Left - Patent Office Drawing) **(Right - Photograph)**

Lombard Log Hauler (1907) 100 H.P. locomotive type boiler. Power transmitted from a horizontal engine on each side through intermediate spur gears. These drive the countershafts, which in turn drive the tracks through chains from sprockets mounted on the rear axle.

Left - "Centiped" Log Hauler (1909) 100 H.P., four cylinder, vertical. Boiler pressure 200 lbs. Link motion. Bevel and spur gear open transmission driving Lombard Patent tracks.

In 1893, Chas. H. Stratton, Moscow, Pa., produced a steam powered traction engine (page 84) designed especially "to travel readily over plowed ground, for cross plowing, and other work." It is propelled by a pair of compactly arranged tracks actuated through gears and pinions from the horizontal engine.

One of the first attempts to manufacture track-type tractors commercially was made by Alvin O. Lombard of Waterville, Maine, in the early '90's. In 1901 he patented one of the first practical track-type tractors (1907 model shown on page 84). Lombard adopted the Linn ball idea, substituting rollers for balls. He built a very workable tractor, and produced and sold a number of machines. The unit was "designed specially for transporting lumber and logs over the rough roads and even cross country in the Maine woods" and embodies "half-track" construction as we know it today. Presently (1954) this Company, now the Lombard Traction Engine Co., Waterville, Maine, manufactures a Diesel powered Tractor-Truck driven by tracks steered with runners in winter and wheels in summer. Another track tractor appearing about this time was the "Centiped" Log Hauler (page 85) manufactured by the Phoenix Manufacturing Co., Eau Claire, Wis. This resembled the Lombard machine, but made use of a vertical engine.

The Stockton Wheel Co. built its first steam traction engine (track type) in 1890 and in 1892 became the Holt Mfg. Co. Due to topography, soil, and the large size of farms and ranches on the Pacific Coast, this type tractor was received more readily there than in the farming sections of the Middle West, East, and South. The reclamation projects (covering thousands of acres of tule land) required track-type machines because wheel tractors in many cases operated unsatisfactorily.

In 1904, on Thanksgiving Day, Benjamin Holt successfully demonstrated his first track-type tractor, the 1904 Holt (page 86) near Stockton, Calif. This demonstration followed considerable experimentation wherein he devised a pair of rough wooden tracks that he installed on a steam engine on which the wheels had been removed.

Three clutches were used — the master clutch for connecting the power source, and the track clutches. By releasing the track clutch on one side, the power applied through the track clutch on the other side caused the tractor to pivot around the declutched track. Application of brakes on the declutched side increased the speed of turning. This method of transmission continues to be used by the Caterpillar Tractor Co. and has been adopted by other track-laying tractor manufacturers.

The Holt Co. had already, in 1891, patented a tractor drive embodying independent friction clutches between the countershaft and each driving wheel, thereby eliminating the need of a differential.

The development of the crawler tractor was the outgrowth of the difficulty the Company had experienced with its wheel-type steam traction engine. The first of these was built in 1890. It soon developed that these wheel machines could not work on the soft peat delta soil without miring down. In an effort to solve the problem the wheels were made wider and wider. In 1903 one Holt outfit was constructed with 18 feet of wheel width on each side, and then it could hardly maneuver (page 12).

In 1889, Daniel Best sold his first steamer, a three wheeler with a vertical boiler. Apparently he had the same difficulty as did the Holt Company—bogging down in the soft peat soil. He also made his drive wheels wider and wider. One machine built shortly after 1900 boasted side wheels of 15 feet width. (page 000)

Of the Holt steamers only 8 of the track-type (page 86, 1906 Holt steam and 1908 Holt steam) were built. Experiments had already been made to replace steam power by gasoline and one model track-type tractor was produced in 1907 (page 87 - Holt gasoline).

The year 1908 marked the sale of the Best facilities to Holt and the end of an intense Best-Holt rivalry which had developed through the years. This was not for long however, as in 1910 C. L. Best, son of Daniel Best, organized the C. L. Best Gas Traction Co., Elmhurst, Calif., for the manufacture of wheel type tractors powered with Buffalo gasoline engines. After 3 years, in 1913, he marketed his first crawler the Best "75". In 1919 the Best "Sixty" was brought out and six years later, in 1925, the Holt and Best organizations combined to form the Caterpillar Tractor Co. The trade mark "Caterpillar" which originated about this time soon came to apply to most products of the Holt line.

During World War I the Holt Caterpillar Track-type Tractor played an important role as an artillery and supply tractor. In addition Holt's crawler was the inspiration of Britain's General E. D. Swinton, the inventor of the tank which created such havoc among the enemy troops and installations.

In 1910, the Bullock Tractor Co., Chicago, Ill., introduced its first track-laying type tractor, the Creeping Grip—12-20 H.P. (page 87). This model, after a few years of limited production, was discontinued in 1918 and in 1919 a larger model, 18-30 H.P., was produced. The following year the Company was bought by the Franklin Tractor Co., Greenville, Ohio, which placed the Franklin 15-30 on the market—patterned somewhat after the Bullock but with a number of improvements in design, such as radiator in place of tubular tank, cylinders cast en bloc in place of in pairs, removable cylinder heads and improved steering.

In 1911, the Yuba Ball Tread idea (page 87) came into being. According to the late B.D. Moses, "C.A. Henneuse a former shop forman of the C.L. Best Tractor Co., Elmhurst, Calif. conceived the idea of mounting the track of a crawler type machine on balls that rolled in a race. He, with a Mr. Frame of Benicia and W.J. Benson of San Jose engaged a man by the name of A.C. Johnson of Winters, Calif. to design the transmission. It contained 2 planetary marine type clutches, one for each track."

In 1912, the Ball Tread Company was formed and the first tractor, containing 2 tracks and a front steering wheel and delivering 12 H.P. at the drawbar was built in Detroit. It was called Model 12-25. In 1914 the Yuba Construction Co., later called the Yuba Mfg. Co., purchased the patent and manufacturing rights of the Ball Tread Co. and went into production. During this same year a larger model along the same lines was built, developing 18 H.P. at the drawbar.

Tractors were a side line with the Company whose chief interest was in gold dredgers and dredger equipment. Tractors were manufactured by this Company, however, until about 1928. According to H. A. Sawin, Sales Engineer of the Yuba Mfg. Co., "this was about the time that Yuba's two plants, one at Marysville, Calif. and one at Benicia, Calif. were combined at Benicia and the Marysville plant sold."

The Leavitt, a gas crawler tractor plow, was produced by H. W. Leavitt, Waterloo, Iowa also in 1912. This tractor had two steering wheels in front, a single track in the rear, and a 3-furrow plow mounted under the frame just back of the front axle and between the steering wheels, with a fourth plow at the rear (page 89). He also produced a track "catenate creeper wheel" which he claimed would "go on any wheeled tractor."

In 1913 the C. L. Best Gas Traction Co. of Stockton, California entered the track-type tractor field after having built round-wheel gas tractors, powered with Buffalo gas engines, beginning about 1908.

In 1914 the Killen-Strait Mfg. Co. of Appleton, Wis., built the Strait's track layer (page 89), and the Phoenix Mfg. Co. of Eau Claire, Wis., the Centiped (page 89) under Lombard patents.

Between 1910 and 1920 Holt, Best, Monarch, Cleveland, Bullock, and Yuba were the principal manufacturers of track-type tractors. While these were developed primarily for agricultural work, they proved useful for such industrial work as road building, logging, oil field work, etc.

In 1911, Rollin H. White, manufacturer of the White steam car and the White internal combustion engine, in cooperation with his brother, Clarence White, developed the Cleveland motor plow, which was the forerunner of the tractor later manufacturered by the Cleveland Tractor Co. These tractors had a planetary system of gears whereby steering was accomplished by holding one or the other of the main or sun gears by a brake band which caused the one track to pivot around the other. This was known as differential steering.

In 1916, the Cleveland Motor Plow Co. was incorporated and began producing their 10-drawbar-horsepower Model R tractor. In 1917 the Company changed its name to the Cleveland Tractor Co., and in 1918 devised as their trade-mark the name "Cletrac." The Tom Thumb (page 88) track-type tractor was also introduced this year as was also the Bates Steel Mule, Model C (page 88).

The Buckeye Mfg. Co., of Anderson, Ind., after 20 years' experimental work in developing a tractor, incorporated in 1896. In 1897 the Company produced a standard 4-wheel tractor, in 1911 a three-wheeler—(front wheel steering and rear wheel driving), in 1913 a three-wheeler (two front wheels steering and single driver in rear), in 1914 and 1915 the driving wheel of the previous model was replaced by a track, and in 1916 the "Trundaar" was produced with two driving tracks (page 89).

The Dayton-Dick Co. (about 1920 called the Dayton Dowd) manufactured the track-type "Leader" (page 88) in 1916, and the "Bear" (page 88) of the Bear Tractors Inc., New York, N.Y. appeared a year later.

In 1917, the design of the Bates Steel Mule was changed into a type having two tracks for driving and two front wheels for steering. The tractor was not marketed until 1921 (page 90).

In 1918 the Cleveland Tractor Co. of Cleveland, Ohio, produced their Model "H" (page 91). The Oliver "B" (page 90) was put on the market by the Oliver Tractor Co. of Knoxville, Tenn., as was also the Austin (page 91), by the F.C. Austin Co. of Chicago, Ill. The Monarch Tractor Co., Watertown, Wis., was producing three track-type models, the 12-20 (page 90) and 18-30 Neverslip and the 6-10 Lightfoot. This Company was organized in 1914, started producing tractors in 1917 and opened a branch factory in Brantford, Ontario, Canada, in 1918. In 1919 it was reorganized and operated under the name of General Tractors through 1921 and then in 1928 became the property of the Allis-Chalmers Co. The Bean Track Pull was produced by the Bean Spray Pump Co., San Jose, Calif., the Belt Rail by the Belt Rail Tractor Co., St. Paul, Minn., the J.T. (page 90) by the J.T. Tractor Co., Cleveland, Ohio, and the Best 25 (page 91) by the C.L. Best Traction Co., San Leandro, Calif. Some other track-type tractors which appeared before 1920 were the Sullivan, Gile, Hackney, Linn, Pan, Homer-Laughlin, Franklin, Hicks, Key Stone, and Union.

By the end of 1919 approximately 10 companies were manufacturing track-type tractors which were being sold not only in various sections of the United States but also in many foreign countries.

Some Early Models of Track-Type Tractors

1904 Holt (steam)

40 H.P. steam traction engine with track substituted for standard wheels. Used Cornish type return flue boiler, single cylinder engine. 10-1/4 inch bore x 12 inch stroke. Track frames rigidly attached to pair of frame cross members - with rollers and drive sprockets journaled to track frame. Adjustable track guide wheel at front of each frame. Tracks driven by long exposed chain. Two clutches.

1906 Holt (steam)

"Paddle wheel" traction engine, the wheels of which are encased in an endless chain of plates or paddles forming an endless track. Power furnished by a 50 H.P. steam engine. Built 6 steamers on tracks then experimented with gasoline powered crawler.

1907 Holt (gasoline)

40 H.P. at 500 r.p.m. - four cylinder vertical engine, valve in head, with open timing gears. Magneto and dry battery high tension ignition. Combination gravity and splash lubricating system. Track drive, sprockets carried entirely on the main frame and idlers adjustably mounted on forward end of frames. Truck frames bolted rigidly to transverse members - did not oscillate. Two-speeds forward, 2 and 4 m.p.hr. and planetary reverse.

1908 Holt (steam)

Holt steam crawler #122. Used on the Los Angeles Aqueduct project. Track referred to as 'platform wheel.''

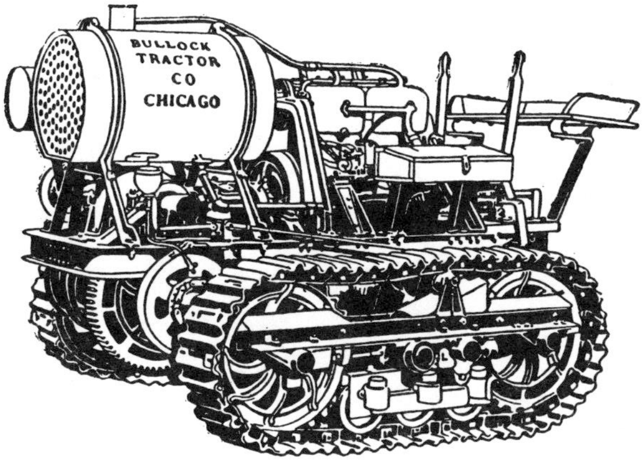

1910 Bullock Creeping Grip

12-20 H.P., 900 r.p.m., 4-cylinder Waukesha engine. Cooling, pump, fan, and tubular open tank; ignition, high-tension magneto with impulse starter; automatic splash lubrication; cone clutch; bevel-gear speed change; chain final drive to track drive sprocket shaft; one forward speed, 2-1/4 m.p.hr.

1911 Yuba Model 12 (gasoline)

12-25 H.P. at 750 r.p.m., 4-cylinder vertical Continental engine. High tension magneto ignition with automatic spark advance. Pump fan and radiator cooling; special carburetor for vaporizing low grade fuels such as distillate. Gasoline to start. Two speed transmission and independent clutch for each tread. Each of the two tracks run on two sets of large balls running in a steel ball race. Tiller-wheel provided to aid in steering.

1912 Holt (gasoline)

30 H.P. Built low for orchard work. Both sprockets and idlers journaled on full-width transverse axles. Track frames pivoted to the main frame at a point midway between track and frame ends. Steering accomplished by hand wheel and worm gear arrangement.

1912 Leavitt (gasoline)

Some Early Models of Track-Type Tractors (Gasoline)

1914 Killen-Strait

25-40 H.P., 650 r.p.m., Waukesha 4-cylinder vertical engine—cylinders cast in pairs. Cooling, closed system-pump, fan, and tank with horizontal tubes; Remy magneto high-tension ignition; bulged-tread endless chain for front wheel (steering) and rear (driving).

1914 Centiped

50 B.H.P., 700 r.p.m., 4-cylinder vertical engine. Double ignition—burned gasoline or motor spirits; frame, spring mounted; track frames oscillated on shaft independently of each other; sliding gear transmission; three forward speeds, 1-1/2, 3, and 5-1/2 m.p.h.

1916 Trundaar

20-35 H.P., 900 r.p.m., Waukesha Buckeye 4-cylinder vertical motor. Ignition, high-tension magneto with impulse starter; master lug at center of each grouser plate tool drive from hardened steel rollers in tread drivers mounted on rear axle; drivers incorporated individual multiple-disc clutch; front wheel smaller than driver, and built with double rim which acted as idler; front axle pivoted at center of main frame and each end rested on heavy spring mounted on main frame; expanding type brake in drum on each clutch housing; main clutch ran in oil; two forward speeds and one reverse.

1916 Tom Thumb

12-20 H.P., 800 r.p.m., 4-cylinder vertical engine. Cooling, pump, fan, and radiator; ignition, high-tension magneto with impulse starter; splash lubrication; friction-disc clutch; enclosed chain final drive to rear track drive sprockets, three forward speeds, 1-1/2 to 3-1/2 m.p.h.

1916 Bates Steel Mule C

13-30 H.P. at 900 r.p.m., Erd 4-cylinder vertical motor mounted lengthwise; cooling, pump, fan and tubular radiator; Dixie magneto with impulse starter; Bennett carburetor equipped with air cleaner; Pickering governor; two front steering wheels and one rear track for driving, forward speed 2-1/4 to 3-1/2 m.p.h.

1916 Leader

25-40 H.P., 750 r.p.m., 4-cylinder engine cast en bloc. Ignition, high-tension KW magneto; cooling, centrifugal pump, fan, and radiator; sliding-gear transmission, enclosed, running in oil; chain drive to overhead sprocket in tracks; two large and two small idlers with track adjustement mounted in frame which pivoted on axle; two forward speeds, 2-1/4 and 3-3/4 m.p.h.

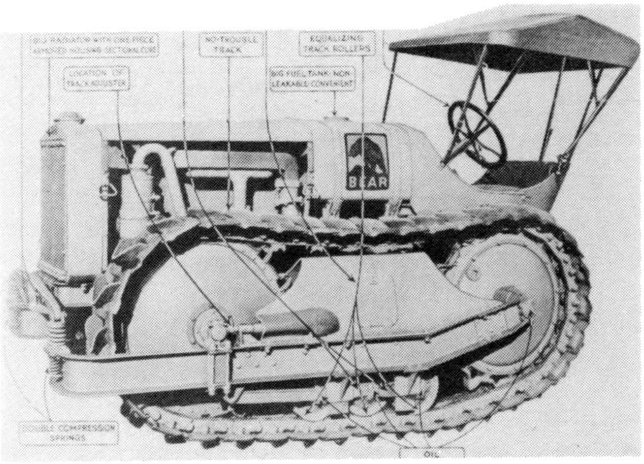

1917 Bear

25-35 H.P., 1,290 r.p.m., 4-cylinder vertical engine cast en bloc. Cooling, centrifugal pump, fan, and radiator; ignition, high-tension magneto with impulse starter; Pomona (Vortox oil type) air cleaner; lubrication, all rotating parts ran in oil bath with combination pump and splash for engine; oscillating 3-point front support of engine; multiple dry-disc clutch with master clutch and steering clutches; sliding gear transmission; three forward speeds, 2.35, 3.80 and 6.15 m.p.h.

1918 Bates Steel Mule D

12-20 H.P., 800 r.p.m., Erd 4-cylinder vertical motor. Cooling, centrifugal pump, fan, and radiator; Bennett carburetor and air cleaner, centrifugal air type; ignition, Eisemann high-tension magneto; two crawlers mounted independently flexible and each with two idlers; clutch, dry-plate disc type; transmission, spur gear; enclosed, running in oil; forward speeds, 2-1/4 to 4 m.p.h.

Some Early Models of Track-Type Tractors (Gasoline)

1918 Monarch

Three sizes, all 4-cylinders—the "Lightfoot" 6-10 H.P., 900 r.p.m., the "Neverslip" 12-20 H.P., 800 r.p.m., and 18-30 H.P., 750 r.p.m. Ignition, high-tension magneto and impulse starter. Track driving and idler sprockets were double-tooth designed to eliminate side play. The center track wheels in the larger two sizes were mounted in a track assembly built around angle irons with roller attached above to minimize sag. Braking and steering were accomplished by the use of a brake band and drum mounted on small sprocket shaft. The two tracks were pivoted at their center on the main axle. Forward speeds 2-1/4 m.p.h. for "Lightfoot" and 1-3/4 m.p.h. and 2-1/4 m.p.h. for "Neverslip."

1918 Oliver

Model A - 15-30 H.P., 900 r.p.m., 4-cylinder engine. Dixie high-tension magneto ignition with impulse starter; Borg and Beck clutch; own make of transmission; two tracks each driven by a cone clutch; steering accomplished by the clutches; two speeds forward 1-3/4 and 2-1/3 m.p.h. Model B - 12-20 H.P., 900 r.p.m. engine. Dixie magneto ignition with impulse starter; Borg and Beck Clutch; Foote-Strige transmission; two front wheels and two tracks, with steering accomplished by front wheels; brake on differential; two forward speeds, 1-3/4 and 2-1/3 m.p.h.

1918 J. T. Model "N"

35 H.P., 800 r.p.m., and 40 H.P., 1,000 r.p.m. 4-cylinder vertical engines. KW high-tension magneto ignition with impulse starter; sliding-gear transmission; double crawler.

1918 Best Model 25

25 H.P., 4-cylinder vertical engine. Cooling, pump, fan and radiator; ignition, high-tension magneto with impulse starter; lubrication splash system; sliding-gear transmission; final drive, internal gear to crawler sprockets; forward speeds, 2 to 3 m.p.h.

1918 Bean Track Pull

6-10 H.P., 4-cylinder vertical Le Roi engine. Bosch magneto high-tension ignition; water cooled with centrifugal pump and fan; combination pump and splash lubrication; spur-gear transmission from engine to internal gear inside driving sprocket which drove track; steering accomplished by hand wheel and chain which actuated a bevel pinion and bevel gear at pivot point; bevel gear was attached to steel casting on which engine, transmission, and traction units were mounted; forward speed, 2 m.p.h.

1918 Belt Rail Model B

12-20 H.P., Waukesha 4-cylinder vertical engine. Cooling, pump fan and Withington radiator; ignition, high-tension magneto splash lubrication; single crawler drive in rear; gear-drive transmission; forward speed, 2-1/4 to 3-1/2 m.p.h.

1918 Austin

20-40 H.P., 4-cylinder vertical engine. Cooling, centrifugal pump, fan, and radiator; lubrication, force feed and splash; ignition, KW high-tension magneto with impulse starter; sliding-gear transmission mounted in front of engine; final drive, chain from countershaft sprocket to rear drive sprocket; three forward speeds.

1918 Cleveland "H"

12-20 H.P., 1,200 r.p.m. Weidley 4-cylinder engine. High-tension magneto ignition; fan, pump, and radiator cooling; steering by planetary compensating differential gears; clutch, double plate; sliding-gear transmission; three forward speeds, 1 to 3-1/2 m.p.h.

THE AGRICULTURAL TRACTOR:
1855-1950

PART II
1920-1950

Introduction

NINETEEN hundred and twenty marked the culmination of a continuous annual upswing in tractor production from 2,000 units in 1909 to 203,207 units in 1920 (table 1). Census figures showed 246,000 tractors on farms at the beginning of 1920.

Table 1.[1] Number of tractor companies,[2] number of tractors produced,[3] and number of horses and mules on farms, 1904-1920

Year	Tractor companies	Tractor production	Horses and mules (000 omitted)
1904	6	—	19,494
1905	6	—	19,947
1906	9	—	22,123
1907	8	—	23,564
1908	6	—	23,861
1909	9	2,000	24,693
1910	15	4,000	25,163
1911	20	7,000	24,600
1912	31	11,500	24,871
1913	39	7,000	24,953
1914	58	15,000	25,411
1915	61	21,000	25,674
1916	114	29,670	25,752
1917	124	62,742	25,933
1918	142	132,697	26,428
1919	164	164,590	26,436
1920	166	203,207	26,104

[1]*Data from September 15,1937, issue of Farm Machinery and Equipment.*
[2]*Inclues only companies listed in Implement Blue Book who had actually built 1 or more tractors.*
[3]*Includes wheel, track-type, and garden tractors. (Bureau of Census — Manufacture and Sale of Farm Equipment.)*

From the standpoint of development, the 1920 tractor, taken collectively, embodied fundamental engineering and designing found perhaps in more refined form in tractors of the present day (1950). The one-piece cast-iron frame, replaceable wearing parts, force-feed and pressure-gun lubrication, enclosed transmissions, carburetor manifolding, air cleaners, electric lighting and starting, the high tension magneto ignition with impulse starter, enclosed cooling systems, antifriction bearings, alloy and heat-treated steels, and the power takeoff had all been introduced and some experiments had been made with rubber tires. The lightweight low-price tractor had been designed and widely accepted, and several fairly successful motor cultivator type units were on the market.

The postwar depression that swept the country after World War I was reflected in a lowering of tractor production from 1921 to 1924, inclusive. The huge effort to produce farm power equipment incident to World War I, and the effort that went into producing food and carrying out other programs to make for good business several years after the war were diverted to the more sane business of incorporating new features into tractor designing. In the keen competition that followed, the quality of the tractor was steadily improved, although a number of companies had to discontinue business. Much effort was being centered on the development of a general-purpose tractor one that not only would plow but, with proper attachments, would plant and cultivate row crops, mow and do related work.

The motor cultivator was a step in the right direction, but few farmers could afford to invest in a unit that could cultivate only, so these machines were soon to be discarded in favor of the more versatile unit the general-purpose tractor.

The first tractor test at Lincoln, Nebr., postponed in 1919 because of snow, was begun on March 21, 1920, and completed April 4 (see page 000). The Nebraska tests for tractors have since become significant not only in Nebarska and the Nation as a whole, but in a score of foreign contries as well. These tests have provided standards for rating tractors, have speeded up improvements on many of them, and have eliminated from the field many others of inferior design.

Of the 69 makes and models of tractors entered in 1920 at Lincoln, 64 were tested in accordance with the recently enacted Nebraska State tractor-test law. These tractors (page 2) included many makes and models that had appeared on the market during the previous years, as well as models appearing the current year for the first time.

Tractor Developments in the 1920's

1920

The Parrett Motors Corp., Chicago, Ill., was organized by Dent and Henry Parrett to build and market a 2-row motor cultivator. The company was entirely independent of the Parrett Tractor Co., which was manufacturing tractors in 1916 in the same city.

The Twin City 20-35 (rated 27-44 in 1926) tractor was produced by the Minneapolis Steel & Machinery Co., Minneapolis, Minn., models of the Electric Wheel Co.'s (Quincy, Ill.) "All Work" tractor "C", "J" (page 3), and "II" 2 models of the Advance-Rumely Thresher Co.'s (La Porte, Ind.) Oil Pull 12-20 and 20-40 and the "Webfoot" (page 3) by the Blewet Tractor Co. of Tacoma, Wash., were introduced. The Best 60 (page 3) tractor was introduced by the C. L. Best Tractor Co., San Leandro, California. About this time both the Best and Holt Companies dropped the tiller (front) wheels from their track-type models.

The 15-30 model of the Keck-Gonnerman Co., Mt. Vernon, Ind. to replace the previous 12-24 model, was produced. Case introduced the 22-40 (page 3), which was rerated at 25-45 in 1924. A larger version of the 15-27 but with heavy channel frame, was also produced by the J. I. Case Threshing Machine Co., Racine, Wis. The Samson M (page 3) 2-plow tractor, manufactured by the Samson Tractor Co., Janesville, Wis., appeared. In general appearance this machine resembled the Fordson, but the Samson M provided a 6-inch gap between the water and fuel tanks.

The John Lauson Co., New Holstein, Wis., introduced its 15-30 hp. tractor, the Lauson Full Jewel; and Kinnard & Sons Mfg. Co, Minneapolis, Minn., produced 4 models the 14-24, 20-35, 30-50, and the 40-70 hp., a couple of which had been on the market several years.

Summary of Results of the Nebraska Tractor Tests, 1920, Conducted By The University of Nebraska—Department of Agricultural Engineering—Lincoln, Nebraska

TRACTOR / MOTOR / BELT TESTS — RATED LOAD, TWO HOURS

No. of Test	NAME	Rating	Weight	Cyls.	Size	R.P.M.	Make	Av'ge H.P.	Average R.P.M.	Fuel Kind	Fuel Gallons perHour	Fuel H.P.H'rs per Gal.	Water Gallons perHour	Temperature Air	Temperature Rad.
1	WATERLOO BOY	12–25	6,183	2	6½x7	750	Own	25.51	771	Kero.	3.28	7.78	0.46	62	177
[illegible]	CASE	10–18	3,760	4	3⅞x5	1,050	Own	18.41	1,058	Kero.	2.95	6.24	0.50	61	180
[illegible]	CASE	15–27	6,460	4	4⅛x6	900	Own	27.52	914	Kero.	2.68	10.27	0.12	65	185
5	CASE	22–40	9,940	4	5½x6¾	850	Own	40.08	856	Kero.	4.12	9.72	0.74	61	180
6	CASE	10–20	5,080	4	4⅛x6	900	Own	20.19	911	Kero.	2.40	8.42	2.41	70	204
7	CASE	20–40	13,780	2	8¾x9	475	Own	40.29	482	Kero.	7.12	5.66	5.54	64	163
8	OIL PULL "E"	30–60	26,000	2	10x12	375	Own	60.20	375	Kero.	7.73	7.79	8.40	68	166
9	OIL PULL "H"	16–30	9,506	2	7x8½	530	Own	30.50	532	Kero.	3.07	9.94	3.28	68	197
10	OIL PULL "K"	12–20	6,638	2	6x8	560	Own	20.06	564	Kero.	1.85	10.82	4.81	58	200
11	OIL PULL "G"	20–40	12,969	2	8x10	450	Own	39.90	462	Kero.	4.29	9.30	4.63	76	193
12	HUBER LIGHT 4	12–25	5,550	4	4½x5¾	1,000	Waukesha	25.28	1,003	Kero.	4.16	6.08	2.63	86	180
13	MINNEAPOLIS	12–25	6,600	4	4¾x7	750	Own	25.10	748	Kero.	3.74	6.71	2.50	85	172
14	MINNEAPOLIS	22–44	12,410	4	6x7	700	Own	44.15	705	Kero.	5.97	7.40	3.74	78	173
15	MINNEAPOLIS	35–70	22,500	4	7¼x9	550	Own	70.31	556	Kero.	8.83	7.96	5.00	73	203
16	HEIDER "C"	12–20	6,200	4	4½x6¾	900	Waukesha	20.18	902	Kero.	2.96	6.81	0.19	77	199
17	HEIDER "D"	9–16	4,100	4	4½x5¾	1,000	Waukesha	15.96	1,001	Kero.	2.18	7.32	0.30	70	183
18	FORDSON	18	2,710	4	4x5	1,000	Own	18.16	993	Kero.	2.48	7.32	4.00	91	210
19	TWIN CITY	12–20	5,000	4	4⅛x6	1,000	Own	20.32	1,009	Kero.	2.58	7.88	0.00	84	179
20	EMERSON-BRAN	12–20	4,400	4	4¾x5	900	Own	20.27	906	Kero.	2.40	8.45	0.13	79	202
21	WISCONSIN "E"	16–30	6,060	4	5x6½	900	Climax	30.24	901	Kero.	5.54	5.46	0.50	87	189
22	GRAY	18–36	6,510	4	4¾x6¾	950	Waukesha	30.46	958	Gaso.	5.22	5.83	2.25	100	209
23	TITAN	10–20	5,708	2	6½x8	575	Own	20.18	577	Kero.	2.57	7.86	6.00	74	210
24	INTERNATIONAL	15–30	8,990	4	5¼x8	575	Own	30.73	580	Kero.	4.47	6.88	4.50	73	212
25	INTERNATIONAL	8–16	3,660	4	4¼x5	1,000	Own	16.06	1,007	Kero.	2.19	7.32	3.00	77	211
26	HART-PARR	30	5,450	2	6½x7	750	Own	30.22	755	Kero.	3.76	8.04	2.50	75	183
27	SAMSON "M"		3,300	4	4x5½	1,100	Own	17.29	1,111	Kero.	2.32	7.46	0.44	86	197
28	BEEMAN "G"	2–4	550	1	3½x4½	230–1,500	Own	2.40	1,036	Gaso.	0.44	5.43	0.63	76	210
29	LA CROSS "G"	12–24	4,670	2	6x7	900	Own	24.23	899	Kero.	3.69	6.56	0.19	84	201
30	AULTMAN-TAYLOR	30–60	24,450	4	7x9	550	Own	60.29	553	Kero.	6.92	8.71	6.00	80	190
31	AULTMAN-TAYLOR	15–30	8,240	4	5x6½	900	Climax	30.40	906	Kero.	4.08	7.44	0.38	83	191
32	AULTMAN-TAYLOR	22–45	12,880	4	5½x8	600	Own	44.92	600	Kero.	5.36	8.38	5.50	80	202
33	MOLINE "D"	9–18	3,590	4	3½x5	1,800	Own	18.27	1,840	Gaso.	2.24	8.15	1.00	93	212
35	COLEMAN "B"	16–30	5,100	4	5x6½	900	Climax	30.27	914	Kero.	5.67	5.34	3.50	100	211
37	PARRET "K"	15–30	5,250	4	4½x6	1,000	Own	31.31	1,015	Kero.	3.46	9.04	0.50	79	196
39	AVERY "C"		3,164	6	3x4	1,250	Own	14.08	1,252	Gaso.	2.33	6.03	0.25	95	209
40	AVERY		3,450	6	3x4	1,250	Own	15.54	1,243	Gaso.	2.50	6.22	0.16	87	202
41	AVERY	12–20	5,500	4	4⅜x6	800	Own	20.18	804	Kero.	2.25	8.98	5.50	72	212
42	AVERY	14–28	7,540	4	4⅜x7	900	Own	28.16	901	Kero.	3.42	8.24	10.00	77	212
43	AVERY	25–50	12,500	4	6½x8	700	Own	50.40	702	Kero.	6.37	7.91	19.50	88	212
44	AVERY*	40–80	22,000	4	7¾x8	600	Own	65.73	612	Kero.	9.06	7.26	17.00	88	212
45	CLETRAC "W"	12–20	3,300	4	4x5½	1,265	Weidely	20.14	1,276	Kero.	2.42	8.33	0.50	97	194
46	FRICK "C"	15–28	6,100	4	4¾x6	900	Beaver	28.26	906	Kero.	4.55	6.21	0.75	91	211
47	FRICK "A"	12–20	5,800	4	4x6	900	Erd	22.40	906	Kero.	3.35	6.69	0.08	84	196
48	TWIN CITY	40–65	25,550	4	7¼x9	535	Own	65.96	534	Kero.	7.87	8.38	10.50	85	212
49	WALLIS	15–25	3,500	4	4¼x5¾	900	Own	25.19	918	Kero.	3.25	7.74	1.12	92	210
50	FLOUR CITY	18–35	10,000	4	5½x6	800	Own	35.23	805	Kero.	4.59	7.68	2.00	64	207
51	LAUSON	15–30	6,500	4	4¾x6	950	Beaver	30.14	960	Kero.	3.68	8.19	1.81	109	212
52	FLOUR CITY	40–70	21,000	4	7½x9	575	Own	70.81	558	Kero.	9.35	7.57	9.00	91	212
53	ALL WORK	14–28	5,000	4	5x6	900	Own	28.41	915	Kero.	4.21	6.75	0.51	100	209
54	ALLIS-CHALMERS	6–12	2,500	4	3⅛x4½	1,000	Le Roi	12.08	1,011	Gaso.	1.70	7.11	0.00	77	204
55	ALLIS-CHALMERS	18–30	6,000	4	4¾x6½	830	Own	30.58	839	Kero.	3.29	9.30	0.67	83	179
56	MONARCH	18–30	8,100	4	4¾x6	950	Beaver	30.55	952	Kero.	3.62	8.43	1.00	84	199
57	AVERY CULTI	5–10	2,650	4	3x4	1,200	Own	10.11	1,246	Gaso.	1.86	5.45	0.00	74	190
58	AVERY	18–36	9,250	4	5½x6	800	Own	36.70	812	Kero.	5.81	6.31	14.37	71	212
59	HOLT "T-11"*	25–40	9,400	4	4¾x6	1,050	Own	30.24	1,070	Gaso.	3.81	7.94	0.25	90	191
60	BATES-S-MULE	15–22	4,600	4	4¼x6	1,100	Erd	22.66	1,110	Kero.	3.18	7.12	0.37	81	206
61	HOLT "T-16"*	40–60	18,500	4	6½x7	750	Own	55.25	763	Gaso.	7.72	7.16	2.25	91	197
62	INDIANA	5–10	2,400	4	3⅛x4½	1,000	Le Roi	10.20	1,024	Gaso.	1.20	8.52	2.41	92	209
63	TOWNSEND	15–30	7,000	2	7x8	525	Own	28.35	526	Kero.	3.02	9.38	8.00	90	210
64	UNCLE SAM	20–30	4,500	4	4¾x6	1,000	Beaver	30.75	1,025	Kero.	6.34	4.85	0.50	90	181
65	TORO	6–10	2,600	4	3⅛x4½	1,200	Le Roi	10.56	1,208	Gaso.	1.25	8.48	0.48	102	208
66	SQUARE TURN*	18–35	7,800	4	5x6½	850	Climax	30.35	854	Kero.	4.75	6.39	0.74	84	206
67	TWIN CITY	20–35	8,100	4	5½x6¾	900	Own	35.22	905	Kero.	4.20	8.39	0.12	78	184
68	BATES-S-MULE	15–22	4,600	4	4½x5¼	1,100	Midwest	22.20	1,112	Gaso.	3.00	7.39	0.39	69	161
69	PORT HURON	12–25	6,750	4	4¾x6	900	Chief	25.14	893	Kero.	4.63	5.43	0.39	61	182

*Rating changed, see text

BELT TESTS — MAXIMUM LOAD, ONE HOUR / VAR. L. 1 HOUR

No. of Test	H. P.	R.P.M.	Fuel Kind	Fuel Gallons perHour	Fuel H.P.H'rs per Gal.	Water Gallons perHour	Temperature Air	Temperature Rad.	Var. L. R.P.M. High	Var. L. R.P.M. Low
1	25.97	724	K	3.80	6.83	0.83	61	178	874	713
[illegible]	18.15	1,043	K	2.75	6.60	0.34	53	180	1,244	1,003
[illegible]	31.23	924	K	3.16	9.90	0.45	60	203	993	846
5	49.97	867	K	6.90	7.24	1.86	62	176	926	787
6	22.81	895	K	2.61	8.75	1.99	71	204	980	881
7	42.80	473	K	7.68	5.57	3.72	62	165	496	467
8	75.60	378	K	10.73	7.05	10.48	70	158	420	374
9	33.52	537	K	5.96	5.62	7.37	68	164	568	529
10	25.87	566	K	3.25	7.96	3.69	69	184	587	560
11	46.19	465	K	5.71	8.10	6.31	63	165	482	458
12	25.70	1,005	K	4.45	5.77	2.77	71	162	1,286	999
13	26.24	760	K	4.24	6.20	2.41	82	161	943	749
14	46.04	701	K	6.74	6.83	3.68	88	178	746	701
15	74.01	556	K	9.97	7.42	4.50	65	203	625	547
16	24.24	902	K	4.34	5.59	0.44	68	173	910	874
17	19.54	1,003	K	2.89	6.75	1.58	73	186	1,004	988
18	19.15	1,014	K	3.00	6.38	3.35	91	209	1,007[1]	996
19	27.93	1,017	K	3.18	8.77	0.23	90	202	1,077	1,003
20	25.90	916	K	3.53	7.29	2.50	90	212	941	905
21	31.50	914	K	5.64	5.58	0.36	92	200	986	893
22	32.20	958	G	4.83	6.67	5.50	103	210	1,452	948
23	28.15	580	K	4.64	6.06	7.00	63	212	591	572
24	36.98	577	K	7.29	5.07	5.00	76	212	641	564
25	18.52	1,007	K	3.24	5.71	1.00	77	210	1,083	1,002
26	31.37	756	K	4.74	6.62	3.00	76	168	811	747
27	19.39	1,111	K	2.79	6.94	2.62	94	203	1,215	1,100
28	2.37	1,023	G	0.42	5.60	0.75	80	210	1,093	951
29	24.94	902	K	4.36	5.72	0.25	90	192	1,225	896
30	75.49	562	K	10.12	7.46	9.00	94	174	584	532
31	34.37	898	K	5.79	5.94	0.13	73	180	968	904
32	46.66	607	K	6.23	7.49	5.00	82	177	720	602
33	27.45	1,788	G	3.09	8.88	7.00	97	215	1,910	1,827
35	30.41	907	K	5.72	5.32	7.00	103	212	1,125	907
37	31.79	1,010	K	3.96	8.03	0.50	82	198	1,109	994
39	14.62	1,246	G	2.39	6.12	0.75	97	208	1,435	1,250
40	15.77	1,240	G	2.54	6.22	0.17	90	202	1,950	1,259
41	24.26	800	K	3.80	6.38	6.25	81	212	847	796
42	31.83	898	K	4.17	7.63	10.69	88	212	958	874
43	56.68	712	K	7.04	8.05	25.00	99	212	893	691
44	69.23	597	K	8.23	8.41	18.75	90	216	678	615
45	24.94	1,321	K	2.90	8.61	2.00	109	212	1,459	1,303
46	29.72	917	K	5.21	5.71	1.00	98	212	1,199	895
47	22.31	903	K	3.39	6.59	0.08	88	197	966	904
48	65.53	530	K	7.88	8.32	6.00	80	212	752	531
49	27.57	907	K	4.01	6.88	0.33	84	199	1,224	919
50	35.46	807	K	4.68	7.58	2.50	71	211	920	785
51	32.46	980	K	4.49	7.23	0.75	94	212	1,008	956
52	72.52	563	K	9.62	7.54	17.00	94	212	857	562
53	28.86	915	K	4.95	5.83	0.26	101	207	1,009	906
54	12.37	1,008	G	1.75	7.05	0.00	81	205	1,214	1,005
55	33.41	842	K	4.61	7.24	1.00	88	164	1,000	834
56	31.40	957	K	3.94	7.97	1.00	86	202	1,056	950
57	11.14	1,228	G	1.86	5.99	0.00	74	188	1,309	1,222
58	44.50	812	K	6.84	6.50	16.00	80	212	837	809
59	35.52	1,066	G	4.38	8.11	0.32	87	201	1,186	1,068
60	24.84	1,116	K	3.76	6.61	0.49	77	192	1,212	1,088
61	57.21	772	G	8.94	6.41	2.00	86	206	900	760
62	11.34	1,023	G	2.21	5.13	0.33	83	210	1,129	1,001
63	29.51	533	K	3.90	7.58	7.00	79	203	582	501
64	32.20	1,041	K	6.41	5.02	0.14	88	192	1,099	952
65	13.31	1,219	G	1.73	7.70	1.27	79	210	1,326	1,197
66	32.19	848	K	5.34	6.02	0.29	85	205	1,020	844
67	46.88	905	K	6.00	7.81	0.93	83	203	925	878
68	29.78	1,108	G	3.59	8.30	0.42	73	187	1,136	1,061
69	28.46	895	K	4.26	6.69	0.98	61	194	1,069	905

[1]Hand controlled

DRAWBAR TESTS — RATED LOAD, TEN HOURS / MAXIMUM / MISCELLANEOUS

No. of Test	H. P.	Pull Lbs.	Miles per Hour	Engine Speed R.P.M.	Wheel Slip %	Gear Used	Lugs Used	Fuel Kind	Fuel Gallons perHour	Fuel H.P.H'rs per Gal.	Water Gallons perHour	Temperature Air	Temperature Rad.	Maximum Pull Pounds	Maximum Miles per Hour	Total No. Hours	Lub. Oil Gal.
1	12.10	1,982	2.29	778	11.84	Low	Angle	K	2.86	4.23	0.19	49	161	2,900	2.07	44	3¾
[illegible]	9.61	1,540	2.35	1,149	17.60	Low	Spade	K	1.35	5.20	0.20	66	178	1,730	2.41	40	8
[illegible]	15.76	2,700	2.19	927	14.16	Low	Spade	K	2.50	6.30	0.22	56	175	3,440	2.05	37	11½
5	23.51	4,180	2.11	852	13.53	Low	Spade	K	3.84	6.13	0.69	60	168	4,965	2.19	53	29
6	11.63	1,904	2.29	919	12.66		Spade	K	1.77	6.57	0.41	67	175	2,631	2.18	30	8½
7	21.48	3,987	2.02	485	4.76	Low	Mall.	K	6.35	3.89	4.44	57	160	5,537	1.67	40	12¾
8	27.91	5,509	1.90	380	5.22		Mall.	K	6.52	4.28	6.48	74	130	10,025	1.87	35	9
9	16.68	3,036	2.06	540	8.90	Low	Angle	K	2.66	6.27	2.75	74	186	4,674	1.84	35	7
10	13.46	2,399	2.10	588	12.08	Low	Angle	K	2.49	5.41	2.73	74	169	2,780	2.02	44	5
11	20.87	3,913	2.00	455	9.09	Low	Angle	K	4.58	4.56	4.80	79	163	6,365	1.77	36	9½
12	15.06	1,976	2.86	989	7.19	Low	Spade	K	3.44	4.37	1.11	68	181	2,505	2.50	45	8½
13	12.33	2,150	2.15	759	10.31	Low	Angle	K	3.25	3.80	1.05	79	182	2,852	2.14	32	6
14	25.35	3,523	2.70	701	6.98		Mall.	K	5.51	4.60	3.90	67	167	5,104	2.44	33	6½
15	38.67	6,910	2.10	556	7.40		Cleats	K	7.27	5.32	1.91	68	175	10,998	1.79	44	6¼
16	13.20	2,135	2.32	906	15.60		Cone	K	3.04	4.35	0.95	86	200	2,223	2.26	36	2½
17	9.88	1,472	2.52	977	8.50		Angle	K	2.18	4.53	0.93	92	202	1,900	2.32	33	4¾
18	6.06	886	2.57	1,060	23.80	Int.	Angle	K	2.48	2.45	2.10	89	209	1,428	2.17	34	3¾
19	13.86	1,772	2.93	988	10.70	High	Angle	K	2.77	5.01	0.08	86	192	2,061	2.92	32	3
20	15.54	2,037	2.86	944	6.76	High	Spade	K	2.71	5.73	0.03	69	180	2,368	2.72	30	3½
21	15.60	2,500	2.34	899	16.80	Low	Angle	K	5.38	2.90	0.34	70	173	3,426	2.43	34	3
22	19.64	2,703	2.73	925	9.91	High	Spade	G	4.39	4.48	3.53	89	212	3,044	2.24	43	2¾
23	9.94	1,827	2.04	573	11.70	Low	Angle	K	3.01	3.31	5.84	67	207	2,660	1.93	30	8
24	15.82	2,405	2.47	594	9.30	High	Angle	K	5.13	3.08	1.72	70	208	4,210	2.31	31	5½
25	8.08	1,120	2.71	1,026	16.50	Int.	Angle	K	2.53	3.19	0.26	72	209	1,532	2.60	32	4¾
26	15.37	2,073	2.78	752	11.10	High	Angle	K	3.24	4.74	1.53	71	173	2,788	2.64	30	2
27	9.32	1,252	2.94	1,084	14.50	High	Angle	K	2.27	4.33	1.32	85	202	1,670	2.60	37	9¾
28	1.08	188	2.17	1,334	22.25		Angle	G	0.29	3.78	0.31		204	187	1.56	40	¾
29	13.63	1,859	2.75	821	7.90		Spade	K	2.86	4.76	0.30	87	198	2,155	3.10	30	4
30	35.09	5,184	2.54	552	2.00		Spade	K	7.06	4.97	5.61	91	176	9,160	2.38	44	17½
31	15.73	2,497	2.36	816	5.40		Spade	K	4.02	3.91	0.00	77	175	2,838	2.80	35	5
32	24.44	3,723	2.46	626	3.60	Low	Spade	K	4.13	5.92	1.53	77	184	4,986	2.11	36	16¾
33	9.61	1,041	3.46	1,695	4.50		Spade	G	2.08	4.63	0.14	85	212	1,778	3.67	31	4¾
35	15.30	1,961	2.92	898	8.70	High	Spade	K	5.60	2.73	3.08	92	210	1,802	3.30	32	6
37	15.99	2,197	2.73	1,002	8.10	Low	Spade	K	3.10	5.15	0.61	85	193	2,988	2.63	54	5½
39	7.03	1,216	2.17	1,156	13.60	Int.	Spade	G	2.52	2.79	0.19	80	209	1,270	2.34	44	4½
40	8.51	1,213	2.63	1,291	10.55	Int.	Angle	G	2.54	3.34	0.06	83	203	1,423	2.26	32	3¾
41	13.98	2,238	2.34	820	12.00	Low	Spud	K	3.29	4.25	6.44	81	211	2,608	2.53	30	6¼
42	17.68	2,646	2.50	787	8.30	Low	Angle	K	3.57	4.94	10.35	87	210	3,049	2.65	33	7¾
43	27.29	3,778	2.71	694	8.30	Low	Spud	K	6.42	4.25	17.81	87	208	4,415	2.67	32	10½
44	46.93	7,812	2.25	634	5.00	Low	Uni.	K	11.85	3.96	11.37	81	210	8,475	2.21	39	12
45	13.47	1,395	3.62	1,386	4.30			K	2.67	5.04	0.51	85	195	1,734	3.36	30	5¾
46	15.24	2,372	2.41	910	7.60	Low	Spade	K	3.41	4.46	0.34	88	203	3,264	2.24	32	4¾
47	12.10	1,900	2.39	900	7.50	Low	Spade	K	2.97	4.07	0.20	88	191	2,340	2.30	30	5
48	41.19	7,736	2.00	548	9.80		Angle	K	7.18	5.73	3.94	80	201	10,820	1.72	28	15½
49	16.06	2,362	2.55	928	16.35	Low	Spade	K	4.54	3.54	0.06	81	190	2,782	2.50	32	4¼
50	18.29	3,784	1.81	752	6.30	Low	Spade	K	4.57	4.00	3.65	78	212	4,116	1.71	30	12¾
51	17.68	2,512	2.64	963	6.76	High	Pyram.	K	3.63	4.89	0.21	76	196	3,115	3.09	31	4½
52	42.76	6,869	2.33	564	8.25		Pyram.	K	7.74	5.52	8.19	77	208	8,404	2.37	38	11½
53	15.24	2,133	2.68	903	7.20	High	Pyram.	K	4.57	3.33	0.07	81	178	2,442	2.79	35	9
54	6.29	1,046	2.28	1,029	14.60		Angle	G	1.61	3.99	0.09	75	205	1,142	2.06	29	2
55	20.55	2,704	2.85	879	12.70	High	Spade	K	3.47	5.93	0.05	72	182	2,450	2.95	33	7
56	19.28	3,443	2.10	946	3.20	Int.		K	3.72	5.18	0.72	79	182	4,670	1.69	31	3½
57	6.00	914	2.46	1,223	8.70	Int.	Angle	G	1.83	3.27	0.09	75	192	854	2.36	30	1
58	20.58	3,202	2.41	811	9.05	Low	V-Cleats	K	5.96	3.45	8.47	71	210	4,590	2.25	30	6
59	26.54	3,336	2.98	1,050	1.90	Int.		G	4.15	6.39	0.75	79	188	3,546	3.53	35	7¾
60	16.99	2,566	2.48	1,203	2.60	Low		K	3.11	5.46	0.52	74	202	2,996	2.59	29	4¾
61	42.76	4,963	3.23	779	0.60	Int.		G	8.70	4.91	0.60	85	173	5,250	3.69	33	9
62	5.00	849	2.21	1,046	10.30		Spade	G	1.35	3.71	0.39	84	209	1,189	1.79	28	2
63	15.26	2,559	2.24	544	15.40		Angle	K	3.23	4.73	6.83	81	202	2,681	2.50	30	2
64	21.98	2,954	2.79	1,027	18.20	Low	Spade	K	5.74	3.83	0.23	71	185	3,264	2.57	34	4½
65	6.32	893	2.66	1,215	8.04	Low	Spade	G	1.47	4.31	0.11	50	206	1,310	2.84	30	1½
66	19.13	2,904	2.47	913	7.41		Angle	K	4.90	3.91	0.16	75	190	3,090	2.85	39	5
67	20.11	2,591	2.91	977	5.09	High	Spade	K	4.42	4.55	0.35	75	174	4,179	2.99	32	8½
68	16.17	2,558	2.37	1,130	2.67	Low		G	2.66	6.07	0.19	69	179	3,100	2.81	32	4½
69	14.36	2,320	2.32	1,197	10.55		Angle	K	3.84	3.74	0.12	44	162	3,427	2.19	32	9¼

Nebraska Tractor Test Summary, 1920

While some tractors appearing before 1920 may not have been tested until 1921 and still others appearing then or since were not tested the year they first made their appearance, it is felt, that a reasonable appearance date would be the year the model first appeared on record in the Nebraska test report, in a tractor field book or other recognized periodical.

Some 1920 Models

Samson M (2 plow)

Case 22-40

Webfoot 28-53

All Work J
(shielded for orchard use)

Minneapolis 22-44

Best 60

1921

A nationwide depression in 1921 reduced the demand for tractors, which occasioned price-cutting on well-known models of all makes. To stimulate the tractor trade a power-farming demonstration was held at Fargo, N. Dakota, with 26 manufacturers competing. As an added attraction eleven teams of horses were also entered with 5 to 8 horses to a team. Unfortunately, however, before the demonstration was over a number of horses had succumbed to the extreme heat. This indicated among other things, the need of tractor power to meet such conditions.

Although business had already started on the downgrade, several new tractors of special interest were produced. In January the United States Tractor & Machinery Co. of Menasha, Wis., announced its new Uncle Sam 12-20 kerosene-burning tractor, a smaller version of the 20-30 produced about 1918. In this model (12-20) the entire drive train behind the engine was made up of roller chains and sprockets instead of gears, as used in the 20-30 model.

The McCormick-Deering 15-30 (page 5) was introduced by the International Harvester Co. of Chicago, Ill. This tractor was of one-piece frame construction, with a 4-cylinder vertical engine fitted with a ballbearing crankshaft, removable cylinder sleeves, and a power take-off incorporated as standard equipment. This model replaced International 15-30, which was powered with a 4-cylinder horizontal double twin engine.

Four styles of Model F Cletrac (page 5) were put on the market by the Cleveland Tractor Co. of Cleveland, Ohio — the Standard with 25-inch tread and 8-inch clearance and 3 others, each with 36-inch tread and 18-inch clearance. This model was characterized by the use of a floating roller chain.

The Toro Mfg. Co., Minneapolis, Minn., offered its first Push Type Toro Tractor with Pennsylvania mowers (page 4). The Light 4 tractor of the Huber Mfg. Co.'s line (Marion, Ohio) was supplemented by another one of the same type but of greater horsepower — the "Super Four" (page 5). Also of special interest was the appearance of the 16-30 hp. steam-tractor engine (page 5) manufactured by the A. D. Baker Co., Swanton, Ohio. This tractor had a cross compound engine fitted with a condenser and stoker.

Other tractors that appeared this year included the Allis-Chalmers 12-20, manufactured by the Allis-Chalmers Co., Milwaukee, Wis., Avery 12-25 and 8-16, manufactured by the Avery Co., Peoria, Ill., Eagle 16-30 and 12-22, by the Eagle Mfg. Co., Appleton, Wis., Minneapolis 17-30, by the Minneapolis Threshing Mach. Co., Hopkins, Minn., Russell Giant A, 30-60, by Russell & Co., Massillon, Ohio, and the Wetmore 12-25, by the Wetmore Co., Sioux City, Iowa.

Two new air cleaners were produced, the Pomona "Vortox" and the United. The Pomona, manufactured by the Vortox Mfg. Co., Pomona, Calif. was a centrifugal and filtering type. Air entered through a tangential inlet and was thus given a swirling motion that caused the dust particles to collect close to the outer wall, and as the communicating passage between the upper and lower chambers was at the center of the cylindrical chamber, the dust fell to the bottom into a removable cup containing used crankcase oil. In the upper part of the cylindrical chamber the air had to pass through a filter of fibrous material on its way to the carburetor, and the passage between the lower and upper parts was of such design that it gave a further centrifugal motion to the air. Thus, the air was subjected to two centrifugal cleanings before it was passed into the filter, the claim being made that this prevented rapid clogging of the filter.

The United produced by the United Mfg. and Distributing Co., Chicago, Ill., made use also of the centrifugal motion of the air. As the air entered the opening at the top, it was caused to abruptly change direction and double back, thereby throwing the dust particles out against the sides and dropping the particles into a removable container at the bottom.

The following 14 makes and models of tractors were tested at Lincoln, Nebr., in 1921.

Test No.	Tractor model and horsepower
70	Minneapolis 17-30
71	Avery 12-25
72	Avery 8-16
73	Wetmore 12-25
74	Huber-Super 4
75	Lauson "S"
76	Best "A" 60 (35-55)
77	Best "S" 30 (18-30)
78	Russell-Giant "A"
79	Hart Parr 20
80	Eagle 16-30
81	Eagle 12-22
82	Allis-Chalmers 12-20 (14-27)
83	Allis-Chalmers 18-30 (22-38)

Some 1921 Models

Toro Push Type Tractor

McCormick-Deering 15-30

Cletrac F. 9-16

Huber Super Four 18-36

Baker - Steamer 16-30

1922

A large number of companies discontinued making tractors, and many reorganizations took place. Among the important companies to discontinue tractor manufacture was the General Motors Corp. This firm entered the farm equipment field in 1917 when it purchased the business of the Samson Sieve Grip Tractor Co. of Stockton, Calif. In the following year it bought the business of the Janesville Machine Co., Janesville, Wis., where later the Samson general-purpose tractor was manufactured. The executive offices were moved to Pontiac, Mich., to become part of the General Motors Truck Co. The Samson name was retained, and tractor production continued in the Stockton plant. During 1918 a number of these tractors were built in a plant leased from Monroe Motor Co. of Pontiac, Mich. Production facilities were expanded in 1919, equipment from the Monroe plant moved to a new plant in Janesville, and the Samson Tractor Co. became a separate division of the corporation. The Janesville Machine Co., a manufacturer of farm machinery, was acquired to serve as source of supply for units used with the Samson tractor. This company later became part of the Samson Tractor Co. Adverse business conditions caused the company to discontinue tractor production, and in 1922 the plant at Janesville was converted into a Chevrolet assembly plant.

Tractor prices that were quoted during the year did not permit a profit to the seller. Fordson tractors were selling at $395—a 35 percent drop from the price in effect in 1921; the Moline tractor, which was selling for $1,325 in 1920 was reduced to $650; and another prominent make of 3-plow tractor was selling $700 under its top price of 1920.

This tractor price war resulted in placing good equipment in the hands of many farmers who otherwise would not have been able to take advantage of power farming. The trend was, however, away from the heavy-type tractors and for light-weight equipment adapted to general usage envisaged for the future market.

The Parrett, 10 Brake hp. "Cul-tor"—a cultivating-type tractor - was produced by the Parrett Motors Corp., Chicago, Ill. This represented an extreme swing in the light-tractor direction and was designed mainly to pull horse cultivators that might be available on the farm. The power plant was a 2-cylinder Mar-tan air-cooled engine of the motorcycle type with a ball-bearing crankshaft and roller bearings for the connecting rods.

A new style wheel of pressed steel construction was introduced with the Case 12-20 (page 7). Other features, such as the 1-piece cast frame, forced feed lubrication, and renewable cylinder sleeves, had appeared on the Case 15-27 produced in 1919.

The 10-20 McCormick-Deering, a smaller version of the 15-30 McCormick-Deering was placed on the market by the International Harvester Co., Chicago, Ill.

Another development of the year was that of the Bryan Light Steam-Tractor produced by the Bryan Harvester Co., Peru, Ind. (page 7). This tractor (15-30) weighed 5,500 pounds and was powered by a steam engine using high-pressure (600 pounds) superheated steam generated in a water tube boiler. Apparently the advent of this outfit was the result of considerable experimentation with a previous model built about

1920, with the same size engine but rated at 26-70 hp. and with a weight, fully equipped, of only 3,900 lb.—a rather light tractor to deliver this amount of power.

Production was started on the Wilson 4-wheel drive (page 7) tractor, developed by Wilson Bros. and marketed through the Wilson Tractor Manufacturing Co., Ottumwa, Iowa. The power was taken from the engine through planetary gears always in mesh, thence through double worms and worm gears for the main reduction, with final drive to rear wheels by bull gears and pinions.

The Avery Co. of Peoria, Ill., entered the track-laying field by introducing the Avery Track Runner (page 6). This 3-4 plow tractor had 2 crawlers in the rear and 2 nondrive wheels in front.

During the period 1922-1924 at Iowa State College, Ames, Iowa, two students, E. D. Gordon and D. B. Lucas, under the supervision of J. B. Davidson and E. V. Collins, designed an experimental manless plow consisting of a small 2-wheel tractor provided with a single cylinder 4-hp. engine. To the frame were attached right and left plow bottoms, so placed that while 1 plow was in the ground the other was out. After the initial furrow was plowed—hand steered—1 driving wheel followed in the furrow and the machine shuttled back and forth across the field on its own. A reversing arm mounted at each end of the tractor changed its direction upon coming in contact with a fence or other obstruction. The reaction lifted the plowing unit out of the ground, swung the other into position and reversed the direction of the drive wheels.

The Holt Mfg. Co. of Peoria, Ill., produced 3 models of Caterpillar tractors—the 2-ton, (page 8), the lightest track-type tractor under its Caterpillar trade name, designed to furnish power for maintaining roads as well as to perform operations on the farm; the 5-ton, a 4-plow; and the 10-ton, a 6-plow size.

The Dayton-Dowd Co. of Quincy, Ill., was producing 2 models of the Leader, each 16-32 hp., one being a crawler type and the other a wheel type. The Yuba Co., Marysville, Calif., introduced its 25-40 Ball Tread (page 6) a larger edition of its previous models. Steering was accomplished by the front wheel actuated by a cable and winding drum on the steering column and independent clutches for each track. The Allis-Chalmers Co. added the 20-35 hp. model to its line.

It had long been known that dust entering the tractor engine from the intake unduly damaged the working parts— particularly the pistons, rings, and cylinders, the extent of damage depending on the nature and amount of the dust. To improve this situation many makes and types of air cleaners were appearing on the market, differing in their ability to remove dust from the entering air, in the degree to which their use imposed vacuum or choking on the carburetor intake, and in their effect on the maximum possible power to be obtained from the engine. To determine the dust separation efficiency, vacuum imposition and effect on power, a long series of tests was made at the Branch (Agricultural Engineering) of the College of Agriculture, Davis, Calif., by A. A. Hoffman under the chairmanship of L. J. Fletcher. Twenty-six cleaners were tested in 1922 and it was found that their dust-separating efficiency varied from a low of 42.7 percent to a high of 99.8 percent. There were 15 cleaners with more than 95 percent efficiency. The test results were published in 1923 as California Agricultural Experiment Station Bulletin 362, "Dust and the Tractor Engine." These studies had a wholesome effect on this segment of the industry and marked improvements in air cleaners resulted.

The following five makes and models of tractors were tested at Lincoln, Nebr. in 1922.

Test No.	Tractor model and horsepower
84	Rogers Four Wheel Drive
85	Cletrac "F" (9-16)
86	Holt 2-ton (15-25)
87	International (15-30)
88	Case (12-20)

Some 1922 Models

Avery Track-Runner (3-to 4-plow)

Yuba Ball Tread 25-40

Wilson Four Wheel Drive 12-24

Bryan Light Steamer (insert shows water-type boiler)

McCormick-Deering 10-20

Case 12-20

1923

With increased tractor production of about 25,000 units in 1922, and an increase of over 30,000 in 1923, manufacturers felt more encouraged about future tractor business and a general refining and improving move that had started a year or two before got under way. One firm announced that while no fundamental change in tractor design had been made, the weight of its product had been materially decreased and yet strengthened structurally, and some 22 refinements and improvements were adopted. Hence, the effects of both the keen trade competition and the depression of 1921 were definitely reflected in improved machine design and operation. While the changes were largely individualistic in character, there was a general tendency toward the more approved engineering practices and designing. The use of antifriction bearings gained considerable headway. One firm announced the use of ball and roller bearings at 28 points on the tractor, including the 2-bearing (ball) crankshaft with lifetime guarantee of both crankshaft and bearings. Pressure lubricating systems with "Alemite" pressure lubrication were widely adopted for bearings not served by forced oil feed. Thermosiphon cooling systems were being replaced by more positive ones; and with the use of improved radiators together with improved designing of water pumps, water jacketing, and increased fan capacities, a reduced requirement for cooling water was possible. Enclosing and dustproofing gear, bearing, and pushrod and rocker-arm assemblies; the hardening of parts subjected to hard wear; and many other features were included in the refinements.

The John Deere Model D (15-27) tractor (page 9) was put on the market. This followed the discontinuance of the "Waterloo Boy" first produced in 1912 by the Waterloo Gasoline Engine., taken over in 1918 by the John Deere Tractor Co. of Moline, Ill. The model D was destined to become one of the company's chief products and subsequent models were modernized in keeping with progressive engineering.

Models 12-24 and 40 were offered by the Hart-Parr Co., Charles City, Iowa, and an experimental high pressure steam farm tractor (page 9) was built by the International Harvester Co. This steam tractor had a wheelbase of 86 inches, weighed 5,300 pounds, and pulled a 3-bottom 14-inch plow. The steam was generated in a special water tube boiler at a pressure of 600 pounds per square inch. The boiler was heated by a burner of the vaporizing type using distillate fuel.

The Frick Co. of Waynesboro, Pa., produced a 36- and a 50-hp. single-cylinder model steam traction engine (page 9), and the Russell Co. of Massillon, Ohio, came out with a single-cylinder steam-traction engine (page 8).

At this time steam power was seldom used for farm drawbar work. There was still a demand, however, for the smooth steady power delivered by the steam tractor for operating large threshing machines—a function with which steam rigs were originally identified. Hence, it was not uncommon to find some of the older established firms hesitant to discontinue a once lucrative line, especially one pioneered by them. The Russell Co. of Massillon, Ohio, for example, had started producing steam tractors in 1879 and continued the line long after the internal combustion tractor was considered a success. This company was established in 1842 and was a pioneer producer of threshing machinery.

The Avery Co., Peoria, Ill., produced its 20-35 hp. model (page 8), typified by sliding frame gear shift, 2 cylinder horizontally opposed engine, and a round vertical tube radiator.

The following 12 makes and models of tractors were tested at Lincoln, Nebr., in 1923.

Test No.	Tractor model and horsepower
89	Avery Track Runner 15-24
90	Case 40-72
91	Case 12-20
92	Wallis 15-27
93	Russell-Little Boss 15-30
94	Russell-Big Boss 20-40
95	McCormick-Deering 10-20
96	Avery 20-35
97	Hart Parr Co. 40
98	Best "60" 40-60
99	Best "30" 20-30
100	Bear "B" 25-30

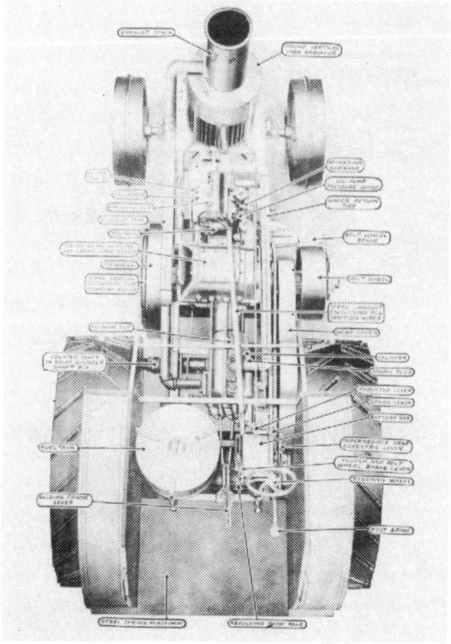

Avery 20-35

International Harvester Company experimental high pressure steamer

Some 1923 Models

Russell Steamer (single cylinder)

Caterpillar 2 Ton

Frick Steamer (single cylinder)

John Deere D

1924

The following list shows changes reported in Bulletin 200 "Nebraska Tractor Tests," published August 24 by the Nebraska Agricultural Experiment Station.

Test No. 3. Case 10-18. "Not being manufactured. Supplanted by Case 12-20." (Test No. 91.)
Test No. 6. Case 10-20. "Not being manufactured. Supplanted by 15-27." (Test No. 4.)
Test No. 7. Case 20-40. "Not being manufactured. Supplanted by 22-40." (Test No. 5.)
Test No. 22. Gray 18-36. "Manufacture of this model tractor discontinued."
Test No. 23. Titan 10-20. "No longer in production. Superceded by McCormick-Deering 10-20." (Test No. 95.)
Test No. 24. International 15-30. "No longer in production. Superceded by McCormick-Deering 15-30." (Test No. 87.)
Test No. 25. International 8-16. "Manufacture discontinued."
Test No. 27. Samson "M". "Taken out of production.
Test No. 33. Moline Universal "D." "Manufacture discontinued."
Test No. 35. Coleman "B" 16-30. Discontinued.
Test No. 37. Parrett "K" 15-30. Discontinued.
Test No. 39. Avery 6 Cylinder Model "C". "Discontinued production."
Test No. 41. Avery 12-20. "Discontinued production."
Test No. 42. Avery 14-28. "Replaced by 20-35." (Test No. 96.)
Test No. 44. Avery 40-80. "Rating changed to 45-65."
Test No. 45. Cletrac "W" 12-20. Now uses engine of own make which has not been tested in Nebraska.
Test No. 49. Wallis "K" 15-25. "Closing out—and working into the Model "O.K." 15-27—test No. 92."
Test No. 50. Flour City 18-35. Engine size is 5 1/4 inches x 6 inches.
Test No. 55. Allis Chalmers 18-30. "Obsolete."
Test No. 57. Avery Single Row Cultivator 5-10. "Discontinued Production."
Test No. 58. Avery 18-36. "Discontinued production."
Test No. 60. Bates Steel Mule "D" 15-22. "Obsolete."
Test No. 64. Uncle Sam 20-30. "Company inoperative."

The manufacture of tractors was discontinued by the Moline Plow Co. of Moline, Ill., the Bates Machine and Tractor Co. of Lansing, Mich., and the Gray Tractor Co. of Minneapolis, Minn.

The Farmall, (page 10), probably the first successful attempt at building a genuine all-purpose tractor of the tricycle design, was placed on the market. This was the result of several years of patient effort by Bert R. Benjamin. Up to this time the conventional 4-wheel tractor was the generally accepted type for farm work, notwithstanding the fact that it failed to meet the problem of rowcrop cultivation. The new Farmall tractor had high rear-axle clearance, gear drive, automatic braking of either wheel for quick and short turns, small closely spaced front wheels designed to run between crop rows, vertical steering column with steering rod over the top of the tractor, vertical steering wheel, 3 forward and 1 reverse speeds, means of mounting cultivator and other attachments, power take-off working off the transmission and delivering power at the center rear, and belt pulley.

The Advance-Rumely Co. of La Porte, Ind., bought the Aultman-Taylor Co., Mansfield, Ohio (established 1865), manufacturer of tractors, steam engines, and threshing machinery.

The Hart-Parr Co., Charles City, Iowa, put its Model 22-40 in production.

The U.S.S.R., which started importing American tractors in 1923, expanded its imports to approximately 6,000 units.

The "Allwork" 20-35 model D tractor, a larger edition of and similar in appearance to the 1917 "Allwork" 14-29, manufactured by the Electric Wheel Co., appeared in late 1924.

The New Wallis OD—15-27 hp. (page 10) was placed on the market by the J.I. Case Plow Works Co., Racine Wis. This tractor embodied many features of previous models, including the one-piece boiler plate unit frame construction.

The Monarch D-6-60 (page 10) was offered to the trade by the Monarch Tractor Co., Watertown, Wis. This machine was similar in appearance to the models produced in 1918 but with considerably more power.

Russell and Co., Massillon, Ohio, produced its Russell 20-40 (page 10), and the J.I. Case Threshing Machine Co., Racine, Wis., started production on its 18-32 hp. model (page 10) of the same general appearance as the tractors this company was currently producing.

The Nichols-Shepard Co., Battle Creek, Mich., added a 20-42 hp. oil-gas tractor (page 10) to its Red River Special line. This oil-gas machine was powered by a 2-cylinder twin horizontal engine mounted on a riveted steel frame. The Minneapolis Threshing Machine Co., Hopkins, Minn., offered it 22-44 oil tractor (page 10), powered by a 4-cylinder horizontal engine.

The following 10 makes and models were tested at Lincoln, Nebr., in 1924.

Test No.	Tractor model and horsepower
101	Bear "B" (25-35)
102	John Deere D (15-27)
103	Oil Pull "S" (30-60)
104	Best "S-30" (25-30)
105	Best "A-60" (50-60)
106	Hart Parr (16-30)
107	Hart Parr (12-24)
108	Monarch "D-6-60" (40-60)
109	Case (18-32)
110	Case (25-45)

Some 1924 Models

Nichols-Shepard 20-42

Case 18-32

New Wallis O.K. (Case) 15-27

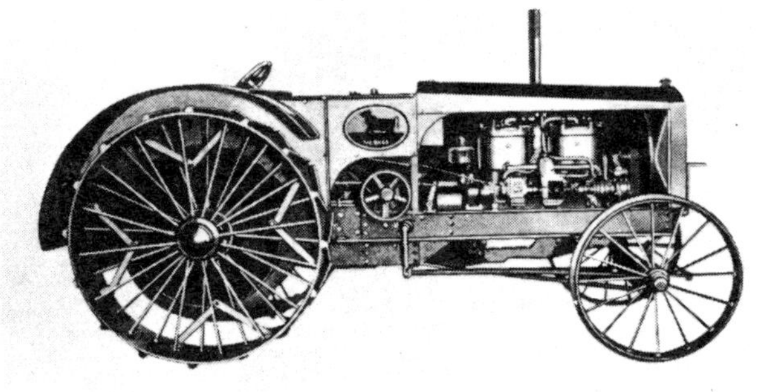

Russell 20-40

Monarch D, 6-60

McCormick-Deering Farmall (2-plow)

1925

In 1925 tractor production exceeded 167,000 units.

The Caterpillar Tractor Co., Peoria, Ill., was formed with $12,300,000 by a merger of the C.L. Best Tractor Co. of San Leandro, Calif. and the Holt Manufacturing Co., which had plants at Stockton, Calif., and Peoria, Ill. The latter firm was the originator and holder of the "Caterpillar" trade mark. The Best Company was organized in 1910 and the Holt Company in 1892, a successor to earlier Holt firms dating back to 1869. Benjamin Holt built the first practical crawler designed tractor in 1904 at Stockton, Calif. In 1908 The Holt Manufacturing Co. purchased the interests of Daniel Best and the latter retired from business. In 1910 C.L. Best, a son of Daniel Best, formed his company and in 1913 started producing "Tracklayers" in his father's old plant at San Leandro. Stiff competition ensued between the two interests. In 1928 the Caterpillar Tractor Co. acquired the Russell Grader Manufacturing Co. of Minneapolis, a manufacturer of road-building machines for more than 20 years. The products of the Caterpillar Tractor Co. then included track-laying tractors, combined harvesters, stationary engines, and a varied line of road-building equipment.

A new Gray Tractor Co. took over the old firm of the same name and added a larger model, the Canadian Special 22-40, to its line of wide rear-wheel-drive tractors. A combination tractor, scarifier, grader, and roller known as the Gray Giant was also put out by this company.

The Square Turn Tractor Co., of Norfolk, Nebr., makers of a popular tractor before World War I, was disposed of at a sheriff's sale for $15,000.

The International Harvester Co. bought the Moline tractor factory at Rock Island, Ill., home of the Moline Universal tractor.

The Russell Co. of Massillon, Ohio, produced its 15-30 hp. 4-cylinder tractor, and the Keck-Gonnerman of Mount Vernon, Ind., its 19- and 22-hp. single-cylinder rear-mounted steam traction engines (page 11).

The Cleveland Tractor Co., Cleveland, Ohio, added Cletrac model K-20 (page 12) to its line. A 3-forward and 1-reverse transmission, a rear belt pulley, and power take-off were incorporated in the tractor.

The Electric Wheel Co., Quincy, Ill., early in the year added the Allwork 25-35 track-type tractor (page 11) to its line. This was designated as the 5-ton crawler-type tractor the following year.

The Oil Pull, 20-35 hp. (page 11), was produced by the Advance-Rumely Thresher Co., LaPorte, Ind., and the Rock Island Plow Co. added the Heider C 15-27 (page 12) to its line. This latter tractor was powered with a 4-cylinder engine, and the friction transmission permitted 7 speeds forward and 7 reverse. The Mead-Morrison Mfg. Co., East Boston, Mass., started production on its M-M 55 crawler (page 12). This succeeded the Bear 25-35, which was formerly (about 1924) manufactured by Bear Tractors Inc., New York City.

The following 8 makes and models of tractors were tested at Lincoln, Nebr., in 1925.

Test No.	Tractor model and horsepower
111	Oil Pull "M"
112	Oil Pull "L"
113	Monarch "C"
114	Heider (15-27)
115	Shawnee "30"
116	Oil Pull "R" (25-45)
117	I. H. C. Farmall
118	The Minneapolis (17-30)

Some 1925 Models

Keck-Gonnerman Steamer 22

Allwork 25-35 (5 ton)

Rumely Oil Pull 20-35

Cletrac K-20

Heider C 15-27

Mead-Morrison MM 55

1926

In June 1926 the new Farmall Works of the International Harvester Co. at Rock Island, Ill., was opened. This marked the beginning of a new period in farm tractor progress, characterized by the wide acceptance of a practical general-purpose tractor, the Farmall, which first appeared in 1924 (page 10). This was later to be designated as the F-20.

The design of the Huber Super 4 (page 13) was changed to the conventional style, the Minneapolis Model B-17-30, Emerson-Brantingham, E-B 15-25, Twin City 27-44, Hart-Parr 12-24 (page 13) and 18-36 were in production or were being groomed for the purpose.

The A.D. Baker Co., Swanton, Ohio, which had been building steam tractors, introduced its first gasoline tractor, the Baker 25-50 hp. (page 12). It was more or less of the conventional 4-wheel type of the period, with channel frame construction, 4-cylinder engine with cylinders cast in pairs, and chain and roller steering.

The Monarch Model F, a 10-ton crawler tractor, was added to the line of the Monarch Tractor Co. (page 13).

The Wizard Tractor Corp., Los Angeles, Calif., placed its 4 Pull 20-35 hp. 4-wheel drive model (page 13) on the market. The final drive was effected by enclosed chains running in oil.

The Avery Power Machinery Co., Peoria, Ill., started production on its 45-65 hp. model (page 13) powered with its 4-cylinder so called "Draft Horse" engine. This was one of the company's earlier models to discard its sliding frame method of gear shift.

Further engineering refinements were apparent, including dash control of some of the fundamental equipment such as magneto, carburetor, and governor; the use of nickel iron for cylinder liners and chrome-nickel steel for crankshaft and studs; and the more general use of standard equipment and accessories.

The following 12 makes and models of tractors were tested at Lincoln, Nebr., in 1926.

Test No.	Tractor model and horsepower
119	Cletrac "K" (15-25) Kero.
120	Cletrac "K" (15-25) gas
121	Twin City (17-28)
122	Twin City (27-44)
123	Huber 4 (18-36)
124	Fordson (20)
125	Cletrac "A" (30-45)
126	Huber (20-40)
127	Twin City (21-32)
128	Hart-Parr (18-36)
129	Hart-Parr (12-25)
130	McCormick-Deering (15-30)

Some 1926 Models

Baker 25-50

Huber New Super 4

Hart-Parr 12-24

Wizard 4 Pull 20-35

Monarch F - 10 ton

Avery 45-65

1927

In January 1927, U.S.S.R. bought 5,000 American tractors. With this purchase the total tractors shipped to U.S.S.R. during the previous 2 1/2 years had reached 27,000. More than 98 percent of the tractors in use in the U.S.S.R. were then of American manufacture.

According to a trade bulletin issued by the Agricultural Implements Division of the United States Department of Commerce, "the supremacy of the American tractor throughout the world is undisputed. Last year (1926) manufacturers shipped to foreign markets 50,000 wheel and more than 1,200 track-laying tractors which represented an increase in business of 11 percent over 1925—a record year. The most important markets were Canada, Russia, Australia, Argentina, France, and Italy. From time to time American tractors have encountered more or less strong competition in foreign markets. At the present time (1927) only a relatively small number of tractors is made in these countries and these are limited to just a few makes. European tractors have been generally of a heavier type unsuitable for ordinary agricultural uses, and their cost has been too high for extensive sales."

Power take-off standards prepared by a committee of the American Society of Agricultural Engineers and approved by representative engineers of the leading tractor manufacturers were adopted. This was an important step toward clarifying a much confused development caused by lack of standardization. The power take-off standards specified the type of fitting and size, speed (535 rpm $\pm$ 10) and direction of shaft rotation. The recommended practice covered such features as vertical distance of end of power take-off shaft from ground, horizontal position of end of shaft, guards, etc. Up to this time p.t.o. shafts were of different sizes, ran at different speeds and rotated in clockwise or anticlockwise direction. In many cases it was a big job to hitch to a power driven machine.

Russell and Co. of Massillon, Ohio, one of the old steam-tractor builders that started the manufacture of gas tractors at an early date, closed out its entire stock.

Advance-Rumely Co., LaPorte, Ind., bought from the Toro Manufacturing Co. of Minneapolis, Minn., exclusive rights to manufacture and market the Toro combination tractor and row-crop cultivating machine for agricultural use. The Toro Co. was identified with some of the pioneer work in attempting to evolve a general-purpose tractor. The Advance-Rumely Co. also announced its new models "W" 20-30 Oil Pull (page 14); Oil Pull X, 25-40; and the Oil Pull Y, 30-50. A feature was the vacuum system for crank case ventilation. This system reduced the temperature of the lubricating oil and the parts inside the crank case, thereby preventing condensation and minimizing crank case dilution.

The Massey-Harris Co. of Racine, Wis., placed on the market its Wallis "Certified" 20-30 tractor (page 14). This resembled the Wallis lines established when the company first introduced the unit boiler plate frame construction about 1913.

The Nichols and Shepard Co., Battle Creek, Mich., introduced two Lauson built models, the Lauson 16-32 and Lauson 20-40 (page 14); the Huber Co., Marion, Ohio, its 40-62 model; the Minneapolis Threshing Machine Co., Minneapolis, Minn., its A 17-30 model (page 14), the Nichols and Shephard Co., Battle Creek, Mich., its oil-gas 25-50 tractor (page 14); and Waterloo Machinery Co., Waterloo, Ontario, 3 sizes of the Waterloo Eagle twin cylinder tractors 13-25, 16-30 (page 15), and 20-40. (Not to be confused with either the Waterloo Boy—Deere Co., or the Eagle, Appleton, Wisconsin).

The following 17 makes and models of tractors were tested at Lincoln, Nebr., in 1927.

Test No.	Tractor model and horsepower
131	Lauson (16-32)
132	Lauson (20-40)
133	Hart-Parr (28-50)
134	Wallis (20-30)
135	Huber (40-62)
136	Shaw "T25" (-1)
137	Shaw "T45" (-1 1/2)
138	Cultor "32"
139	Monarch "10 ton"
140	Hart-Parr (28-50)
141	Oil Pull "W" (20-30)
142	McCormick-Deering (10-20)
143	Oil Pull "X" (25-40)
144	Rock Island (18-35)
145	Oil Pull "Y" (30-50)
146	John Deere (15-27)
147	Monarch "6 ton" (40)

Rumely Oil Pull "W" 20-30

Wallis (certified) 20-30

Minneapolis A 17-30

Some 1927 Models

Nichols-Shepard 25-50

Lauson 20-40 (by Nichols-Shepard)

Waterloo Eagle 16-30 (Waterloo, Ontario)

1928

The Massey-Harris Co., Racine, Wis., took over the J.I. Case Plow Works also of Racine, Wis. Operation of the newly purchased plant was continued. The acquisition of the Case business gave the Massey-Harris Co. a well-known line of farm implements including the Wallis tractor. By agreement between the Massey-Harris Co. and the J.I. Case Threshing Machine Co., Racine, Wis., the latter acquired the exclusive right to use the names "Case" and "J.I. Case" on farm equipment.

The Allis-Chalmers Manufacturing Co. of Milwaukee, Wis., purchased the business of the Monarch Tractors Corp., Springfield, Ill., manufacturers of track-laying tractors used largely in the industrial field.

The J.I. Case Threshing Machine Co. bought the farm implement business of the Emerson-Brantingham Corp., Rockford, Ill. This transaction marked the passing of a farm equipment business that had been in continuous operation since 1852. It was founded by John Manny, one of the early reaper manufacturers, who in 1853 admitted to partnership Ralph Emerson and Wait Talcott. The E-B Corp. in the 1928 reorganization retained its industrial products division and continued to manufacture malleable castings, certain automobile parts, enameled and metal products.

In July the International Harvester Co. manufactured 200 tracks for its 10-20 (page 18) and 15-30 wheel-type tractors. These crawler-fitted tractors were called TracTracTors.

The Bates Tractor Co., Lansing, Mich., sold its entire assets to Foote Bros. Gear & Machine Co., Chicago, Ill., which operated the tractor activities under the name of the Bates Manufacturing Co. Divison. Around 1935 the Foot Bros. Gear & Machine Co. was reorganized and the tractor was sold to a Chicago firm which continued making tractors for several years, but selling most of them to export markets.

The J.I. Case Threshing Machine Co., Racine, Wis., substituted letters for horsepower ratings—the 12-20 became Model A, the 18-32 became Model K, and the 25-45 Model T.

The year (1928) was productive of great diversity in tractor designing. Track-laying models ranged from the 20-25 horsepower Caterpillar 20 (page 18) produced by the Caterpillar Tractor Co. of Peoria, Ill., to the 12-ton Model 75, Cletrac 40 and Cletrac 100 (page 18) produced by the Cleveland Tractor Co. of Cleveland, Ohio and the 3 1/2-ton "Creeping Grip"—a product of the Bullock Tractor Co., Chicago, Ill. Among new developments in wheel-type tractors were the Rock Island 18-35, the Lauson 25-45 hp., 6-cylinder tractor (page 18), the Wallis 12-20, and 2 general purpose tractors produced by the Advance-Rumely Co. and the John Deere Co., respectively.

These two general purpose models were of special interest in that they were radically different in design to that of the International Harvester Co.'s Farmall. The Rumely DoAll (page 18), when used for plowing or other general farm work, functioned the same as any conventional 4-wheel tractor; when fitted for cultivating, the front wheels and axle were removed and the cultivator attachment with caster wheel was substituted. The John Deere 10-20 (page 18) general purpose tractor was practically a small Model D (the standard 15-27 Deere product) but with an arched front axle. In cultivating, the tractor straddled 1 row and hence planted or cultivated 3 rows at a time; high clearance and other adaptations for row-crop work were designed in the outfit. It is of interest to note that this model was fitted with the first power lift (mechanical) and added a fourth way of applying tractor engine power by applying power for lifting drawn machines off the ground. Up to this time there had been 3 methods—the belt pulley, the drawbar, and the power takeoff.

Another model, the Baker 25-50, was placed on the market by the A.D. Baker Co., Swanton, Ohio, that resembled the 22-40 introduced in 1926.

Although previous attempts had been made to use rubber tires on tractors (see Part I, page 4), it was not until about 1928 that successful use was made of them. According to L.B. Sperry, tractor engineer for the International Harvester Co., it was another case of "necessity is the mother of invention." Mr. Sperry (footnote #8, page 73, Part I) went on "a peculiar use of rubber had made its appearance in 1928 on tractors in Florida. Operators in the orange groves there noting that the steel lugged-tractor wheels damaged the roots of their orange trees, undertook to make use of discarded tire casings to overcome the difficulty. They accordingly altered the rear steel tires of their tractors to accommodate the use of 8 x 40 casings (without inner tubes) and placing two or three side by side on each wheel secured both flotation and traction. Because this construction proved so successful, it was not long until discarded 8 x 40 truck tires were shipped to Florida in quantity.

"This use of rubber attracted the attention of the rubber tire manufacturers and early in 1931 the B.F. Goodrich Co. brought out their 'zero pressure' tire for agricultural tractors. It consisted of a rubber arch built on a perforated steel base for attachment to standard steel tire drive wheels. The piers supporting the arch were of sufficient size to provide ample load capacity. The all-rubber arch or tread was made flexible to give full tread contact under the comparative light weight of tractors and through this flexibility and full contact a traction exceeding that of either solids or pneumatics was claimed. The tire was not under air pressure consequently was not subject to puncture.

"In 1932 several companies were experimenting with rubber-tired tractors with encouraging results. For instance, the Firestone Rubber Co. fitted an Allis-Chalmers tractor with pneumatic tires and began active development work with low-pressure airplane tires, and announced the same year an air-tired farm tractor with air tires as standard equipment. The use of lug-type chains on low-pressure air tires was found to expedite traction under adverse field conditions. As incentive to the general move, plowing matches and road hauling demonstrations were staged. At a plowing match at Big Rock, Ill., it was reported that a McCormick Deering tractor equipped with Firestone rubber tires "Stole the show." Later the same year an Allis-Chalmers model "U" air-tired tractor pulling a trailer loaded with farm implements was driven to its destination at the rate of 17 1/2 miles per hour for 5 hours official time."

It might be recalled, in connection with this development, that Barney Oldfield driving a pneumatic-tired tractor through a small town in Indiana, received a "ticket" for exceeding the speed limit. He was probably traveling at the rate of 10 to 15 m.p.h. (Mr. Oldfield had "raced" tractors at 40 m.p.h.). This was no doubt in the early days of the "speed trap."

In 1933 certain tractor models of many of the leading companies, including Allis-Chalmers, Case, Deere, Ford, Huber, International Harvester, Massey-Harris, Minneapolis-Moline, Oliver Farm Equipment, were mounted on rubber.

It is significant to note that rubber tires for agricultural tractors were developed during a depression period. In 1930 about **200,000** tractors were produced and in 1932 approximately 19,000 were sold. The success of the development doubtless had a marked influence on tractor demand which in 1935 resulted in over 160,000 being produced. Of the wheel-type tractors produced 14 percent were mounted on rubber in 1935, 31 percent in 1936, 43 percent in 1937, 95 percent by 1940, and almost 100 percent by 1950.

In a study made in 1933 by the University of Nebraska at Lincoln (Bulletin No. 291), of the comparative tractor performance on (a) steel tires and (b) low-pressure rubber tires, the following were among the findings established:

1. Rubber tires equipped with lug-type chains will pull as much as steel wheels and lugs on dry soil, and more under muddy conditions.

2. That a creeping oscillating movement of the lug-type chain will keep them clean in soil conditions that will fill up the steel wheel with fixed lubs.

3. Maximum drawbar horsepower developed with rubber tires was in high gear (4 to 5 mph). The maximum with steel wheels was in low gear.

4. Rubber tires showed the better fuel economy based on drawbar horsepower. High gear with rubber showed the maximum fuel economy.

The advantages of pneumatic over steel tires in fuel economy and in rate of performance on everyday field operations is shown on page 17, which summarizes 24 tests made in 1933 in this connection. The tests were made by the Department of Agricultural Engineering of the Experiment Station of the University of Nebraska in cooperation with several of the tractor and 4 commercial rubber companies. In making these experiments 2 tractors of the same make and model were run in the same field at the same time, each pulling as nearly as possible an equal load. One tractor was equipped with pneumatic tires and the other with steel wheels and space lugs. When half of the work was finished the wheel equipment from one tractor was exchanged with that of the other and the work completed.

With many of the small comparatively lightweight tractors in general use at this time, it was necessary to provide wheel weights of several hundred pounds so that the maximum power of the engine could be utilized. The weights were designed to be hung on the wheels; with faster speeded rubber-tired tractors the wheel weights became both cumbersome and dangerous especially on the road at high speeds.

It was found after much experimentation that water could be inserted in air tires in sufficient quantity to act as a substitute for the conventional metal wheel weights. The idea of weighting tractor wheels by use of a liquid agent was not original with pneumatic tires. The principle was incorporated in the metal wheels of the Allen Water Ballast Tractor of 1917. (see page 50, Part I).

After a year's trial, the B.F. Goodrich Co. found that water in tires resulted in no detrimental effects upon the tires and that there were certain advantages arising from the practice. Aside from the low cost of the agent, water inflation provided normal cushioning without the disadvantage of rebound or bounce. The "shock-absorber" action made tractor operation more satisfactory where ground conditions were rough. Since no traction problem was involved in the front wheels of the tractor weighting here was not necessary.

To facilitate filling tires with water an "adaptor" was developed, one end of which was fastened to one end of the garden hose and the other end to the tire valve. With water as the weighting agent, an antifreezing solution had to be provided when operating the tractor during freezing weather. Commercial calcium chloride at a cent or two per pound was found economical and satisfactory for the purpose.

In wheel-weighting, studies showed that carrying capacities and inflation pressures were not altered by the addition of water. It was, therefore, necessary to compensate for the added weight of the water by increasing the inflation pressure up to the recommended standard the same as when wheel weights are used.

Proper size and type of air tire was an important consideration, and there were doubtless many misfits and disgruntled customers before this phase became standardized.

The Fordson played an important part in tractor development during its manufacture in this country (1918-28). It entered the field at a most opportune time, first furnishing thousands of units for World War I purposes after which it predominated, for several years, the power equipment on American farms. Starting with a production of over 34,000 in 1918, or over 25 percent of the 132,700 tractors produced by 142 companies , the Fordson enjoyed its normal share of the increasing business for the next two years with practically 100 percent Fordson increase over 1918. It was in 1921, however, when the post-World War I depression was beginning to be felt, that the low-priced

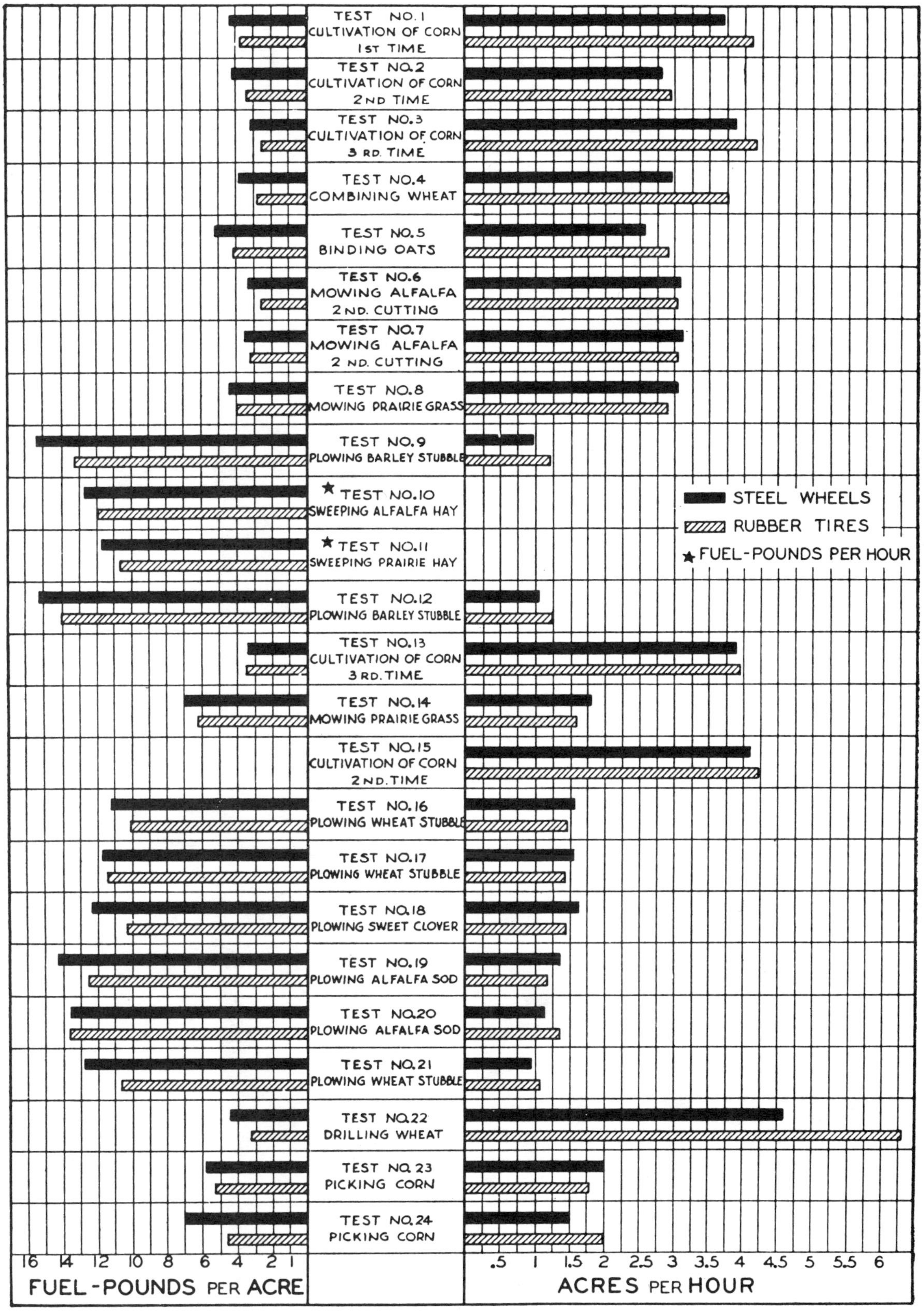

A comparison of pneumatic-tired tractors with those having steel wheels and lugs, in respect to fuel consumption and acres per hour, as shown by 24 tests. (from Bulletin 291, September 1934, University of Nebraska, Lincoln, Nebraska)

Fordson represented about 50 percent of the total tractor production; and in 1923 and 1925, over 100,000 Fordson units were produced each year, or 60 to 75 percent of the tractors produced by all companies. After 1925 with returning normalcy augmented by the influence of the general purpose tractor, Fordson production decreased and its manufacaturer was finally discontinued in this country in 1928. Strangely enough the general purpose tractor the following year (1929) took up, in production, where the Fordson left off (1928) at about 40,000 units, only soon to experience the ravages of a period of financial recession similar to the one of a decade before.

The following 7 makes and models of tractors were tested at Lincoln, Nebr., in 1928.

Test No.	Tractor model and horsepower
148	Lauson "S 12" (20-35)
149	Cletrac "40" (40-55)
150	Caterpillar "20" (20-35)
151	Allis-Chalmers (20-35)
152	Twin City (21-32)
153	John Deere (gen. purpose) (10-20)
154	Advance-Rumely Do-All

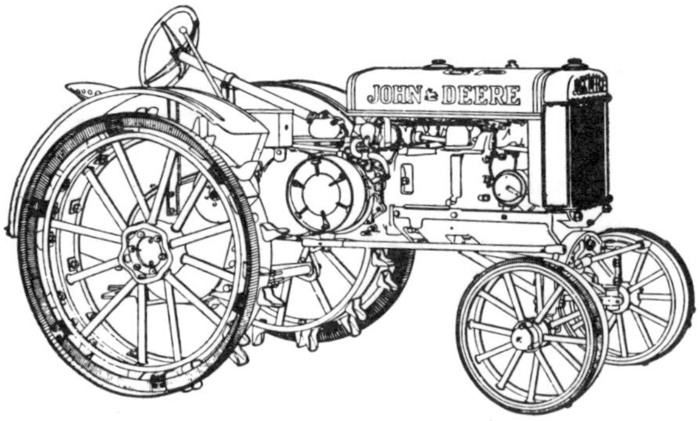

John Deere General Purpose 10-20

Lauson 25-45

Some 1928 Models

Rumely DoAll

I.H.C. TracTracTor T20

Cletrac 100

Caterpillar 20

1929

In 1929 the Farm Machinery Blue Book listed 47 tractor manufacturers; in 1921 there were 186. Production in 1929 however, reached an all time peak of 229,000 tractors.

One of the interesting developments of the year was the competition that the American tractor industry was beginning to encounter from Fordsons built by the Ford Motor Company's plant in Cork, Ireland. The United States Department of Commerce Circular 256 reported the first importations from Ireland as 1,517 in November and 1,955 in December. Fordson tractors were imported into the United States duty free under Paragraph 1604 of the Tariff Act.

The imported Fordson was a much improved machine over the one formerly produced in the United States. The engine power had been increased by adding 1/8-inch to the bore; a high-tension ignition system replaced the low-tension flyweel magneto with its troublesome high-tension vibrating coils, and the thermosiphon cooling system was replaced by pump circulation. The transmission was built heavier and taper roller bearings were used, the front end was spring mounted, fenders were standard equipment, while the muffler and lighting generator and equipment were optional.

The Oliver Farm Equipment Co. was organized February 25, 1929. It was a merger of the Oliver Chilled Plow Works, South Bend, Ind., the Hart-Parr Co., Charles City, Iowa, the American Seeding Machinery Co., and the Nichols & Shepard Co., Battle Creek, Michigan. The Oliver Chilled Plow Works was founded in 1855 by James Oliver, the inventor and pioneer manufacturer of chilled plows. The Nichols & Shepard Co., founded 1848, was one of the early concerns to manufacture grain separators. The Hart-Parr Co. was organized in 1905 and was credited as being one of the first firms to manufacture farm tractors on a commercial basis.

The Minneapolis-Moline Power Implement Co. was incorporated in Delaware. The company was a consolidation of the Moline Implement Co., the Minneapolis Steel & Machinery Co. (builders of the Twin City tractors) and the Minneapolis Threshing Machine Co. (builders of Minneapolis tractors).

The Electric Wheel Co. abandoned production of its "Allwork" tractors.

The United Tractor & Equipment Corp., Chicago, Ill., was organized by 32 independent makers and distributors of tractors, farm implements, and industrial equipment. The combined assets of the companies in this cooperative manufacturing and marketing movement was in excess of $125,000,000. Plans called for a full line of farm and industrial tractor equipment designed to be operated by their own tractor, the "United." The tractor was in full production under contract by the Allis-Chalmers Manufacturing Co. of Milwaukee, a member of the corporation. The first public showing of the "United" (later renamed Allis-Chalmers Model U, (page 20) was at the Southwest Tractor and Road Show in Wichita, Kan., March 1929.

Further competition in the agricultural wheel tractor trade was indicated by the announcement of the light Caterpillar "15" (page 20), by the Caterpillar Tractor Co., to sell for about $1,500. The addition of this size gave the company a complete "light line" of "10", "15", "20", "30" and "60" models.

The J.I. Case Threshing Machine Co. of Racine, Wis., announced a change in its corporate name. Hereafter the organization was to be officially known as the J.I. Case Co., Inc. With the company handling a full line of farm operating equipment, use of the word "threshing" in the corporate name no longer indicated adequately the class of machines manufactured. The same year the Case Model L, 26-40 (page 20), was produced. Later the same year the Case Model C, a 2- to 3-plow conventional 4-wheel tractor designed along the same general lines as the Model L was produced, also a Model CO orchard tractor, and the CI and LI industrial tractors. All were built with 3-bearing crankshafts, force-feed lubrication, and removable cylinder barrels.

The Huber Super Four, 40-62 model (page 20), appeared. This was a larger version of the super 4 (18-36) produced in 1926.

The Four Drive "Cat," 15-30 hp., Model "E" (page 20) was added to the line of the 4-wheel drive tractors produced by the Four Drive Tractor Co., of Big Rapids, Mich. (the word "Cat" was discontinued in Sept., 1929.) This was powered with a 4-cylinder vertical Waukesha engine in place of the Climax engine that powered some of the earlier models designated as Fitch Four Drive.

The Rock Island Plow Co., Rock Island, Ill., furnished to the trade the new Heider 18-35 (page 20). This was a larger version of the Heider 15-27 appearing in 1925, but the 1929 tractor was fitted with sliding gear transmission in place of the friction drive used in their earlier models.

Up to the year 1930 internal combustion tractor engine design was greatly influenced by the type of fuel used, and efforts toward engine improvement were being directed along two independent lines: (a) designing for the use of the more highly refined fuels, and (b) designing for the use of the lower grades of crudes. The advocates of gasoline-burning equipment had anticipated a more abundant, and, through improved refining processes, a more economical use of high-grade fuels. The diesel-tractor engine that was being tried out in this country about this time gave food for further consideration on the use of crudes. This compression-ignition engine, although burning a low-priced oil, required the use of precision-made parts for straining and metering the fuel and the special dsigning of parts to withstand the extremely high pressures, thus entailing a relatively high first cost. The economy realized with such equipment was dependent upon a high-annual use—at least 1,000 hours.

The following 17 makes and models of tractors were tested at Lincoln, Nebr., in 1929.

Test No.	Tractor model and horsepower
155	Case "L" (26-40)
156	McCormick-Deering (22-36)
157	Rock Island "G 2" (15-25)
158	Rock Island "G 2" (18-30)
159	Caterpillar "15" (10-15)
160	Caterpillar "10" (10-15)
161	Baker (43-67)

162	Minneapolis (27-42)
163	Minneapolis (39-57)
164	Wallis (12-20)
165	Four Drive "E" (15-25)
166	No entry
167	Case "C" (17-27)
168	Huber "Light Four" (20-36)
169	Case "CC"
170	Allis-Chalmers (United)
171	Monarch "35" (30)

Some 1929 Models

Allis-Chalmers Model U (United)

Case L (26-40)

Fitch Four Drive E - 15-30, The Cat

Caterpillar 15

Huber Super Four 40-62

Heider 18-35

1930

In 1930 tractor production dropped to 196,297 units, or 12 percent under the peak production of the previous year (Table 2).

The 1930 figures show an increase of 82 percent above those of 1925 and an increase of 275 percent for the 10-year period. Illinois retained first place in number of tractors on farms. Kansas shows the largest increase in units, Iowa second, Illinois third, and Ohio fourth. Some of the less important farm tractor states show largest percentages of increase; Wyoming led this field with an increase of 211 percent from 1925 to 1930; Mississippi was second with 196 percent increase; Montana third with 188 percent; and Idaho fourth with 143 percent increase.

Table 2. Number of tractor companies, number of tractors produced, and number of horses and mules on farms, 1921-1930.

Year	Tractor companies	Tractor production	Horses and mules (000 omitted)
1921	186	68,029	25,138
1922	116	98,794	24,587
1923	93	131,908	24,018
1924	64	116,838	23,273
1925	58	164,097	22,558
1926	69	178,074	21,970
1927	61	194,913	21,169
1928	51	171,469	20,415
1929	47	223,081	19,699
1930	38	196,297	19,050

During the year general-purpose tractors were being produced by International Haravester Co., Oliver Farm Equipment, Massey-Harris, Minneapolis-Moline, Case, and Allis-Chalmers. This does not imply that the entire field had gone "general purpose," but that the large potential market for this equipment was realized and that something was being done about it. At the same time tractors of 3- to 5-plow and larger capacity were being further developed to handle the new heavy machinery and implements either drawn or operated by power take-off from tractor, such as corn pickers, potato diggers, combines, wheatland plows, power sprayers, etc.

Early in 1930 the Oliver Farm Equipment Co. announced its new tricycle "Row Crop" (page 22) tractor with its "tiptoe" drive wheels, which had been developed as a result of the work carried on by the Oliver and Hart-Parr divisions of the company. The drive wheels were essentially steel rings—flat vertically—that were solely for the purpose of attaching the spokes and to provide an attachment point for the 20 lugs of each wheel. Two lugs were generally in contact with ground, although in soft mellow ground three lugs were in contact. Also their Standard tractors "80" and "90-99" appeared.

A 4-wheel drive general purpose tractor was put out by the Massey-Harris Co. (page 22). This was made in 4 widths of tread to conform to different row spacings. With the 2 hand-operated brakes operative on both wheels on one side or the other, it was claimed the tractor could be turned within a 6-foot circle. The rear axle assembly was free to oscillate about a horizontal axis, thereby giving flexibility in traveling over uneven ground.

The Twin City 11-20 hp. KT (Kombination Tractor) (page 22) was announced in June by the Minneapolis-Moline Power Implement Co. of Minneapolis, Minn. "Kombination" was inspired by the fact that the tractor was equipped to handle any combination of power jobs, such as plowing, cultivating, belt work, and hauling. A power take-off available as extra equipment further added to its versatility.

The Case Company made formal announcement of their general-purpose Model CC (page 22) in September, although a number of tractors of this model had been working under actual field conditions some months before.

The Allis-Chalmers "All Crop" (page 24) general-purpose tractor was announced in the fall of 1930 as was also the 6-cylinder Rumely 6 (page 24). Six forward speeds (ranging from 2 1/2 to 4 3/4 miles per hour) and 2 reverse speeds were provided.

Table 3. U. S. Census of Tractors on Farms, 1920, 1925, 1930.

State	April 1 1930	Jan. 1 1925	Jan. 1 1920
Illinois	69,628	43,325	23,102
Kansas	66,275	31,171	17,177
Iowa	66,258	37,230	20,270
Ohio	52,974	30,905	10,469
Wisconsin	50,173	29,651	9,407
Minnesota	48,457	26,739	15,503
California	44,437	29,948	13,852
Indiana	41,979	23,567	9,230
Nebraska	40,729	18,765	11,100
New York	40,369	25,681	7,497
North Dakota	37,605	17,483	13,006
Texas	37,348	16,780	9,048
Michigan	34,579	19,217	5,884
South Dakota	33,837	17,426	12,939
Pennsylvania	33,513	19,764	5,697
Oklahoma	25,962	10,950	6,210
Missouri	24,999	12,745	7,889
Montana	19,031	6,602	7,647
Colorado	13,334	6,693	4,990
North Carolina	11,426	7,909	2,277
Oregon	9,838	5,768	3,070
Virginia	9,757	6,750	2,544
Washington	8,388	4,490	2,635
New Jersey	8,088	4,419	946
Kentucky	7,322	4,994	2,029
Maryland	7,208	4,026	1,525
Tennessee	6,865	4,767	1,872
Georgia	5,870	4,145	2,252
Arkansas	5,684	3,476	1,822
Florida	5,618	2,777	680
Mississippi	5,542	1,871	667
Louisiana	5,016	3,482	2,812
Idaho	4,691	1,927	1,587
Alabama	4,664	2,465	811
Wyoming	4,110	1,323	1,075
Massachusetts	3,921	2,212	592
South Carolina	3,462	2,906	1,304
Maine	3,410	1,774	635
West Virginia	2,792	1,860	572
Connecticut	2,667	1,441	440
Arizona	2,558	1,239	930
New Mexico	2,497	1,080	491
Vermont	2,426	1,568	444
Delaware	1,600	694	239
Utah	1,426	850	583
New Hampshire	1,096	531	207
Rhode Island	589	318	79
Nevada	360	221	210
	920,378	505,925	245,370

In August the Eagle Manufacturing Co. of Appleton, Wis., announced a new 6A 3-4 plow tractor (page 24) powered by a 6-cylinder Hercules vertical L head 4 x 4 1/2 inch engine mounted on a 2-piece frame. Wheel tread was adjustable from 40 inches to 46 inches by reversing the wheels.

An Allis-Chalmers Monarch Diesel "75" tractor equipped with a Monarch-Atlas Diesel engine especially designed for this tractor was being tried out on the ranch of the Campbell Farming Corp., in Montana where it was found economical for plowing and heavy-duty construction jobs. In April the Allis-Chalmers Manufacturing Co. took over the Advance-Rumely Co. of La Porte, Ind. The consolidation gave the purchasing company a full line of threshers and combines, the Do-All and Rumely Six tractors, and other power machinery.

The Cletrac "30" (page 24) was added to the Cleveland Tractor Co. line.

Between 1926 and 1929 track-type tractor production in the U.S. had increased threefold. Even during 1930, which showed a decline in wheel-type tractor production, track-type tractor production held up due to its wide use in industry.

The manufacture of tractors in eastern and western Europe began to threaten both domestic and export business of American manufacturers. In the Putilovetz factory at Leningrad (which had been operating before the revolution) 10,000 units were built during 1929-30. The Ford factory in Ireland was employing 2,700 workers 3 shifts a day in the beginning of 1930. During that year Ford built 15,196 tractors, of which 6,815 were shipped to the United States.

The following 10 makes and models of tractors were tested at Lincoln, Nebr., in 1930.[1]

Test No.	Tractor model and horsepower
173	Fordson "F" (11-21) K
174	Fordson "F" (14-26) G
175	Twin City K T (11-20)
176	Oliver Row Crop (18-27)
177	Massey-Harris G.P. (15-22)
178	(no test made)
179	Monarch "50" (42-50)
180	Oliver Hart-Parr (18-28)
181	(no test made)
182	Cletrac "80-60" (60-80)
183	Oliver Hart-Parr (28-44)
184	Eagle "6A" (22-37)

On page 23 is a summary of the Nebraska tests, which includes only tractors on the market January 1, 1930.

[1]*Beginning with 1931, the names of the tractors tested at Nebraska will no longer be listed at the end of the text for that year.*

Oliver Tricycle Row Crop

Twin City Kombination Tractor - KT

Some 1930 Models

Case CC - General Purpose

Massey-Harris Four Wheel Drive

SUMMARY OF RESULTS OF THE NEBRASKA TRACTOR TESTS

CONDUCTED BY THE UNIVERSITY OF NEBRASKA, DEPARTMENT OF AGRICULTURAL ENGINEERING, LINCOLN, NEBRASKA

INCLUDES ONLY TRACTORS ON THE MARKET JANUARY 1, 1930

Tractor, Engine and Belt Tests

TEST NO.	TRACTOR: NAME	RATING	WEIGHT AS TESTED	ENGINE: CYL'S.	BORE AND STROKE	RATED R.P.M.	MAKE	FUEL USED	TOTAL HRS. RUN	BELT: RATED LOAD ~ TWO HRS.: H.P.	AV. ENG. R.P.M.	FUEL: GALS. PER HOUR	H.P. HRS. PER GAL.	LBS. PER H.P. HR.	WATER GALS. PER HR.	TEMP. °F ATM.	TEMP. °F RAD.	MAXIMUM LOAD ~ ONE HR.: H.P.	AV. ENG. R.P.M.	FUEL: GALS. PER HR.	H.P. HRS. PER GAL.	LBS. PER H.P. HR.	WATER GALS. PER HR.	TEMP. °F ATM.	TEMP. °F RAD.	VARYING LOAD ONE HOUR R.P.M.: HIGH	LOW
	1	2	3	4	5	6	7	8	9	10	11	12	13	14	15	16	17	18	19	20	21	22	23	24	25	26	27
	~1920~																										
22	GRAY	18-30	6,510	4	4¾×6¾	950	WAUKESHA	GAS	43	30.46	958	3.22	5.85	1.05	2.25	100	209	32.20	958	4.83	6.67	0.92	5.50	103	210	1452	948
28	BEEMAN "G"	2-4	550	1	3½×4½	1800	OWN	GAS	40	2.40	1036	0.44	5.45	1.13	0.63	76	210	2.37	1025	.42	5.60	1.09	0.75	80	210	1093	951
29	LA CROSS "G"	12-24	4,670	2	6×7	900	OWN	KERO.	30	24.23	899	3.69	6.56	1.03	0.19	84	201	24.94	902	4.36	5.72	1.18	0.25	90	192	1225	896
43	AVERY	25-50	12,500	4	6½×8	700	OWN	KERO.	32	50.40	702	6.37	7.91	0.85	12.50	88	212	56.68	712	7.04	8.05	0.84	25.00	99	212	893	691
44	AVERY	45-65	22,000	4	7¾×8	600	OWN	KERO.	39	65.73	612	9.06	7.26	0.83	17.00	88	212	69.23	597	8.23	8.41	0.80	18.75	90	216	678	615
45	CLETRAC "W"	12-20	3,500	4	4×5½	1265	WEIDELY	KERO.	30	20.14	1276	2.42	8.33	0.81	0.50	97	194	24.94	1321	2.90	8.61	0.78	2.00	109	212	1459	1505
65	TORO	6-10	2,600	4	3⅝×4½	1200	LeROI	GAS	30	10.56	1208	1.25	8.48	0.75	0.48	102	208	15.31	1219	1.75	7.70	0.81	1.27	79	210	1326	1197
68	BATES~S~MULE	15-22	4,600	4	4⅛×5¼	1100	MIDWEST	GAS	32	22.20	1112	3.00	7.39	0.84	0.39	69	161	29.78	1108	3.59	8.30	0.75	0.42	75	187	1136	1061
	~1921~																										
73	WETMORE	12-25	3,000	4	4×5½	1265	WEIDELY	KERO.	52	25.27	1253	3.65	6.95	0.98	0.00	72	195	27.94	1264	3.46	8.08	0.84	0.64	70	206	1280	1246
	~1924~																										
101	BEAR "B"	25-35	7,000	4	4¾×6½	1290	STEARNS	GAS	40	36.53	1293	4.53	8.06	0.77	0.18	75	180	33.36	1295	8.43	6.59	0.94	0.00	68	180	1399	1286
104	BEST "S~30"	25-30	9,065	4	4¾×6½	850	OWN	GAS	36	30.24	853	3.26	9.28	0.67	0.00	74	175	37.83	850	4.91	7.71	0.81	0.00	79	198	977	840
105	BEST "A~60"	50-60	20,000	4	6½×8½	650	OWN	GAS	34	60.00	651	7.81	7.68	0.81	0.00	87	196	72.51	653	11.36	6.38	0.97	0.00	87	195	696	652
110	CASE	25-45	10,610	4	5½×6¾	850	OWN	KERO.	36	45.18	854	5.13	8.80	0.77	1.17	73	183	52.59	852	5.75	9.14	0.74	2.37	64	180	909	849
	~1925~																										
117	I.H.C. FARMALL	—	3550	4	3¾×5	1200	OWN	KERO.	39	18.03	1196	1.87	9.64	0.70	1.88	91	210	20.05	1201	2.14	9.39	0.72	1.10	77	210	1288	1197
	~1926~																										
119	CLETRAC "K"	15-25	4,775	4	4×5½	1375	OWN	KERO.	45	25.38	1384	2.80	9.08	0.74	0.00	67	180	28.44	1378	3.10	9.18	0.74	0.26	85	205	1598	1572
120	CLETRAC "K"	15-25	4,775	4	4×5½	1575	OWN	GAS	39	25.16	1382	3.38	7.44	0.83	0.00	87	185	30.15	1380	4.27	7.07	0.87	0.00	81	182	1646	1551
121	TWIN CITY	17-28	5,895	4	4¼×6	1000	OWN	KERO.	39	28.40	1007	2.96	9.58	0.71	0.19	67	194	30.91	1005	3.16	9.79	0.69	1.40	89	210	1105	993
122	TWIN CITY	27-47	10,657	4	5½×6¾	900	OWN	KERO.	38	44.04	900	4.69	9.38	0.72	2.90	95	210	49.05	899	5.53	8.89	0.76	0.89	85	209	953	895
123	HUBER SUPER 4	18-36	8,595	4	4¾×6½	1000	STEARNS	GAS	43	36.18	999	4.58	7.90	0.78	0.00	88	197	43.15	998	5.86	7.37	0.83	0.00	83	195	1175	995
125	CLETRAC "A"	30-45	7,225	6	4×5	1577	WISCONSIN	GAS	36	45.19	1582	5.64	8.01	0.77	1.42	96	210	48.62	1575	5.93	8.20	0.75	0.96	89	210	1722	1575
126	HUBER	20-40	9,336	4	5⅛×6½	1100	STEARNS	GAS	47	40.23	1102.5	4.89	8.25	0.75	0.00	75	167	50.05	1104	7.08	7.07	0.87	0.00	83	175	1257	1098
128	HART PARR	18-36	7,325	2	6¾×7	800	OWN	DIST.	37	36.53	802	3.75	9.74	0.71	1.77	75	167	42.85	795	5.03	8.52	0.81	1.97	71	178	835	799
129	HART PARR	12-25	5,440	2	5¾×6½	850	OWN	DIST.	34	24.09	852	2.38	10.12	0.68	0.49	73	198	31.99	856	3.90	8.19	0.84	2.56	65	179	889	849
	~1927~																										
134	WALLIS	20-30	4,525	4	4⅜×5¾	1050	OWN	DIST.	38	30.24	1052	2.94	10.28	0.67	0.95	70	210	35.29	1050	4.23	8.34	0.83	0.00	63	183	1234	1043
135	HUBER	40-62	9910	4	5½×6½	1100	STEARNS	GAS	48	62.42	1100	6.45	9.39	0.65	0.00	79	190	69.76	1100	8.66	8.06	0.76	0.00	77	178	1295	1095
136	SHAW "T25"	—1	400	1	2½×2½	2200	BRIGGS&STR.	GAS	41	1.02	2241	0.31	3.30	1.86	0.00	65	—	1.08	2215	0.29	3.76	1.63	0.00	61	—	2400	2177
137	SHAW "T45"	—1½	457	1	2⅞×3	1600	"	GAS	31	1.51	1611	0.32	4.73	1.30	0.00	81	—	1.68	1610	0.35	4.80	1.28	0.00	80	—	1595	1519
138	CULTOR "32"	—	1840	4	3¾×4	1050	FORD "T"	GAS	29	—	—	—	—	—	—	—	—	—	—	—	—	—	—	—	—	—	—
139	MONARCH "10 TON"	—	21,700	4	6½×7	850	BEAVER	GAS	37	—	—	—	—	—	—	—	—	—	—	—	—	—	—	—	—	—	—
140	HART PARR	28-50	10,394	4	5¾×6½	850	OWN	DIST.	38	50.35	855	4.69	10.73	0.64	3.38	77	210	64.56	851	7.94	8.15	0.84	5.75	89	201	951	856
141	OIL PULL "W"	20-30	6776	2	5½×7	850	OWN	KERO.	41	30.24	848	3.14	9.62	0.69	2.47	64	180	35.36	850	4.48	7.89	0.84	4.11	76	160	866	841
142	McCORMICK DEERING	10-20	5028	4	4¼×5	1025	OWN	KERO.	35	20.46	1028	2.04	10.05	0.68	0.00	63	206	24.81	1025	2.72	9.12	0.75	0.35	64	208	1176	1022
143	OIL PULL "X"	25-40	9440	2	6½×8¼	725	OWN	KERO.	34	40.50	724	3.63	11.15	0.61	2.90	80	201	50.26	725	5.77	8.71	0.78	5.43	76	174	745	720
144	ROCK ISLAND	18-35	5740	4	4½×6	1100	BUDA	DIST.	37	35.52	1094	3.92	9.02	0.76	0.96	76	186	36.50	1094	4.74	7.70	0.89	1.15	74	177	1169	1073
145	OIL PULL "Y"	30-50	13,207	2	7½×9½	635	OWN	KERO.	33	50.56	632	4.67	10.79	0.63	3.29	69	217	63.32	635	7.16	8.85	0.77	5.96	80	200	657	626
146	JOHN DEERE	15-27	4917	2	6½×7	800	OWN	KERO.	35	27.45	802	2.56	10.74	0.64	0.97	92	210	36.98	801	3.98	9.29	0.74	3.68	91	204	843	796
147	MONARCH "6 TON"	40	15,735	4	5⅛×6½	1000	STEARNS	GAS	24	—	—	—	—	—	—	—	—	—	—	—	—	—	—	—	—	—	—
	~1928~																										
148	LAUSON "S12"	20-35	6,228	4	4½×6	1100	LeROI	GAS	46	35.24	1099	4.419	7.97	0.778	0.22	61	191	40.88	1113	4.629	8.83	0.702	0.24	60	191	1159	1071
149	CLETRAC "40"	40-55	12,038	6	4½×5	1575	WISCONSIN	GAS	37	55.24	1577	7.816	7.07	0.874	0.00	67	178	63.00	1579	8.447	7.46	0.829	0.00	67	174	1724	1565
150	CATERPILLAR "20"	20-25	7,822	4	4×5½	1100	OWN	GAS	41	25.23	1102	3.036	8.31	0.744	0.00	86	181	29.49	1102	3.333	8.85	0.699	0.00	82	181	1185	1096
151	ALLIS CHALMERS	20-35	7,095	4	4¾×6½	950	OWN	GAS	50	35.29	933	4.226	8.35	0.741	0.00	86	179	44.29	951	5.018	8.83	0.701	0.00	85	177	1001	925
152	TWIN CITY	21-32	6,463	4	4½×6	1075	OWN	GAS	44	32.37	1075	3.539	9.15	0.678	0.00	73	192	39.14	1075	3.984	9.82	0.631	0.00	68	191	1110	1070
153	JOHN DEERE GEN. PURP.	10-20	4265	2	5¾×6	950	OWN	KERO.	42	20.20	948	2.362	8.55	0.793	0.93	62	197	24.97	948	2.721	9.18	0.741	1.54	62	199	1021	940
154	ADVANCE RUMELY DO-ALL	—	3702	4	3¾×4½	1400	WAUKESHA	GAS	40	19.46	1400	2.335	8.33	0.744	0.00	56	206	21.61	1400	2.374	9.10	0.681	0.00	47	206	1644	1380
	~1929~																										
155	CASE "L"	26-40	5307	4	4½×6	1100	OWN	KERO.	58	40.22	1102	4.123	9.76	0.701	0.00	62	183	44.01	1099	4.461	9.87	0.693	0.00	61	190	1134	1095
156	McCORMICK DEERING	22-36	7486	4	4¾×6	1050	OWN	KERO.	48	36.15	1055	3.898	9.27	0.740	1.80	82	185	40.66	1050	4.070	9.99	0.687	2.56	76	177	1187	1062
157	ROCK ISLAND "G2"	15-25	5070	4	4¼×5¾	1100	WAUK.	DIST.	30	25.36	1102	3.014	8.41	0.825	0.00	55	161	29.00	1100	3.620	8.01	0.866	0.00	54	162	1130	1084
158	ROCK ISLAND "G2"	18-30	5380	4	4⅜×5¾	1100	WAUK.	GAS	25	30.16	1102	3.264	9.24	0.666	0.00	67	168	33.69	1102	3.906	9.14	0.675	0.00	66	175	1113	1089
159	CATERPILLAR "15"	15-20	5951	4	3¾×5	1250	OWN	GAS	37	20.04	1250	2.626	7.63	0.806	0.00	72	202	24.77	1249	3.005	8.24	0.748	0.00	71	182	1305	1239
160	CATERPILLAR "10"	10-15	4375	4	3⅜×4	1500	OWN	GAS	37	15.26	1504	1.925	7.93	0.779	0.00	67	194	18.10	1501	2.107	8.59	0.718	0.00	64	176	1586	1498
161	BAKER	43-67	10375	4	5½×7	1100	LeROI	GAS	35	67.37	1101	7.236	9.31	0.661	0.00	80	206	73.88	1101	8.267	9.18	0.670	0.58	74	209	1423	1079
162	MINNEAPOLIS	27-42	8375	4	4⅝×7	925	OWN	GAS	39	42.30	926	4.240	9.98	0.617	0.00	66	165	48.42	925	5.133	9.43	0.653	0.00	58	187	991	920
163	MINNEAPOLIS	39-57	9695	4	5½×6½	1000	STEARNS	GAS	40	57.39	997	6.445	8.90	0.692	0.00	72	183	64.55	1004	6.367	10.14	0.608	0.00	72	197	1151	995
164	WALLIS	12-20	3895	4	4⅜×5¾	1000	OWN	DIST.	48	20.32	1001	2.277	8.92	0.778	0.00	72	205	24.16	999	2.467	9.79	0.709	1.06	90	212	1093	990
165	FOUR DRIVE "E"	15-25	6305	4	4⅛×5¼	1100	WAUK.	KERO.	46	25.29	1105	3.398	7.03	0.976	0.00	73	179	28.42	1100	3.945	7.20	0.952	0.00	75	190	1151	1101
167	CASE "C"	17-27	4155	4	3⅞×5½	1100	OWN	KERO.	35	27.53	1100	2.472	11.06	0.621	0.76	90	212	29.81	1102	2.624	11.36	0.604	0.89	89	213	1214	1101
168	HUBER "LIGHT FOUR"	20-36	6725	4	4⅝×6¼	1150	WAUK.	GAS	36	36.05	1148	4.625	7.79	0.793	0.00	95	182	42.26	1152	5.207	8.12	0.761	0.00	96	194	1214	1142
169	CASE "CC"	—	4240	4	3⅞×5½	1100	OWN	KERO.	40	27.37	1098	2.601	10.52	0.652	0.00	70	202	28.97	1101	2.880	10.06	0.682	0.00	79	207	1203	1099
170	ALLIS CHALMERS (UNITED)	—	4821	4	4¼×5	1200	CONT.	GAS	35	30.27	1200	3.671	8.25	0.756	0.00	65	185	35.04	1200	4.181	8.38	0.724	0.00	65	182	1255	1187
171	MONARCH "35"	50	10,680	4	4⅞×6¼	930	OWN	GAS	27	NO BELT PULLEY			—	—	—	—	—	—	—	—	—	—	—	—	—	—	—

Drawbar Tests

TEST NO.	RATED LOAD ~ TEN HOURS: DRAW BAR H.P.	DRAW BAR PULL LBS.	MILES PER HR.	AV. ENG. R.P.M.	SLIP OF DRIVERS PER CENT: POINT LUGS	FACE RIM	TYPE OF LUG	GEAR USED	FUEL: GALS. PER HR.	D.B.H.P. HRS. PER GAL.	LBS. PER D.B.H.P. HR.	WATER GAL. PER HR.	TEMP. °F ATM.	TEMP. °F RAD.	MAXIMUM LOAD: LOW GEAR: DRAW BAR H.P.	DRAW BAR PULL LBS.	MI. PER HR.	INT. GEAR: DRAW BAR H.P.	DRAW BAR PULL LBS.	MI. PER HR.	HIGH GEAR: DRAW BAR H.P.	DRAW BAR PULL LBS.	MI. PER HR.
	28	29	30	31	32	33	34	35	36	37	38	39	40	41	42	43	44	45	46	47	48	49	50
	~1920~																						
22	19.64	2703	2.73	925	9.91	—	SPADE	HIGH	4.39	4.48	1.57	3.53	89	212	19.15	3390	2.12	—	—	—	18.21	3044	2.24
28	1.08	187.8	2.17	1334	22.25	—	ANGLE	—	0.29	3.78	1.62	0.51	—	204	0.78	187	1.56	—	—	—	—	—	—
29	13.63	1859	2.75	821	7.90	—	SPADE	—	2.86	4.76	1.41	0.30	87	198	—	—	—	—	—	—	17.83	2155	3.10
43	27.29	3778	2.71	694	8.30	—	SPUD	LOW	6.42	4.25	1.59	17.81	87	208	51.47	4415	2.67	—	—	—	32.62	3225	3.79
44	46.93	7812	2.25	634	3.00	—	UNI.	LOW	11.85	3.96	1.70	11.37	81	210	49.97	8475	2.21	—	—	—	45.59	5020	3.24
45	13.47	1395	3.62	1386	4.30	—	—	—	2.67	5.04	1.34	0.51	85	195	—	—	—	—	—	—	13.32	1734	3.36
65	6.32	893	2.66	1215	8.04	—	SPADE	LOW	1.47	4.31	1.44	0.11	50	206	9.92	1310	2.84	—	—	—	—	—	—
68	16.17	2558	2.37	1130	2.67	—	—	LOW	2.66	6.07	1.03	0.19	69	179	23.19	3100	2.81	—	—	—	22.98	2103	4.10
	~1921~																						
73	13.28	1477	3.37	1255	14.00	5.60	ANGLE	INT.	2.88	4.61	1.48	0.04	62	176	11.36	2525	1.69	16.10	2260	2.67	—	—	—
	~1924~																						
101	22.79	2855	3.68	1285	2.91	—	ANGLE	INT.	4.90	5.49	1.09	0.06	61	165	51.55	6805	1.74	44.64	4863	3.44	41.28	2708	5.72
104	26.56	3796	2.62	852	1.33	—	—	INT.	3.40	7.82	0.80	0.00	79	181	30.20	7563	1.50	53.21	4823	2.58	52.02	3343	5.59
105	51.61	7654	2.53	649	1.45	—	—	INT.	8.49	6.08	1.02	0.00	77	175	61.53	12560	1.86	61.27	9155	2.51	60.22	6240	3.62
110	26.09	2966	3.30	856	6.69	-2.49	SPADE	HIGH	4.53	5.76	1.17	0.54	37	193	32.96	5750	2.15	—	—	—	34.65	3969	3.28
	~1925~																						
117	9.35	1114	3.15	1207	10.17	-5.71	ANGLE	INT.	1.79	5.23	1.29	1.16	91	210	12.70	2727	1.74	13.27	1625	3.06	10.41	867.5	4.50
	~1926~				SLIPPAGE																		
119	15.30	2703	2.12	1426	4.08		ANGLE	LOW	2.64	5.80	1.16	0.02	58	174	25.42	4373	2.00	—	—	—	20.12	1790	4.21
120	15.31	2763	2.08	1377	2.44		ANGLE	LOW	2.82	5.44	1.13	0.03	78	164	24.53	4590	2.00	—	—	—	20.73	1840	4.22
121	17.30	2927	2.22	998	9.40		SPADE	LOW	2.70	6.40	1.06	0.03	71	190	22.50	3778	2.25	—	—	—	21.47	2548	3.17
122	27.17	4368	2.33	904	7.66		SPADE	LOW	4.37	6.22	1.09	0.59	82	194	34.65	5640	2.30	—	—	—	32.63	3992	3.06
123	18.86	3198	2.21	1011	12.17		CONE	LOW	3.48	5.42	1.14	0.19	93	186	27.02	5420	1.87	—	—	—	30.33	2997	3.80
125	29.16	4520	2.42	1581	2.56		ANGLE	LOW	4.71	6.19	1.00	0.29	81	206	38.38	6170	2.34	—	—	—	36.43	2799	4.88
126	22.12	3233	2.57	1111	5.57		CONE	LOW	4.01	5.51	1.12	0.17	85	186	40.79	5953	2.57	—	—	—	37.78	3931	3.60
128	18.10	2248	3.02	756	4.94		SPADE	HIGH	2.78	6.50	1.06	1.22	62	177	30.17	5535	2.04	—	—	—	32.25	4075	2.97
129	12.11	1274	3.57	848	6.26		SPADE	HIGH	2.63	4.60	1.50	0.35	59	172	21.78	2950	2.77	—	—	—	20.43	2193	3.49
	~1927~																						
134	19.80	2452	3.03	1048	9.30		SPADE	LOW	2.58	7.68	0.90	0.00	46	173	27.01	3409	2.98	—	—	—	26.36	2625	3.77
135	40.26	4442	3.40	1113	11.44		SPADE	HIGH	6.12	6.58	0.93	0.14	84	184	49.44	7607	2.43	—	—	—	50.03	5355	3.50
136	0.55	123	1.66	2214	25.71		SPADE	HIGH	0.27	2.06	2.98	—	76	—	0.46	176	0.99	—	—	—	0.73	140.5	1.94
137	0.53	145	1.37	1627	28.64		SPADE	HIGH	0.30	1.77	3.47	—	82	—	0.55	222	0.93	—	—	—	0.71	159.5	1.65
138	4.95	591	3.13	1040	5.03		SPADE	—	1.21	4.06	1.53	0.42	84	204	8.41	1050	3.01	—	—	—	—	—	—
139	59.56	9023	2.48	855	1.43		ANGLE	INT.	8.64	6.89	0.90	0.05	77	176	74.85	20,050	1.40	78.17	11,960	2.45	65.97	7550	3.28
140	28.52	2967	3.60	855	2.96		SPADE	HIGH	4.79	5.95	1.15	1.92	89	194	45.38	7547	2.22	—	—	—	46.03	4940	3.49
141	20.96	2455	3.20	844	6.93		SPADE	INT.	3.46	6.06	1.09	2.29	63	183	24.89	3995	2.34	26.10	3007	3.25	25.04	2315	4.06
142	10.90	1196	3.42	1026	4.19		SPADE	INT.	1.78	6.11	1.12	0.05	61	204	19.60	2955	2.49	17.92	1937	3.47	19.08	1486	4.98
143	26.42	3276	3.02	724	6.24		SPADE	INT.	3.34	7.91	0.86	1.97	56	182	37.79	6385	2.22	38.66	4787	3.02	34.58	3477	3.73
144	18.77	1928	3.65	1105	2.82		SPADE	LOW	2.66	7.07	0.97	0.00	72	174	30.42	3190	3.58	—	—	—	28.12	1995	5.29
145	30.53	3504	3.25	634	4.19		SPADE	INT.	4.49	6.75	1.01	3.45	76	204	47.18	6965	2.54	4410	5252	3.15	42.85	4025	3.99
146	15.38	1577	3.66	796	3.85		SPADE	HIGH	2.30	6.69	1.02	0.64	78	210	28.53	4462	2.40	—	—	—	27.77	3090	3.37
147	40.59	5450	2.79	982	1.48		ANGLE	INT.	5.50	7.38	0.84	0.08	41	154	50.55	10,557	1.80	49.07	6580	2.80	47.53	4402	4.05
	~1928~																						
148	20.04	2201	3.41	1101	3.37		SPADE	HIGH	3.939	5.09	1.219	0.25	63	183	28.53	3140	2.08	—	—	—	29.63	3259	3.41
149	40.74	4317	3.54	1571	0.62		ANGLE	INT.	6.231	6.54	0.944	0.05	88	193	55.51	9725	2.14	51.67	3457	3.56	39.61	2637	5.63
150	20.83	2556	3.06	1099	0.45		ANGLE	INT.	2.957	7.04	0.878	0.00	83	181	26.32	3721	1.725	24.77	3006	3.09	25.15	1860	4.67
151	21.25	2272	3.51	925	3.34		SPADE	HIGH	4.014	5.29	1.169	0.00	84	166	33.20	4400	2.83	—	—	—	32.37	3488	3.48
152	22.58	2499	3.39	1075	5.42		SPADE	INT.	3.440	6.56	0.945	0.09	85	174	29.16	4716	2.32	30.20	3333	3.40	27.73	2176	4.78
153	10.20	1078	3.55	937	1.01		SPADE	INT.	2.061	4.95	1.374	0.16	42	185	17.24	2489	2.60	16.83	1837	3.44	13.93	1060	4.93
154	11.63	1466	2.98	1404	7.79		SPADE	LOW	2.334	4.98	1.244	0.00	55	201	16.32	2012	3.04	—	—	—	13.93	1198	4.36
	~1929~																						
155	26.28	2833	3.45	1095	9.71		SPADE	INT.	4.028	6.52	1.049	0.13	65	185	30.02	4333	2.47	29.66	3427	3.25	30.08	2643	4.26
156	22.78	2258	3.78	1058	4.56		SPADE	INT.	3.688	6.18	1.111	0.64	68	184	30.07	3912	2.88	28.03	2846	3.69	25.46	2190	4.36
157	15.32	2023	2.84	1103	8.55		SPADE	LOW	2.934	5.22	1.329	0.00	67	177	22.08	2880	2.88	—	—	—	18.56	1627	4.28
158	18.68	2389	2.93	1110	6.96		SPADE	LOW	3.262	5.73	1.077	0.00	70	175	25.50	3549	2.69	—	—	—	22.91	2079	4.13
159	15.00	2179	2.58	1255	0.57		—	INT.	2.596	5.78	1.068	0.00	79	195	21.29	4166	1.92	21.79	3175	2.57	19.46	2039	3.58
160	10.10	1466	2.58	1505	0.99		—	INT.	1.726	5.85	1.054	0.00	80	196	14.89	2816	1.98	14.43	2087	2.59	15.38	1455	3.50
161	43.57	4275	3.82	1097	4.60		SPADE	HIGH	7.812	5.58	1.103	0.10	86	190	55.72	7840	2.67	—	—	—	54.61	3324	3.83
162	27.79	2956	3.53	926	4.78		SPADE	HIGH	4.190	6.63	0.929	0.09	94	186	33.99	4918	2.59	—	—	—	34.50	3693	3.48
163	39.84	3723	4.01	992	4.05		SPADE	HIGH	5.878	6.78	0.909	0.19	91	187	47.77	5950	3.01	—	—	—	47.57	4356	4.09
164	12.34	1232	3.76	1007	3.60		SPADE	INT.	2.079	5.94	1.169	0.05	81	192	18.07	3018	2.24	17.25	1784	3.62	16.39	1295	4.81
165	15.18	1639	3.47	1089	2.53		SPADE	INT.	3.607	4.21	1.630	0.08	84	183	18.97	3022	2.35	17.68	1879	3.53	7.76	405	7.17
167	17.41	1856	3.52	1100	8.76		SPADE	INT.	2.479	7.02	0.977	0.00	78	195	19.60	3289	2.20	21.41	2338	3.40	20.20	1525	4.97
168	20.35	2191	3.48	1153	7.06		SPADE	HIGH	4.035	5.04	1.225	0.00	91	166	27.25	4482	2.28	—	—	—	29.18	3298	3.32
169	17.88	1599	4.19	1104	3.01		SPADE	INT.	2.377	7.52	0.912	0.00	72	190	22.70	2950	2.88	20.91	2038	3.85	18.34	1187	5.79
170	19.28	1934	3.74	1192	5.53		SPADE	INT.	3.114	6.19	0.980	0.00	40	186	25.63	3679	2.61	24.62	2466	3.74	20.39	1314	5.82
171	30.08	4126	2.73	929	1.28		—	INT.	4.142	7.26	0.836	0.00	48	172	40.99	8450	1.82	38.63	5337	2.71	37.67	3525	4.01

Summary of Bulletin 242, Nebraska Tractor Tests, March, 1930

Nebraska Tractor Test Summary, 1930

Allis-Chalmers - All Crop

Eagle 6A

Rumely 6

Cletrac 30 (standard tread)

1931

In 1931, as in other periods of depression, mechanical power farming slowed down. This was more general abroad than in the United States, because foreign farmers were confronted by comparatively higher prices for fuel. In some cases this reactionary trend meant farming with hand tools instead of horse-drawn machines. Apparently the European governments such action on the theory that if machine sales could be prevented, men out of work could be given jobs.

Tractor production in the United States fell from 202,458 in 1930 to 71,704 in 1931. The output of tractors at the Ford plant at Cork, Ireland, which was at the rate of 50 tractors a day at the close of 1930, continued steadily during the first 2 months of 1931, but by the end of May production was practically discontinued owing to lack of new orders.

More bids were made for general-purpose tractor trade, the Huber, Caterpillar, and Sears-Roebuck companies competing; also the International Harvester Co. produced its 3-plow "F-30" general-purpose tractor.

The Huber Manufacturing Co. of Marion, Ohio, developed a general-purpose model the Modern Farmer in 2 styles for the 1931 trade, a high-clearance narrow tread 4-wheel machine straddling 1 row and a wide tread tricycle model straddling 2 rows (page 25), both of which were available with steel or rubber tires.

In March the High Clearance 10 (page 25) was announced by the Caterpillar Tractor Co. This company had for several years been producing its model 10, of which a revised model with high ground clearance (22 inches) was made, thus adapting it for row-crop work.

Sears-Roebuck Co. made its entry into the tractor field with a Bradley 2-plow general-purpose tractor (page 25) equipped with a Waukesha motor, Borg-Warner clutch, and other standard parts. This design embodied one of the early variable tread constructions and direct engine driven p.t.o.

The Farmall 30 was produced by the International Harvester Co. (page 25). This was a 3-plow model built closely along original Farmall 20 lines of 1922. The Farmall 30 was intended to meet the demand of larger farms in row-crop territory where 3-bottom plowing was practiced and heavier drawn equipment, such as combines and 2-row pickers, were already being used.

Deere & Co. offered their first general-purpose tricycle type or wide tread tractor. This was a 2-plow tractor capable of cultivating 2 or 4 rows and was fitted with a power take-off for front or rear connection and a mechanical power lift, featured in their G.P. tractor of 1928.

The Caterpillar Tractor Co. produced the Diesel 65 (page 25) in October, the first diesel tractor in the United States to be offered to the trade. The tractor was powered by a 4-cylinder 4-cycle engine governed at 650 r.p.m. A 2-cylinder spark ignition gas engine was provided for a starting unit.

The Cleveland Tractor Co. placed their Cletrac Model 15 (page 25) on the market. This was a 2-plow (15-22 hp.) crawler tractor, with 3 speeds forward and fitted with a standard power take-off.

Some 1931 Models

Cletrac 15

Caterpillar Diesel 65

Caterpillar 10, High Clearance

Huber-Modern Farmer

Bradley GP (Sears, Roebuck)

Farmall 30 (Radio controlled)

1932

In 1932, owing to the continued world-wide depression, tractor sales in the United States dropped to around 19,000 units—the lowest since 1915. Despite this, the year was marked with a number of important tractor developments.

The Minneapolis Moline Power Implement Co., Minneapolis, Minn., produced a new "Universal" tractor (page 26) of the wide tread tricycle type capable of burning gasoline, kerosene, or distillate fuel. This model together with others of conventional type that were soon to follow were equipped, as desired, with high-turbulence cylinder head, and with patented "Hot Spot" manifold attachment for use with different fuels.

Two 6-cylinder tractors—both of the track-laying type—were introduced; the Allis-Chalmers Model L (page 26) by the Allis-Chalmers Co., Milwaukee, Wis., and the McCormick-Deering T-40, TracTractor (page 26) by the International Harvester Co., Chicago, Ill. The latter, fitted with a 6-cylinder gasoline engine, was also available with a 4-cylinder diesel-type engine.

Two other track-laying tractors were announced by the Cleveland Tractor Co., Cleveland, Ohio, and the Caterpillar Tractor Co., Peoria, Ill., respectively. Each machine was a Model "25" (page 26).

In October the International Harvester Co. announced its F-12 tractor (page 27). This was a general-purpose tractor capable of pulling one 16-inch plow and was provided with a power take-off. Of the tricycle type, the tread of the rear wheels could be adjusted from 44 to 78 inches. This was in the $500 price range.

The Lauson Co., New Holstein, Wis., discontinued building tractors.

A new method of surface hardening of metal parts subject to excessive wear was announced by the Ohio Crankshaft Co. of Cleveland. Known to trade as the Tocco process, it was based on the principle of electric induction. A high-frequency electric current at high voltage was transformed into low voltage with high amperage. The current passed into inductor blocks that surrounded but did not actually touch the bearing area it was planned to harden. Thus, a current (which was the heating factor) was induced in the surface of the metal. At the end of the controlled period, the circuit was opened and the heated surface at the same time was quenched by a water-jacket spray. After the bearings had been hardened the entire shaft was drawn at a low temperature to reduce strains. Then a final grind completed the process. The process had already been adopted by some of the leading engine manufacturers.

Some 1932 Models

Allis-Chalmers L

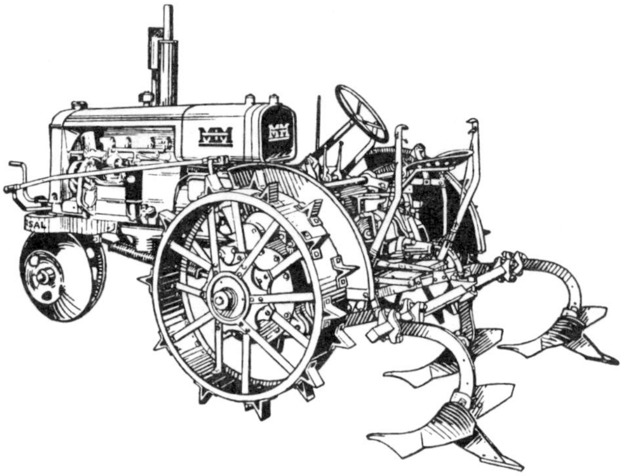

Minneapolis-Moline Universal

McCormick-Deering T40

Cletrac 25

Caterpillar 25

International Harvester F-12

1933

During the year 1933 tractor production was about 80 percent above the preceding year. Chief among tractor developments were diesel engines and rubber tires.

The first production diesel-engine powered crawler tractor model 65, built in 1931 by the Caterpillar Tractor Co. was followed in 1933 by other models of different sizes.

The Cleveland Tractor Co. produced a Cletrac diesel "80" (page 28) having the same chassis as the gasoline 80, both equipped with electric starting.

A 1-plow Light 4 tractor was produced by the New Beeman Tractor Co. of Minneapolis, Minn. This was especially adapted for work in narrow-spaced rows. The rear-axle tread of the tractor was adjustable so that the wheels could be set wide for plowing and narrow for cultivating.

A Massey-Harris Model "25" burning distillate appeared. This was a 3 - 4 plow tractor with Twin Disc clutch and patented Wallis U-frame construction.

At this time some types of farm machinery were being equipped with rubber tires. (Low pressure pneumatic rubber tires were available for tractors in 1932.) Small power take-off combines mounted on rubber were being operated by rubber-tired tractors driven at a maximum of 5 miles per hour.

During the year 1933 about 20 firms engaged in manufacturing farm tractors, of which 90 percent of the total output was produced by 9 companies, namely:

1. International Harvester Co.
   Chicago, Ill.
2. Deere and Co.
   Moline, Ill.
3. J. I. Case Co.
   Racine, Wis.
4. Massey-Harris Co.
   Toronto, Ontario
5. Oliver Farm Equipment Co.
   Chicago, Ill.
6. Minneapolis-Moline Power Implement Co.
   Minneapolis, Minn.
7. Allis-Chalmers Co.
   Milwaukee, Wis.
8. Cleveland Tractor Co.
   Cleveland, Ohio
9. Caterpillar Tractor Co.
   Peoria, Ill.

During 1920-33 the number of tractor manufacturers had decreased from a total of 90 to 9 principal producers, who had either through merger, combination, or reorganization survived the lean business years.

The International Harvester Co., the largest manufacturer of agricultural machinery in the United States, originally incorporated in 1902 and began producing tractors in 1906. Its formation united several large harvesting machine builders of that day, each a well established firm. The company maintained a dozen or more manufacturing plants in this country besides those in Canada and other countries. Tractors were being manufactured at plants located at Chicago, Milwaukee, and Rock Island.

Deere & Co., the largest manufacturer of steel plows, and the second largest manufacturer of agricultural machinery in the United States, bought the Waterloo Gasoline Engine Co. in 1918 and began the manufacture of farm tractors. Deere & Co. had started to distribute tractors manufactured by the Waterloo Gasoline Engine Co. in 1914. In 1926 the name of the gasoline engine company was changed to John Deere Tractor Co. and became the property of Deere & Co. Back of this tractor company was an organization originally established by John Deere in 1837.

The J. I. Case Co., the third largest manufacturer of agricultural machinery in the United States, was incorporated in 1880 as the J. I. Case Threshing Machine Co. In 1929 the name was changed to its present one. The business was originally established by Jerome I. Case in 1842. The J. I. Case Threshing Machine Co. was actively engaged in the building of steam tractors in the nineties and was among the first to turn to the gasoline tractor. In 1912 the plant at Racine, Wis., was enlarged to permit starting tractor production purchasing engines from the Davis Motor Co., Milwaukee, Wis. In 1913 they started building their own engines. In 1919 it merged with the Grand Detour Plow Co. and in 1928 it purchased the implement plant of the Emerson-Brantingham Corp. at Rockford, Ill. At this time the J. I. Case Plow Co. of Racine was taken over by the Massey-Harris Co. of Toronto. It sold the rights to use of the name "Case" and "J. I. Case" to the J. I. Case Threshing Machine Co. of Racine. Thus an end came to years of confusion caused to these two concerns bearing the name J. I. Case, and being in no way connected with each other.

The Massey-Harris Co., an amalgamation in 1891 of the Massey and Harris Companies dating back to 1847, Canadian manufacturers of agricultural machinery and tractors, extended its holdings in this country through the purchase of the J. I. Case Plow Works at Racine, Wis., in 1928. The J. I. Case Plow Works, incorporated in 1919, was a consolidation of the J. I. Case Plow Works Co. and the Wallis Tractor Co., both of Racine, Wis. The plow company, established in 1876, engaged primarily in the manufacture of plows and tillage equipment; the Wallis Tractor Co. was organized in 1912 to manufacture farm tractors.

The Oliver Farm Equipment Co., incorporated in 1929, acquired the business and property of 6 manufacturers to become a full line agricultural implement company. Chief among these were the Oliver Chilled Plow Works of South Bend, Ind., Nichols and Shepard of Battle Creek, Mich., and Hart-Parr Co. of Charles City, Iowa, and the American Seeding Machine Co. of Springfield, Ohio. When this consolidation was effected, efforts were concentrated on the well-known Hart-Parr tractor, and the Nichols and Shepard factory was converted into one handling harvesting and threshing machines only. The tractors were produced under the name of Oliver Hart-Parr. In 1944 the Cleveland Tractor Co. combined with the Oliver Farm Equipment Co. to form the Oliver Corp. and all of their equipment and tractors are produced under the name of Oliver.

Formation of the Minneapolis-Moline Power Implement Co., incorporated in 1929, was the result of a merger of the Minneapolis Steel and Machinery Co., the Minneapolis Threshing Machine Co., and the Moline Implement Co. The Minneapolis Steel and Machinery Co., organized in 1902, had been building Twin City tractors since 1908, and at the time of merger it was engaged in the manufacture of threshing machines in addition to tractors. The principal products of the other two companies included tillage implements, which combined with the tractor and threshing machine companies gave another full line agricultural implement company.

The Allis-Chalmers Manufacturing Co. was incorporated in 1913. The expansion of this company through the acquisition of 8 manufacturing concerns, widened the scope of its business to embrace a diversified line of power machinery. Shortly after the beginning of World War I, this company began to manufacture farm tractors. In 1928 it took over the Monarch Tractor Co. (incorporated in 1918) and has since manufactured the Monarch (a track-laying type) tractor under the name "Allis-Chalmers." When the Advance-Rumely Co. was acquired in 1931, an agricultural implement company long engaged in the manufacture of threshing machines and farm tractors was added.

The Cleveland Tractor Co., incorporated in 1916, was the only tractor manufacturer of any size that has continued to remain a single-line manufacturer up to 1945, when it was merged into the Oliver Corp. In 1911, Rollin H. White, manufacturer of the White steam car and the White internal combustion engine, in cooperation with his brother, Clarence White, developed the Cleveland motor plow, which was a forerunner of the product later turned out by the Cleveland Tractor Co. Their product, a track-laying machine was known under the trade name of "Cletrac."

1934

Although there was an increase of about 40 percent in tractor production in 1934 over 1933, the total number produced was still below 50,000 units. This was about one-fifth of the high of 1929. It was the second year in which improvement had been observed, however, and the general feeling among manufacturers was optimistic as to future business.

Experiments with diesel engines, which had been under way for several years, were beginning to bear fruit and several of the companies were planning soon to place diesel-powered tractors on the market.

While the Allis-Chalmers Co. was experimenting with diesels, it was not quite ready to put them on the market. It did, however, come out with its 2 crawler models powered with spark ignition engines, the Model KO is shown on page 29. The Allis-Chalmers Co. produced their WC 2-plow tractor selling at $675 on steel wheels or $825 on air tires, front and rear (page 29). The rear wheels were reversible, making possible treads of 65 and 76 inches. This was the first general-purpose tractor mounted on rubber to be officially tested at Nebraska.

The Bates Manufacturing Co. of Joliet, Ill., adopted a Waukesha-Hesselman engine for use in the "Bates Steel Mule 35" tractor. The Hesselman was a fuel-injection spark-ignition engine. Air was inducted into the engine and fuel injected into the compressed air near the end of the compression stroke by means of a typical solid-injection system. As part of the mixture passed to the spark plug, it was ignited as in a carburetor engine. The compression ratio was not much different than that of the ordinary spark ignition gasoline engine, but use could be made of refined commercial fuel oil (diesel).

The Cleveland Tractor Co. announced, in the fall, a diesel "35" equipped with a Hercules engine and Bosch fuel pumps (page 29).

An improved Alemite lubricating system was provided by the Alemite Corp. Pressures up to 5 tons per square inch were claimed by the use of new fittings and "hydraulic" pressure gun. This was sufficient to insure the forcing of all of the old grease out of the bearing, while replacing it with a fresh supply.

An important development was the John Deere Model A 2-plow tractor (page 29) capable of burning distillate fuel. This was a tricycle-type as compared to the 2 earlier 10-20 general-purpose models.

Some 1933 and 1934 Models

1933 Cletrac Diesel 80

1934 Allis-Chalmers KO (spark ignition)

1934 Cletrac Diesel 35

1934 Allis-Chalmers WC

1934 John Deere General Purpose A

1935

The year 1935 brought a marked return of business; tractor production snapped back to 161,000 units. Despite decreased purchases of all kinds of farm equipment during the depression, there had been a steady increase of tractors on farms in the United States as shown in the following table, according to estimates of the Research Department of the Farm Equipment Institute.

Year Jan. 1	No. of tractors on farms	Percent gain
1925	505,933[1]	—
1926	584,332	15.5
1927	665,038	13.8
1928	766,332	15.2
1929	826,183	7.8
1930	920,032[2]	11.4
1931	997,943	8.5
1932	1,051,646	5.4
1933	1,077,121	2.4
1934	1,090,203	1.2
1935	1,123,251[3]	3.0
1935	1,174,889	4.6

[1]*United States Bureau of Census.*
[2]*Census reported 920,021 for this date.*
[3]*As of July 1, 1935.*

Expansion in diesels, rubber tires, and general-purpose tractors and the special tractor designing for use of Ethyl gasoline of 70 octane rating (motor method) were given considerable attention.

Case tractors were equipped with a "motor lift" for lifting or lowering implements. By stepping on a trip button the implement was lowered, then by tripping the button again the implement was raised. The lift was driven by engine power through an enclosed worm and gear operated by a starter button near the driver's heel. Among the tools developed for use with the motor lift were 2- and 4-plow corn and cotton cultivators, 2-row potato cultivators, 10-row truck crop seeders and cultivators, 6-row beet planters and cultivators, 4-row corn and cotton planters, 3-row middle busters, 2-row listers and 7-foot mower attachments.

Something new in valve arrangement and manifolding was brought out in the Universal Model J of the Minneapolis-Moline Power Implement Co. The valves were so arranged that the exhaust was in the block and the intake was in the head, so that the engine classed as a "F" type (page 30). It was claimed that this construction provided maximum water jacket space around the valves, and since the exhaust valves were lower down, they got the benefit of cooler water. The drive wheels that were clamped on the rear axle were adjustable for tread widths of 54 to 76 inches. The wheels were also reversible. Its Universal MTA, an improved version of the original M brought out in 1931, was also brought out this year. A high compression cylinder head for 70 octane gasoline was optional, as was also the special high speed gears for speeds up to 10 miles per hour.

The John Deere Co. produced its general-purpose Model B tractor (page 30). This was built along the lines of the Model A produced in 1934 but of smaller, or 1-plow capacity.

Up until about 1933 gasoline regularly available throughout the country varied in antiknock quality from approximately 50 to 65 octane rating. Gasolines were sold by gravity, with no consideration for their knocking characteristics, resulting in varying tendencies of gasolines of the same gravity to knock. Later tetra ethyl lead was introduced in sufficient quantity to insure a standardized antiknock fuel with a 70 octane (Research Rating) rating. In 1935 approximately two-thirds of the gasoline sold was of this quality.

In October the Oliver Farm Equipment Co. produced a 6-cylinder tractor, the 70 (HC) row crop (page 30), so-called because it was especially designed to burn a standard antiknock (68-70) fuel. Later in the year a companion model, the Row Crop 70 KD, was produced that was capable of burning kerosene or distillate fuel.

The Cletrac Model E cultivating track-type tractor (page 30) was produced by the Cleveland Tractor Co. This 3-4 plow tractor was capable of 4-row cultivation; and to take care of variation in row widths, 5 models (E 31, E 38, E 62, E 68, and E 76) were available with treads from 31 to 76 inches. The 31- and 38-inch models were especially designed to go between the rows of row crops when it was not desirable to straddle the rows.

A wheel-type diesel tractor WD-40 (page 30) was put out by the International Harvester Co. The engine was the same as the one that was being used on the TD-40 TracTracTor produced in 1932.

In November the Caterpillar Tractor Co. announced 3 new diesel tractors and its 10,000th diesel engine. The new machines were designated as: RD 8 (95 drawbar hp), RD 7 (70 drawbar hp and RD 6 (45 drawbar hp). In addition the company put out its R4 (30) spark ignition tractor (page 31).

The Allis-Chalmers Co. added model UC, a 3-plow tractor, to its line, a larger version of the 2-plow WC model produced in 1934.

Some 1935 Models

Oliver 70 HC

Minneapolis-Moline Universal J

John Deere General Purpose B

Cletrac E Row Crop

International Harvester WD-40

Caterpillar 30 (spark ignition)

1936

During 1936 tractor production rose to 227,185 units, which was topped only by the alltime high of 229,000 units in 1929.

March 19 the first production of the Co-op tractor (page 32) left the Duplex Machine Co. plant at Battle Creek, Mich., for distribution through affiliates of the Farmers Union Central Exchange, Inc., St. Paul, Minn. The tractor was powered by a 6-cylinder Chrysler-built engine, and the axles, transmission, and differential were supplied by the Clark Equipment Co.

The Caterpillar Tractor Co. introduced its RD4 (35 drawbar hp.).

The Minneapolis-Moline Power Implement Co. produced a model J (page 32) designed for 70 octane gasoline. This was an orchard version of the orginal Model J brought out in 1934. Some of the machines were equipped with electric starters. The Kerosene and distillate-burning tractors were continued in production.

The use of water in pneumatic tractor tires was beginning to come in to use for the purpose of adding needed weight for traction. This system was intended to supplant the iron weights attached to the wheels, which frequently caused trouble especially under high road speeds (see page 16 for discussion of introduction of use of water in rubber tires).

Another 6-cylinder tractor, the Eagle 6-B (page 31), was announced by the Eagle Manufacturing Co. of Appleton, Wis., one of the pioneer builders of 6-cylinder engines. This was of the tricycle-type with adjustable rear wheels, 66 or 77 inches. Two-row cultivator equipment was provided, and rubber or steel wheels were optional.

The J. I. Case Co. produced its general-purpose model RC (page 32), which was intended for jobs of less capacity than that provided in the model CC. This was available either with steel or rubber-tired wheels, and the tread was adjustable from 44 to 80 inches. For use in narrow-spaced crop rows, a single, front wheel could be substituted for the dual narrow-tread wheels otherwise used.

The Massey-Harris Co. produced the Challenger and a 4-wheel drive general-purpose tractor (page 31) available on rubber, an improved model of the original 4-wheel drive of 1930. The Challenger was a 2-3 plow tricycle-type tractor available on rubber or steel wheels and was powered with a Massey-Harris 4-cylinder valve-in-head all-fuel engine. Adjustable tread was provided on rear axle, the rear wheels being reversible.

Manufacturers for some time had been giving considerable attention to means of attaching, operating, and disconnecting machines and implements used in connection with power farming. This was especially needed for general-purpose tractors because of the added planting and cultivating equipment involved. It was claimed by one manufacturer that over a dozen "quick on and off" implements could be set up or disconnected in from 2 to 8 minutes.

The Fate-Root-Heath Co., Plymouth, Ohio, produced the "Silver King" (page 32), a tricycle-type general-purpose tractor. When mounted on rubber (Model R-66), the tractor could pull 2 plows. Rear wheels adjustable at 66 and 70 inches.

Deere and Co. placed their Model G in production.

Some 1936 Models

Eagle 6B (six cylinder)

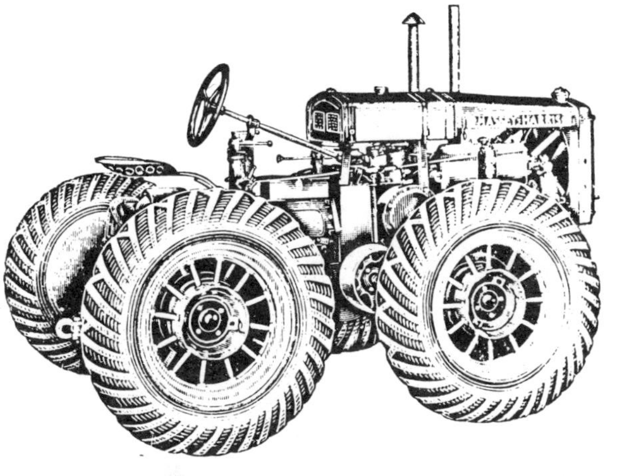

Massey-Harris 4WD

Duplex Machinery Company - Co-op

Minneapolis-Moline J

Case RC

Silver King

1937

About this time the word "streamline" was borrowed from the automobile trade and applied to the new styling of tractors.

The Huber Manufacturing Co. produced 2 general-purpose tractors, a 2-plow model B (page 33) and a 3-plow model LC. Both models were of the tricycle type with rear wheels adjustable; the first model was powered with a 4-cylinder Buda engine and the LC model with a Waukesha.

The Graham-Paige Motors Corp., Detroit, Mich., produced the Graham-Bradley general-purpose tractor (page 33), powered with a 6-cylinder engine. Rubber tires were standard equipment, and a power take-off, hydraulic power lift, and electric starting and lighting were also provided. Tread of rear wheels was adjustable from 56 to 84 inches. Four forward speeds, from 3 to 22 m.p.h., were provided. This tractor was sold exclusively by Sears, Roebuck & Co.

A few years before the U.S.S.R. Government was purchasing American tractors in large numbers. In 1937 the Soviet Government was advertising for export agents to handle 3 sizes of Soviet-built tractors, namely, 48-60, 15-30, and 10-20. An advertisement seeking such agencies appeared in the March number of the Implement and Machinery Review of London.

A new Fordson, the All-Around (page 33), was produced by the Ford Motor Co., Ltd., Dagenham, Essex County, England. This was a 2-plow general-purpose tricycle-type tractor with tread of rear wheels adjustable from 52 to 88 inches.

As an expedient for the steel tracks of track-laying tractors when hauling loads on hard surfaced roads, an endless rubber belt reinforced longitudinally with a number of steel cables, and in which belt were valcanized steel driving members, was produced by the B. F. Goodrich Co. The rubber track was designed to be driven positively by means of sprockets and interlocking guiding links.

The question of tractor fuels was still rampant among manufacturers and was directly reflected for 1937 in the output which included several high-octane jobs duplicated in companion machines for fuel oils, a tractor with adjustable head by which high or low compression could be attained, and a row-crop track-laying diesel tractor of small size.

In July the Universal (general-purpose) "Z" Model tricycle-type tractor (page 33) of the Minneapolis-Moline Co. was announced. A special feature was the detachable part of the head, which could be adjusted to alter the compression pressure. This was on the opposite side to the valves and contained the spark plugs and water jackets. By changing the size of the protrusions on the detachable head the volume of the combustion chamber and hence the compression pressure could be altered for the use of high-octane fuels, or for the burning of distillate fuel. The valves were horizontal and operated by rocker arms.

In October the Oliver Farm Equipment Co. announced a new model 70 (page 33). The first model 70 had been produced 2 years previously and was designed especially for the high octane (70) gasoline then being produced and sold at nonpremium prices. The new 70 was produced in 2 engine types, namely, the HC (high compression) for regular 68-70 octane gasoline, and the KD (kerosene-distillate) for the heavier types of fuels. The engine had 6 cylinders, self-starter, and electric lights and was equipped with either steel "tiptoe" wheels or rubber tires.

In November the Massey-Harris Co. announced its "Twin-Power" Pacemaker high-compression 3-plow tractor (page 33). Twin power involves the power produced by 2 different settings of a lever so that top speed in the one case, for example, was 1,200 r.p.m. and in the other 1,400. Owing to this increase in speed, it was claimed that 5 additional horsepower could be obtained. This tractor was patterned after the original Challenger, but the Pacemaker was designed to burn high octane (68-70) gasoline and was available on steel wheels or rubber tires.

The Cleveland Tractor Co. produced its first small agricultural diesel model ED. This was an adaptation of its row-crop gasoline model E for use with diesel fuel.

Some 1937 Models

Minneapolis-Moline Universal Z

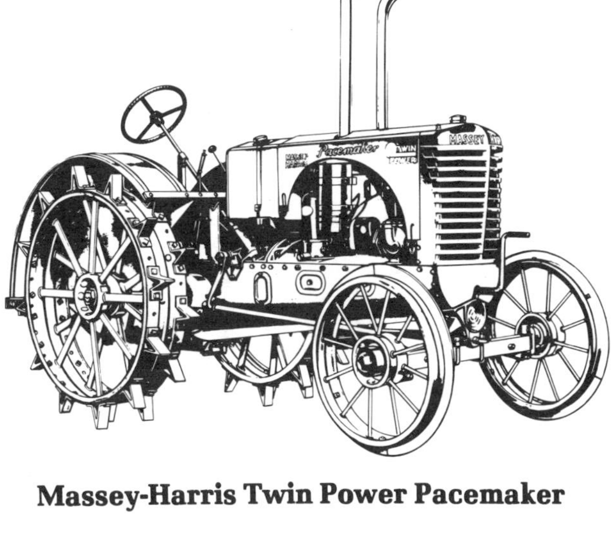

Massey-Harris Twin Power Pacemaker

Huber B

Graham-Bradley

Fordson All Around

Oliver Row Crop 70

1938

The Caterpillar Tractor Co. announced its small diesel D-2 tractor (page 35) especially designed for farm work. It was claimed the new tractor would handle 3- to 4-plow bottoms, one 10-foot disk harrow, a 12-foot combine, 20 feet of grain drills, or similar loads.

The Oliver Farm Equipment Co. produced its Row Crop 80 series (page 34). These 3-plow machines were available to the trade with 2 types of engine: (1) For gasoline fuel (68-70) octane, and (2) for kerosene or distillate. Although this tractor was of the tricycle type, it could be fitted with the conventional 4-wheel wide tread especially adapted for row-crop work on hilly land.

A new general-purpose Massey-Harris 2-plow tractor, the 101 (page 35) was produced embodying "twin power" designing as in the 1937 model. An L-head Chrysler-built 6-cylinder motor burning 70 octane gasoline was used. The rear wheels were provided with adjustable tread that could be varied from 52 to 90 inches on rubber and 52 to 80 inches on steel.

The Minneapolis-Moline Co. put out its M Universal Model U (page 35), a 3- to 4-plow conventional tractor available in either a high (5.33 to 1) or low (4.30 to 1) compression ratio head. Forward speeds of from 2.7 to 20.3 m.p.h. (on rubber) were provided. For short turning, 2 internal-expanding differential brakes were manipulated independently, or for road braking they were operated simultaneously through a third foot pedal.

Considerable restyling was accomplished by Deere & Co., and the following models were offered, all fitted with 2-cylinder twin horizontal engines (opposed crank). (1) The "B" series was capable of pulling 1 16-inch or 2 10-inch plows; the model "B" was a tricycle type (two wheels in front); (2) model "BN," single front wheel; (3) model "BW" (page 35) with adjustable front axle; (4) model "BR" standard tread; and (5) model "BO" for orchard use. The "A" series was a little more powerful, capable of pulling 2 14-inch plows. (1) the model "A" was a tricycle-type (2 wheels in front); (2) model "AN" (page 35) single front wheel; (3) model "AW" adjustable front axle; (4) model "AR" standard tread; and (5) model "AO" the orchard type. Other models were model G, tricycle-type capable of pulling 3 14-inch plows, and the model "D" with standard tread and of 3 to 4 plow capacity.

The fourth size of tractor in the International Harvester Co.'s Farmall series was produced: the F-14 (page 35) designed for 2 14-inch plow bottoms under average conditions. It burned gasoline or tractor distillate. The rear wheels were available with rubber or steel tires, and treads were adjustable from 44 to 78 inches. Weights were provided for use on both front and rear wheels when tractor was mounted on rubber.

At the fall tractor show, two so-called deLuxe tractors were exhibited by the Minneapolis-Moline Co. An all-weather model with cab, embodied all the comforts and equipment of the modern sport coupe; and the same tractor without cab was equipped in a manner comparable, with the modern sport roadster automobile model.

The Avery Farm Machinery Co. of Peoria, Ill., produced a straddle-row tractor that could be readily changed from a tricycle type for cultivating to a standard-tread 4-wheel conventional type. This "Ro-Trak" tractor (page 34) was adapted for cultivation of any crop planted in rows ranging from 10 to 60 inches apart.

The Allis-Chalmers Co. announced a new 1-plow general-purpose 4-wheel tractor, Model B (page 35), weighing less than 2,000 lb., and priced at $495 f.o.b. factory. Rubber tires were standard equipment. This tractor was available with arched front axle and with treads adjustable front and rear.

The Eagle Manufacturing Corp. added the model 6C (page 35) tractor to its line. This was a standard 4-wheel tractor powered with the same size engine as the model 6B tricycle, which appeared in 1936.

Some 1938 Models

Oliver Row Crop 80

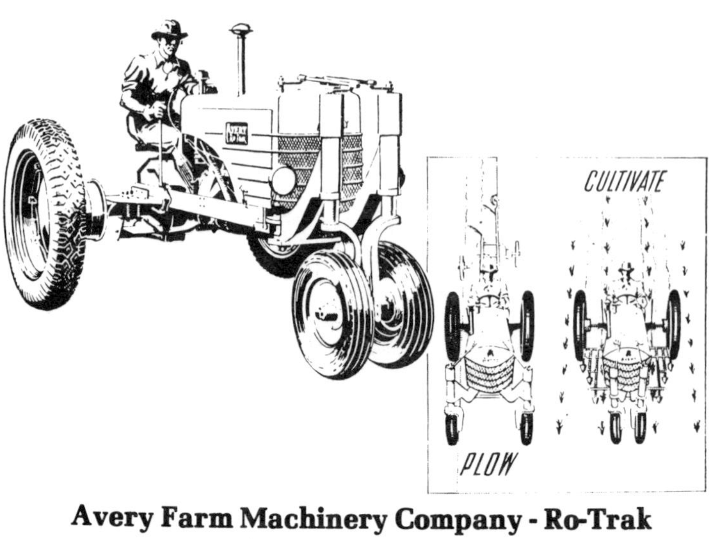

Avery Farm Machinery Company - Ro-Trak

Caterpillar D2

Deere AN

Massey-Harris 101

Deere BW

Minneapolis-Moline Universal U

International Harvester Farmall F-14

Eagle 6C

Allis-Chalmers B

1939

The year 1939 saw several companies introducing new models. The J. I. Case Co. introduced its model "D" series (page 38), a standard 4-wheel design that provided 3-plow power. It was equipped with a belt pulley, power take-off, and all-purpose motor lift. Model DC-3 featured wheel spacing adjustable to various row widths and synchronized steering that provided 2 major steering advantages, quick dodge for close cultivation in uneven rows and high leverage for short turns with full load. "Eagle Eye" visibility was also incorporated in this model. This gave the operator full view of controls, implements, and work. Model DC-4 was similar to model DC-3 except that it provided extra clearance. Model DO was designed for heavy orchard and grove work, while DV took care of vineyards.

The Allis-Chalmers Co. introduced its model C (page 39). This differed from the older model B in that it was a tricycle type designed for 2-row planting and cultivation, and had a slightly more powerful engine. For closely spaced row crops, the model C was available with single front wheel and 6-12 rib-type tire that fitted more easily between narrow-spaced crop rows. For lands and conditions where it was necessary that the front wheels of the tractor track with the rear wheels, the model C tractor could be equipped with adjustable front axle in place of the regular front-wheel equipment.

The Cleveland Tractor Co. introduced the Cletrac model HG (page 38), which was a smaller and more compact tractor than the model E.

Models 60 (page 39) and 900 of the Oliver Farm Equipment Co. went into production. The Oliver 60 was a small tractor weighing 2,100 pounds with "tiptoe" steel wheels and 2,000 pounds with rubber tires, and had crop clearance of 21 1/2 inches. The Oliver 900 was an industrial tractor that did have some agricultural use, particularly in irrigated sections for land preparation such as leveling, ditching, and sub-soiling. During 1939, changes incorporated in its standard models included: (1) The ASAE master power take-off shield and the standardization of the power take-off and drawbar relation were adopted for all agricultural models; (2) wide-base rear tractor tires were adopted for all agricultural models; (3) 4-speed transmissions were applied to the 80 and 90 series tractors and 6 speeds became standard equipment for the 70 tractors. The 60 series started in production with a 4-speed transmission, and this was later changed to 5 speeds as regular equipment; (4) the 60 series was changed to battery ignition, since all tractors were sold with starting and lighting equipment.

The International Harvester Co. introduced several new models mainly of the Farmall class. The Farmall A was a 1-row tractor with Culti-Vision (page 39). By Culti-Vision was meant that the operator in the seat had clear vision ahead. The engine was mounted to the left of center to permit this. Farmall AV, high-clearance tractor, was a modification of the Farmall A and was especially adapted for work in crops planted on high beds and for cultivating such crops as asparagus, the cultivation of which is continued until the plants have reached considerable height. It was powered with the same 4-cylinder engine as the Farmall A and was in all respects the same tractor, but it was modified to give it 6 extra inches of crop clearance. Farmall B, similar to Farmall A but more powerful, cultivated 2 full rows. This model also had Culti-Vision. The Farmall BN, a 3-wheel 1-plow tractor cultivated 2 rows, and was an all-purpose tractor designed for narrower row spacings than the Farmall B. The BN could work 2 rows spaced from 28 to 42 inches and had the power to pull 1 16-inch bottom.

The Farmall H (page 30), larger than Farmall A, was capable of pulling 2 14-inch bottoms. Farmall HV, a cane tractor, a modified Farmall H, provided a clearance of 30 1/2 inches. Farmall M, a more powerful tractor than models A and H, was capable of pulling 3 14-inch bottoms. As with the Farmall H, Farmall M could be equipped with a distillate-gasoline engine or with a high-compression engine for operation on high-test gasoline. Farmall MD was essentially the same tractor as the regular Farmall M but was powered with a 4-cylinder diesel engine. Farmall MV and MDV were also cane tractors, the MV similar to M but with high clearance and the MDV similar to MD but with high clearance.

The latest model put out by this company in 1939 and the largest of its crawler line up to this time was the TD-18 Trac-Tractor (page 38), diesel-powered, which developed 80 hp. on the belt and 70 hp. on the drawbar. As with other International diesels, the TD-18 had a special starting arrangement. By opening a valve an auxiliary combustion chamber for each cylinder, each fitted with a spark plug, was brought into use. This reduced the compression pressure and the engine operated as a gasoline engine. After a short interval the valves were closed and the engine operated as a full diesel. The tractor was available in 2 treads, the wide with 74 inches and the narrow with 62 inches.

The Minneapolis-Moline Power Implement Co. came out with several new developments. They effected standardization of a power take-off shield, and also the drawbar location. Standard factory-built tractors were put out for the burning of LP-gas as tractor fuel (page 39). Starting and lighting equipment became available as standard equipment on all models. The road speeds were reduced to less than 15 miles per hour. The Model R tractors were also introduced. This was a 2-row, 2-plow tractor and was offered in 3 models, 2 front wheels on a pedestal-type mount, with single front wheel, or with standard front wheel assembly.

Kansas State College Engineering Experiment Station Bulletin No. 71, "Liquefied Petroleum Gas for Tractors," by George H. Larson, gives the following history of LP-gas:

"LP-gas is an abbreviated term for 'liquefied petroleum gasses' which include butane (C_4H_{10}), propane (C_3H_8), or any mixture of the two. The early history of LP-gas dates back to about 1890. Before this time, the light ends of the crude oil were largely a waste product and a drug on the market, since most crude oil was then processed for kerosene and lubricating oils. The advent of the automobile opened up a market for gasoline to utilize part of the light ends, but the light materials, butane and propane, were hard to handle and refineries still resorted to the practice of wasting them.

"The first attempts to liquefy these gases were made about 1903. Herman Blau, a German, liquefied a very high pressure gas called 'Blaugas,' which contained ethane, butane, and propane. About 1910, A. N. Kerr began experiments on the compression of these waste gases and bottling them as a fuel. Frank P. Peterson of Tulsa, Oklahoma, who was an early associate of Mr. Kerr, was recognized by the Natural Gasoline Association of America in 1943 as the 'father of LP-gas.' He is claimed to have driven an automobile powered by LP-gas 'up Broadway' in New York City in 1912. The bottleneck was the high pressure required; pressures as high as 700 to 1000 psi were not uncommon. Better processes were developed to separate the very light hydrocarbons. As a result, during the period between 1922 and 1928, the industry standardized on propane and it was refered to as 'bottle gas' fuel. It was and still is widely used for cooking.

"By 1935 the increased use of this fuel called for larger containers or tanks in order to make deliveries economical. These storage tanks were of both the underground and the aboveground types. About this time, the lower pressure gas, butane, was flooding the market, so that butane, propane, or mixtures were made available. The war consumed large quantities of butane for aviation fuel, and this made it scarcer. There are indications that it may continue to be more scarce than propane, since there is a growing demand for it in many synthetic producing industries.

"The physical properties of these fuels are shown in Table I. It will be noted that the boiling of either fuel is well below normal atmospheric temperatures and pressures. At practically all temperatures likely to be encountered in the United States, propane is a gas or vapor but it can be liquified at moderate pressures. Butane boils at about the freezing point of water 32°F. and since its evaporation cools it somewhat, its liquid cannot supply vapor at a satisfactory pressure unless the surroundings of the container are at a temperature considerably above the freezing point of water. For this reason, butane is stored in underground tanks for domestic purposes, whereas propane is generally stored in aboveground tanks. An odorant is added to LP-gases to aid in detecting leaks.

"There are three principal sources for the LP gases: (a) natural gasoline plants where they are recovered from the dry gas and gasoline removed from the crude oil, (b) recycling plants where the wet gas is drawn from natural gas wells, and (c) oil refineries where the crude oil is processed into commercial gasoline, fuel oils, and distillates.

Table I. Physical Properties of Two LP-Gases.[1]

Properties	Propane	Butane
Formula	C_3H_8	C_4H_{10}
Boiling point, °F	44	33
Pounds per gallon of liquid at 60°F	4.24	4.85
Heating value of the gas:	21,680	21,330
Btu per pound	21,680	21,330
Btu per gallon	91,900	103,400
Btu per cu. ft. at 60°, 30 in. Hg.	2,572	3,393
Specific gravity of gas (air = 1)	1.55	2.14
Vapor pressure, psi gage at:		
33°F	54	0
100°F	196	37
Volume of air required to burn 1 volume of gas	23.9	31.0

[1]*E. R. Weaver, Propane, Butane and Related Fuels, Nat'l Bureau of Standards (U.S.) Circular C420, 1939.*

"Table II shows a comparison of the physical properties of common tractor fuels.

Table II. Comparison of Physical Properties of Common Tractor Fuels.[2]

Fuel	EPI Gravity Deg.	Weight lbs. per gal.	BTU per lb.	BTU Octane No. gal.	Average
Propane	145	4.25	21,680	92,140	115
Butane	114	4.08	21,300	102,200	100
Premium-grade gasoline	63	6.06	20,320	123,100	80
Regular-grade gasoline	61	6.13	20,280	124,300	72
Third-grade gasoline	60	6.16	20,260	124,800	60
Kerosene	43	6.76	19,830	134,100	

[2]*E. L. Barger and others, Tractors and Their Power Units, John Wiley and Sons, 1952.*

The 9N Ford tractor incorporating the Ferguson System (page 8) was placed in production in the United States by the Ford Motor Co., Dearborn, Mich. This 2-plow tractor was powered with a 4-cylinder vertical engine and fitted with a 3-point hydraulic lift to which a plow or other equipment could be attached. By a small lever the attached implement could be raised or lowered at will with little effort. The claim was made that the suction of the plow pulled down on the rear wheels thereby increasing their traction.

The Ferguson System consisted of a combined linkage and hydraulic control, applicable to a wide variety of implements and was perfected in 1935 after 17 years of engineering development by Harry Ferguson.

The combined results of the Ferguson implement linkage and hydraulic control mechanism were a revolutionary new ease of implement operation. The whole trend of design of farm tractors and equipment in ensuing years has been strongly influenced by the development of the Ferguson System.

By 1936, Mr. Ferguson had developed a new light tractor incorporating his invention featuring a hydraulic system built into the tractor to operate implements an an integrated unit, and within two years approximately 1260 were manufactured and sold in the British Isles and Norway.

In late 1938, Mr. Ferguson had brought his tractor and equipment to the U.S. and demonstrated it to Henry Ford. A working agreement was subsequently established between Mr. Ford and Mr. Ferguson for mass production of a tractor incorporating the Ferguson System. Under this arrangement more than 300,000 tractors were built between June 1, 1939 and June 30, 1947.

Deere and Co. announced its model L, an all-purpose tractor (page 39) with a 2-cylinder vertical engine, adjustable treads front and rear, and capable of pulling 1 14-inch plow; its model H tricycle type (twin front wheels), with 2-cylinder horizontal engine capable of pulling one 16-inch plow; and the model HN with the same power as the H but with a single front wheel. The power of both the A and B series produced in 1934 and 1935, respectively, was considerably increased.

Some 1939 Models

Case DC

International TD-18

International Harvester Farmall H

Cletrac HG

Ford-Ferguson 9N

Oliver 60

Minneapolis-Moline U for LP-gas

International Harvester Farmall A

John Deere L

Allis-Chalmers C

1940

The Eagle Co. was sold to the Four Wheel Drive Auto Co. of Clintonville, Wis. and the manufacture of tractors was discontinued.

The International Harvester Co. put out 11 new models. Models 0-4 and 0-6 (page 40) were designed for orchard and grove work, 2- and 3- plow size, respectively. Of the standard farm 4-wheel tractors, the W-4, 2-plow size was the smallest. It was available for use with either distillate or with high-compression for use with gasoline. The W-6, similar to the W-4, was 3-plow size. The model W-9, 4-plow tractor, was the largest of the 3 distillate-gasoline standard tractors. Also a group of crawler tractors became available. The T-6, a 3- to 4-plow crawler tractor, was powered by basically the same carbureted engine that was used in the Farmall M and 6 series. The TD-6 had the same chassis as the T-6 but the TD-6 was fitted with the same diesel engine that was used in the Farmall MD and 6 series of diesel farm tractors. Operation of the TD-6 was the same as the T-6, except that a starting change-over lever was furnished on the TD-6. The T-9 rated as a 3- to 4-plow crawler tractor. The TD-9 had the same forward speeds, features, and controls as the T-9, except that it was equipped with the 9 series drawbar capacity for the larger farm operations, was capable of pulling 5 to 8 14-inch plow bottoms. It had a 4-cylinder diesel engine, including the all-weather starting system.

The Caterpillar Tractor Co. increased the horsepower of its model D8 and provided it with an oil-cooling system. The company also announced its new model D7 (page 40) with a maximum belt horsepower of 93 and maximum drawbar horsepower of 81. The D7 was provided with a 2-cylinder gasoline engine starter with cooling system integral with that of the diesel. The company also introduced its first rubber-tired tractor, the 4-wheeled DW 10. This was designed for use with earth-moving equipment and probably marked the beginning of the Caterpillar Tractor Co's entry into the rubber tired equipment business. The rubber tired Auto Patrol had already been on the market for some time.

The Allis-Chalmers Co. brought out 3 models of crawler tractors powered by General Motors 2-cycle diesel engines, the HD 7 (page 40) (3-cylinder vertical), the HD 10 (4-cylinder vertical), and the HD 14 (6-cylinder vertical); 6-, 10-, and 14-plow capacity, respectively. All were equipped with electric starting motors.

Deere & Co. offered its model HN with single front wheel; this model had the same power as the H introduced in 1939.

The J.I. Case Co. added the Case models VC (page 40), V, VI, and VO to its line. The VC was powered with a 4-cylinder vertical engine of 1- to 2-plow capacity, tricycle construction with dual wheels in front, adjustable front and rear wheel treads, and 19-1/2-inch cultivating clearance.

Caterpillar D7

Allis Chalmers HD-7 (2 cycle Diesel)

Case VC (all purpose row crop)

Some 1940 Models

McCormick-Deering (I. H. Co.) Orchard 06

1941

The following summary (pages 41-44), issued by the (page 45) for narrow rows could be equipped with market January 1, 1941.

The J. I. Case Co. presented their new "S" series which is comprised of smaller tractors than the "D" series. Model S brought easily handled power to farms where row-crop cultivation was not a factor. Tractor SC (see page 000) for narrow rows could be equipped with a single front wheel, fitting it for multi-row cultivation. Orchard and grove work was taken care of by Model SO which provided 2-plow power in a compact tractor

SUMMARY OF RESULTS OF THE NEBRASKA TRACTOR TESTS

CONDUCTED BY THE UNIVERSITY OF NEBRASKA, DEPARTMENT OF AGRICULTURAL ENGINEERING, LINCOLN, NEBRASKA

INCLUDES ONLY TRACTORS ON THE MARKET JANUARY 1, 1941

Test No.	TRACTOR: Name (1)	Horsepower Rating (2)	Horsepower Rating (3)	Weight as Tested (4)	ENGINE: Cylinders (5)	Bore and Stroke (6)	Rated Eng. R.P.M. (7)	Make (8)	Fuel Used (9)	Oil: S.A.E. No. (10)	Oil: Gals. to Motor (11)	Oil: Gals. from Motor (12)	Total Hrs. Run (13)
	1920-1935												
185	Huber Super Four	40-62		9,910	4	5½x6½	1,100	Stearns	Gas	....	....	....	48
168	Huber Light Four	20-36		6,725	4	4¾x6¼	1,150	Waukesha	Gas	....	....	....	36
183	Oliver Hart Parr 3-5 Plow [1]	Not Rated	28.54-44.63	6,415	4	4¾x6¼	1,125	Own	Kero.	30	5.929	1.840	60
184	Eagle 6A	Not Rated	22.17-37.09	5,670	6	4 x4½	1,416	Hercules	Gas	40	3.024	1.998	50
186	Bates Steel Mule 35	35-45	32.91-46.90	11,218	6	4¼x4¾	1,500	Waukesha	Gas	40	2.195	1.892	47
187	Bates Steel Mule 45 [2]	45-55	40.07-60.38	13,509	6	4⅝x5½	1,500	Waukesha	Gas	40	4.221	1.055	55
198	Mc'Deering Farmall F-30	Not Rated	20.27-30.29	5,990	4	4¼x5	1,150	Own	Kero.	30	6.949	7.825	66
210	McCormick-Deering W-30	Not Rated	19.69-31.31	5,575	4	4¼x5	1,160	Own	Kero.	40	3.782	4.220	43
216	Allis-Chalmers M	Not Rated	21.81-31.81	6,620	4	4⅜x5	1,200	Own	Gas	40	2.525	1.931	47
219	Massey-Harris 3-4 Plow [3]	Not Rated	26.44-41.01	5,385	4	4⅜x5¾	1,200	Own	Dist.	30	4.889	1.345	43
224	Caterpillar R-5	Not Rated	41.24-54.64	13,675	4	5⅛x6½	950	Own	Gas	40	3.701	3.060	41
226	Caterpillar Twenty-two	Not Rated	18.86-26.75	6,605	4	4 x5	1,250	Own	Dist.	30	2.506	2.485	48
228	Caterpillar Twenty-two	Not Rated	19.34-27.17	6,605	4	4 x5	1,250	Own	Gas	30	2.436	1.988	30
237	Allis-Chalmers U	Not Rated	19.28-30.97	5,030	4	4⅜x5	1,200	Own	Dist.	40	4.660	4.488	82
237	(With Rubber Tires)			5,140									
238	Allis-Chalmers UC	Not Rated	19.12-30.86	5,710	4	4⅜x5	1,200	Own	Dist.	40	5.234	4.069	71
238	(With Rubber Tires)			6,115									
239	Allis-Chalmers M	Not Rated	22.79-31.64	6,855	4	4⅜x5	1,200	Own	Dist.	40	1.942	1.804	50
242	Caterpillar Diesel Forty [4]	....	39.46-50.12	15,642	3	5¾x8	1,000	Own	*F.Oil	40	3.664	3.418	51
243	Caterpillar Diesel Forty [5]	....	33.56-44.08	15,642	3	5¾x8	850	Own	*F.Oil	40	3.480	3.471	26
	1936												
251	Case RC	Not Rated	11.27-17.44	3,350	4	3¼x4	1,425	Waukesha	Gas	20	1.633	0.942	52
260	Cletrac AG	Not Rated	19.28-27.72	7,025	4	4 x4½	1,400	Hercules	**Gas	30	1.703	0.774	50
261	Cletrac E	Not Rated	16.58-25.93	6,100	4	4 x4½	1,300	Hercules	**Gas	30	2.114	1.392	59
262	Cletrac FG	Not Rated	70.50-94.29	26,670	6	5¾x6	1,120	Hercules	**Gas	30	7.350	4.168	49
265	Massey-Harris Challenger	Not Rated	16.12-25.99	4,200	4	3⅞x5¼	1,200	Own	Dist.	40	3.056	3.693	44
266	Massey-Harris Pacemaker	Not Rated	16.02-26.49	4,050	4	3⅞x5¼	1,200	Own	Dist.	40	3.056	3.536	40
267	Oliver Row-Crop 70 KD	Not Rated	16.53-24.49	3,900	6	3⅛x4⅜	1,500	Own	Dist.	30	2.019	1.045	44
269	McCormick-Deering WK-40 [6]	Not Rated	28.34-44.80	7,625	6	3¾x4½	1,750	Own	Dist.	40	2.719	2.046	63
271	Caterpillar Thirty [7]	Not Rated	24.29-32.27	9,975	4	4¼x5½	1,400	Own	Dist.	30	7.597	8.183	72
272	Caterpillar Thirty [7]	Not Rated	26.50-34.71	9,950	4	4¼x5½	1,400	Own	Gas	30	2.972	2.784	47
273	Caterpillar Diesel RD-4 [8]	....	26.76-34.99	10,100	4	4¼x5½	1,400	Own	*F.Oil	30	3.452	3.163	57
274	Co-op No. 3	Not Rated	28.97-37.09	6,065	6	3⅜x4½	1,600	Chrysler	**Gas	10	1.234	1.030	48
275	Co-op No. 2	Not Rated	22.70-29.44	5,000	6	3⅛x4⅜	1,500	Chrysler	**Gas	10	1.485	1.151	54
276	Mc'Deering Farmall F-20	Not Rated	15.98-23.80	4,310	4	3¾x5	1,200	Own	Dist.	30	3.198	4.240	45
	1937												
283	Oliver Standard 70 HC	Not Rated	15.89-24.95	3,500	6	3⅛x4⅜	1,500	Own	**Gas	30	2.272	0.710	61
284	Oliver Standard 70 KD	Not Rated	15.87-23.89	3,500	6	3⅛x4⅜	1,500	Own	Dist.	30	2.503	0.779	49
289	Cletrac CG	Not Rated	34.01-47.08	11,700	6	4¼x4½	1,565	Hercules	**Gas	30	2.517	1.354	49
290	M-M Twin City ZT	Not Rated	15.91-23.43	4,280	4	3⅝x4½	1,500	Own	Dist.	30	3.405	3.780	46
291	Huber LC	Not Rated	23.22-37.42	5,300	4	4½x5¼	1,200	Waukesha	Dist.	30	4.870	5.343	77
291	(With Rubber Tires)			5,500									
292	Huber B	Not Rated	16.28-24.00	3,745	4	3 13/16x4½	1,300	Buda	Gas	30	2.852	2.103	75
292	(With Rubber Tires)			4,465									
293	Massey-Harris TP-CH	Not Rated	19.67-31.60	4,570	4	3⅞x5¼	1,200	Own	**Gas	20	3.534	2.366	74
293	(With Rubber Tires)			5,900									
293	(1400 r.p.m.)		-36.32				1,400						
294	Massey-Harris TP-PA	Not Rated	20.30-31.81	4,090	4	3⅞x5¼	1,200	Own	**Gas	20	4.344	1.965	98
294	(With Rubber Tires)			5,250									
294	(1400 r.p.m.)		-36.78				1,400						
295	John Deere G	Not Rated	20.70-31.44	5,160	2	6⅛x7	975	Own	Dist.	30	2.708	2.340	50

BELT TESTS

Test No.	TEST B—100% MAXIMUM LOAD—TWO HOURS: Belt H.P. (14)	Av. Eng. R.P.M. (15)	Fuel: Gals. per Hour (16)	Fuel: H.P. Hours per Gal. (17)	Fuel: Lbs. per H.P. Hour (18)	Water: Gals. per Hour (19)	Temp. °F. Rad. (20)	Temp. °F. Atm. (21)	Carb. Adj. % of Max. (22)
	1920-1935								
185	....	....	....	....	....	....	......	....	......
168	....	....	....	....	....	....	......	....	98.8
183	51.23	1,128	5.941	8.62	0.791	0.000	174	63	96.0
184	41.67	1,417	4.897	8.51	0.727	0.000	181	77	96.9
186	52.60	1,500	6.466	8.13	0.746	0.000	179	78	100.0
187	66.57	1,502	7.815	8.52	0.713	0.000	193	91	100.0
198	34.36	1,150	3.771	9.11	0.742	0.000	179	70	97.0
210	35.01	1,160	3.884	9.01	0.748	0.000	178	78	95.0
216	35.54	1,200	3.576	9.94	0.608	0.000	199	74	100.0
219	46.04	1,200	5.522	8.34	0.835	1.651	186	77	96.1
224	59.52	950	6.616	9.00	0.686	0.000	188	103	98.9
226	30.41	1,250	3.347	9.09	0.765	0.000	183	65	96.5
228	30.97	1,250	3.833	8.08	0.760	0.000	180	64	99.2
237	34.12	1,200	4.063	8.40	0.824	0.000	193	95	97.2
238	34.09	1,199	4.096	8.32	0.831	0.000	189	92	98.3
239	35.05	1,200	4.164	8.42	0.830	0.000	189	77	98.8
242	56.05	1,000	4.087	13.71	0.510	0.000	167	78	......[4]
243	48.60	850	3.381	14.37	0.486	0.000	173	85	......[5]
	1936								
251	19.80	1,428	2.335	8.48	0.727	0.000	193	60	95.9
260	30.25	1,398	3.460	8.74	0.702	0.000	195	98	99.6
261	28.76	1,300	3.266	8.81	0.697	0.000	196	90	98.4
262	104.16	1,120	11.982	8.69	0.709	0.223	214	93	100.0
265	28.58	1,199	3.184	8.98	0.769	0.000	195	91	95.0
266	29.09	1,200	3.141	9.26	0.745	0.000	191	87	94.6
267	27.15	1,500	3.346	8.11	0.852	0.000	167	80	96.1
269	49.76	1,749	5.557	8.95	0.774	0.000	176	85	97.5
271	36.37	1,400	4.447	8.18	0.849	0.000	203	68	93.8
272	39.15	1,400	4.776	8.20	0.745	0.000	187	65	96.6
273	39.82	1,400	2.885	13.80	0.506	0.000	186	62	......[8]
274	42.29	1,600	4.015	10.53	0.577	0.000	173	61	96.5
275	33.17	1,500	3.522	9.42	0.646	0.000	180	73	95.4
276	27.84	1,201	2.780	10.01	0.689	0.000	198	41	95.8
	1937								
283	27.79	1,499	2.829	9.82	0.624	0.000	169	79	95.6
284	26.75	1,500	3.380	7.91	0.877	0.000	159	75	96.0
289	52.60	1,565	6.563	8.02	0.768	0.000	172	78	100.0
290	26.84	1,498	2.905	9.24	0.750	0.000	189	61	98.9
291	43.24	1,199	4.020	10.76	0.644	0.000	169	60	97.7
292	27.50	1,300	3.024	9.09	0.674	0.000	168	63	100.0
293	34.85	1,200	3.718	9.39	0.649	0.016	204	87	97.2
293 (1400 r.p.m.)	40.78	1,400	4.361	9.34	0.652	0.000	190	73	96.6
294	36.83	1,200	3.616	10.19	0.598	0.000	180	59	97.4
294 (1400 r.p.m.)	42.13	1,401	4.154	10.14	0.600	0.000	185	68	98.5
295	35.91	975	3.928	9.15	0.757	0.025	208	61	94.9

Test No.	TEST C—OPERATING MAXIMUM LOAD—ONE HOUR: Belt H.P. (23)	Av. Eng. R.P.M. (24)	Fuel: Gals. per Hour (25)	Fuel: H.P. Hours per Gal. (26)	Fuel: Lbs. per H.P. Hour (27)	Water: Gals. per Hour (28)	Temp. °F. Rad. (29)	Temp. °F. Atm. (30)
	1920-1935							
185	69.76	1,100	8.656	8.06	0.762	0.00	178	77
168	42.26	1,152	5.207	8.12	0.761	0.00	194	96
183	49.04	1,132	4.906	10.00	0.682	0.000	178	58
184	40.36	1,417	4.300	9.39	0.660	0.000	181	71
186	52.58	1,498	6.395	8.22	0.738	0.000	179	72
187	66.53	1,501	7.815	8.51	0.713	0.000	192	91
198	32.80	1,150	3.414	9.61	0.704	0.000	182	76
210	33.26	1,160	3.270	10.17	0.663	0.000	184	78
216	35.43	1,199	3.575	9.91	0.609	0.000	203	76
219	44.24	1,201	4.647	9.52	0.731	1.364	194	76
224	58.89	950	6.112	9.64	0.640	0.000	199	111
226	29.36	1,252	2.970	9.89	0.703	0.000	187	73
228	30.71	1,250	3.345	9.18	0.669	0.000	180	58
237	33.18	1,199	3.338	9.94	0.696	0.000	202	98
238	33.52	1,203	3.410	9.83	0.704	0.000	203	98
239	34.62	1,201	3.629	9.54	0.733	0.000	186	75
242	56.05	1,000	4.087	13.71	0.510	0.000	167	78
243	48.60	850	3.381	14.37	0.486	0.000	173	85
	1936							
251	18.98	1,425	1.958	9.69	0.635	0.000	193	74
260	30.12	1,399	3.318	9.09	0.675	0.000	194	96
261	28.30	1,298	3.054	9.27	0.663	0.000	205	95
262	104.16	1,120	11.982	8.69	0.709	0.223	214	93
265	27.16	1,202	2.696	10.07	0.685	0.000	206	102
266	27.52	1,199	2.648	10.39	0.664	0.000	182	83
267	26.09	1,500	2.534	10.30	0.671	0.000	173	76
269	48.53	1,751	4.775	10.16	0.682	0.000	188	89
271	34.13	1,400	3.660	9.33	0.744	0.000	206	70
272	37.81	1,401	4.286	8.82	0.693	0.000	187	66
273	39.82	1,400	2.885	13.80	0.506	0.000	186	62
274	40.80	1,600	3.684	11.07	0.549	0.000	170	60
275	31.66	1,501	2.919	10.85	0.561	0.000	177	72
276	26.67	1,201	2.539	10.50	0.657	0.000	193	39
	1937							
283	26.58	1,500	2.439	10.90	0.562	0.000	172	78
284	25.67	1,502	2.419	10.61	0.654	0.000	190	80
289	52.60	1,565	6.563	8.02	0.768	0.000	172	78
290	25.20	1,500	2.495	10.10	0.686	0.000	189	60
291	42.24	1,200	3.756	11.25	0.616	0.000	167	60
292	27.50	1,300	3.024	9.09	0.674	0.000	168	63
293	33.86	1,201	3.238	10.46	0.582	0.015	195	77
293 (1400 r.p.m.)	39.36	1,400	3.846	10.23	0.595	0.000	188	71
294	35.87	1,201	3.356	10.69	0.570	0.000	180	68
294 (1400 r.p.m.)	41.51	1,401	3.847	10.79	0.564	0.000	178	62
295	34.09	975	3.218	10.59	0.654	0.044	209	61

Test No.	TEST D—RATED LOAD—ONE HOUR: Belt H.P. (31)	Av. Eng. R.P.M. (32)	Fuel: Gals. per Hour (33)	Fuel: H.P. Hours per Gal. (34)	Fuel: Lbs. per H.P. Hour (35)	Water: Gals. per Hour (36)	Temp. °F. Rad. (37)	Temp. °F. Atm. (38)	TEST E—VARYING LOAD-TWO HOURS: R.P.M. High (39)	R.P.M. Low (40)
	1920-1935									
185	62.42	1,100	6.645	9.39	0.654	0.00	190	79	1,295	1,093
168	36.05	1,148	4.625	7.79	0.798	0.00	182	95	1,214	1,142
183	44.29	1,125	4.252	10.42	0.655	0.000	186	58	1,240	1,107
184	37.14	1,428	4.047	9.18	0.674	0.000	181	72	1,596	1,347
186	45.18	1,500	5.863	7.71	0.788	0.000	180	65	1,581	1,479
187	55.60	1,501	6.834	8.14	0.746	0.000	174	78	1,645	1,488
198	30.31	1,150	3.327	9.11	0.742	0.000	191	75	1,318	1,140
210	31.62	1,160	3.018	10.48	0.643	0.000	187	79	1,307	1,135
216	32.01	1,202	3.219	9.94	0.607	0.000	183	77	1,365	1,155
219	41.07	1,198	4.365	9.41	0.740	1.224	187	76	1,344	1,184
224	54.83	950	5.684	9.65	0.640	0.000	198	114	997	935
226	26.87	1,250	2.607	10.31	0.674	0.000	189	78	1,310	1,178
228	27.27	1,251	2.992	9.11	0.674	0.000	179	61	1,327	1,194
237	31.07	1,199	3.251	9.56	0.724	0.000	195	100	1,433	1,172
238	30.98	1,199	3.032	10.20	0.678	0.000	196	98	1,392	1,150
239	31.78	1,199	3.247	9.79	0.714	0.000	190	82	1,351	1,169
242	50.43	1,001	3.721	13.55	0.516	0.000	162	76	1,068	979
243	44.14	850	3.112	14.18	0.493	0.000	159	75	921	833
	1936									
251	17.55	1,423	1.778	9.87	0.624	0.000	208	92	1,508	1,394
260	27.89	1,399	3.246	8.59	0.715	0.000	184	90	1,541	1,370
261	25.97	1,300	2.940	8.83	0.695	0.000	197	96	1,456	1,253
262	94.54	1,120	11.984	7.89	0.781	0.000	206	97	1,258	912
265	26.21	1,201	2.632	9.96	0.698	0.000	207	106	1,273	1,191
266	26.69	1,201	2.599	10.27	0.672	0.000	181	85	1,272	1,182
267	24.80	1,502	2.384	10.40	0.664	0.000	182	83	1,636	1,466
269	45.11	1,749	4.570	9.87	0.702	0.000	192	91	1,884	1,718
271	32.64	1,401	3.621	9.01	0.770	0.000	206	70	1,539	1,398
272	34.90	1,400	4.064	8.59	0.711	0.000	183	65	1,518	1,363
273	35.61	1,400	2.574	13.83	0.505	0.000	185	68	1,523	1,363
274	37.36	1,600	3.446	10.84	0.561	0.000	164	60	1,701	1,570
275	29.65	1,500	2.827	10.49	0.580	0.000	168	69	1,618	1,465
276	24.13	1,200	2.297	10.51	0.657	0.000	198	40	1,305	1,184
	1937									
283	25.14	1,500	2.385	10.54	0.582	0.000	177	72	1,548	1,419
284	24.19	1,500	2.308	10.48	0.662	0.000	190	80	1,613	1,468
289	47.28	1,565	6.222	7.60	0.810	0.000	166	76	1,679	1,471
290	23.62	1,502	2.349	10.06	0.689	0.000	190	60	1,621	1,500
291	37.73	1,200	3.646	10.35	0.670	0.000	171	60	1,372	1,145
292	24.15	1,299	2.778	8.69	0.705	0.000	159	61	1,442	1,082
293	31.76	1,200	3.151	10.08	0.604	0.000	188	68	1,304	1,165
293 (1400 r.p.m.)	36.27	1,402	3.732	9.72	0.627	0.000	178	64	1,498	1,349
294	31.94	1,199	3.007	10.62	0.573	0.000	176	62	1,308	1,130
294 (1400 r.p.m.)	36.88	1,402	3.529	10.45	0.583	0.000	172	59	1,508	1,356
295	31.51	975	2.955	10.66	0.650	0.070	208	60	1,069	942

BELT TESTS

Test No.	TRACTOR: NAME	HORSEPOWER RATING	HORSEPOWER RATING	Weight as Tested	ENGINE: Cylinders	Bore and Stroke	Rated Eng. R.P.M.	Make	Fuel Used	OIL: S.A.E. No.	OIL: Gals. to Motor	OIL: Gals. from Motor	Total Hrs. Run	TEST B—100% MAXIMUM LOAD—TWO HOURS: Belt H.P.	Av. Eng. R.P.M.	Fuel: Gals. per Hour	Fuel: H.P. Hours per Gal.	Fuel: Lbs. per H.P. Hour	Water: Gals. per Hour	Temp. °F.: Rad.	Temp. °F.: Atm.	Carb. Adj. % of Max.	TEST C—OPERATING MAXIMUM LOAD—ONE HOUR: Belt H.P.	Av. Eng. R.P.M.	Fuel: Gals. per Hour	Fuel: H.P. Hours per Gal.	Fuel: Lbs. per H.P. Hour	Water: Gals. per Hour	Temp. °F.: Rad.	Temp. °F.: Atm.	TEST D—RATED LOAD—ONE HOUR: Belt H.P.	Av. Eng. R.P.M.	Fuel: Gals. per Hour	Fuel: H.P. Hours per Gal.	Fuel: Lbs. per H.P. Hour	Water: Gals. per Hour	Temp. °F.: Rad.	Temp. °F.: Atm.	TEST E—VARYING LOAD-TWO HOURS: R.P.M. High	R.P.M. Low
	1	2	3	4	5	6	7	8	9	10	11	12	13	14	15	16	17	18	19	20	21	22	23	24	25	26	27	28	29	30	31	32	33	34	35	36	37	38	39	40
	1938																																							
296	Graham-Bradley 503-103	Not Rated	19.87-26.99	4,955	6	3¼x4⅜	1,500	G. Paige	**Gas	30	1.149	1.056	43	30.38	1,502	3.083	9.85	0.618	0.000	157	62	93.1	28.27	1,497	2.721	10.39	0.586	0.000	161	68	27.06	1,498	2.690	10.06	0.605	0.000	165	72	1,709	1,467
297	Mc'Deering Farmall F-14	Not Rated	11.52-15.48	4,900	4	3 x4	1,650	Own	Dist.	30	1.907	1.620	44	17.44	1,649	1.670	10.44	0.662	0.039	199	75	97.5	17.00	1,651	1.566	10.86	0.636	0.097	200	73	15.54	1,651	1.515	10.26	0.674	0.026	200	71	1,732	1,608
300	Oliver Row Crop 80KD	Not Rated	23.27-35.14	4,930	4	4½x5¼	1,200	Own	Dist.	30	2.134	2.490	48	38.78	1,200	4.982	7.78	0.888	0.000	170	81	93.7	36.33	1,200	3.294	11.03	0.626	0.000	171	72	35.24	1,201	3.204	11.00	0.628	0.000	174	72	1,372	1,176
301	Oliver Standard 80KD	Not Rated	22.16-35.16	4,950	4	4½x5¼	1,200	Own	Dist.	30	3.494	1.748	44	39.32	1,200	4.752	8.27	0.835	0.000	173	79	94.2	37.03	1,199	3.281	11.29	0.612	0.000	170	75	35.20	1,199	3.085	11.41	0.606	0.000	175	71	1,308	1,184
302	Allis-Chalmers B	Not Rated	10.16-13.86	2,620	4	3¼x3½	1,400	Own	Dist.	30	1.554	1.685	42	15.68	1,400	1.540	10.18	0.677	0.000	146	62	96.6	15.14	1,400	1.360	11.13	0.619	0.000	155	63	14.00	1,400	1.257	11.14	0.619	0.000	169	61	1,608	1,291
303	Allis-Chalmers WC	Not Rated	14.80-22.98	3,175	4	4 x4	1,300	Own	Dist.	30	3.738	3.842	58	25.45	1,300	2.611	9.75	0.707	0.000	177	81	95.1	24.21	1,300	2.202	10.99	0.627	0.000	184	84	23.04	1,300	2.103	10.96	0.629	0.000	181	87	1,491	1,264
303	(With Rubber Tires)			4,545																																				
304	Allis-Chalmers WC	Not Rated	17.69-26.72	3,165	4	4 x4	1,300	Own	**Gas	30	1.956	1.264	63	29.93	1,300	2.861	10.46	0.589	0.000	166	77	95.0	28.44	1,300	2.455	11.58	0.532	0.000	175	83	26.80	1,300	2.321	11.55	0.534	0.000	171	84	1,444	1,259
304	(With Rubber Tires)			4,535																																				
305	John Deere B	Not Rated	10.76-16.86	3,390	2	4½x5½	1,150	Own	Dist.	30	3.219	1.150	66	18.53	1,150	1.870	9.91	0.695	0.352	208	96	98.8	18.31	1,150	1.734	10.56	0.653	0.384	207	98	16.94	1,149	1.662	10.19	0.676	0.328	207	97	1,230	1,094
305	(With Rubber Tires)			4,360																																				
306	Massey Harris "101"S	Not Rated	23.84-31.31	5,725	6	3⅛x4⅜	1,500	Chrysler	**Gas	10	2.480	2.137	83	35.02	1,500	3.083	11.36	0.543	0.000	192	78	96.7	33.87	1,500	2.857	11.86	0.521	0.000	183	77	31.50	1,500	2.702	11.66	0.529	0.000	183	81	1,697	1,489
306	(With Steel Wheels)			3,805																																				
306	(1800 r.p.m.)		-36.01				1,800							40.04	1,800	3.613	11.08	0.557	0.000	192	82	98.1	39.29	1,800	3.420	11.49	0.537	0.000	188	81	36.15	1,800	3.321	10.89	0.567	0.000	180	76	1,928	1,785
307	Massey Harris "101"R	Not Rated	24.78-31.30	5,400	6	3⅛x4⅜	1,500	Chrysler	**Gas	10	2.625	2.149	85	35.40	1,500	3.105	11.40	0.541	0.000	181	69	96.8	34.10	1,500	2.880	11.84	0.521	0.000	182	71	31.40	1,500	2.775	11.32	0.545	0.000	182	75	1,655	1,424
307	(With Steel Wheels)			3,800																																				
307	(1800 r.p.m.)		-36.19				1,800							40.67	1,800	3.649	11.15	0.553	0.000	183	76	95.0	38.65	1,802	3.313	11.67	0.529	0.000	183	79	36.34	1,800	3.253	11.17	0.552	0.000	184	81	1,948	1,750
308	Case "R"	Not Rated	14.17-18.30	4,140	4	3¼x4	1,425	Waukesha	Gas	20	1.479	1.351	43	20.52	1,425	2.166	9.47	0.649	0.000	198	74	93.8	19.25	1,425	1.836	10.48	0.586	0.000	199	79	18.48	1,423	1.800	10.27	0.599	0.000	198	78	1,609	1,389
309	Case "L"	Not Rated	32.04-42.02	8,025	4	4⅝x6	1,100	Own	Dist.	20	2.986			47.04	1,101	4.462	10.54	0.656	0.000	204	78	95.7	45.02	1,102	3.876	11.62	0.595	0.415	211	90	42.10	1,101	3.758	11.20	0.617	0.000	200	83	1,265	1,102
309	(With Steel Wheels)			5,300						40	1.466	3.266	62																											
310	M-M Twin City UTS	Not Rated	30.86-38.12	7,940	4	4¼x5	1,275	Own	**Gas	20	2.707	2.021	55	42.88	1,274	4.112	10.43	0.589	0.000	182	75	97.2	41.67	1,273	3.725	11.19	0.549	0.000	185	80	38.31	1,275	3.733	10.26	0.598	0.000	178	80	1,364	1,231
311	M-M Twin City UTS	Not Rated	24.54-31.54	7,940	4	4¼x5	1,275	Own	Dist.	20	2.199	1.993	36	36.48	1,274	3.871	9.42	0.734	0.000	186	53	96.5	35.19	1,273	3.408	10.33	0.670	0.000	187	52	31.47	1,273	2.986	10.54	0.656	0.000	187	55	1,325	1,229
312	John Deere "H"	Not Rated	9.68-12.97	3,035	2	3-9/16x5	1,400	Own	Dist.	30	0.991	0.795	43	14.84	1,399	1.305	11.37	0.608	0.000	201	54	95.8	14.22	1,400	1.190	11.95	0.578	0.000	201	56	13.01	1,400	1.110	11.72	0.590	0.000	201	57	1,496	1,321
313	John Deere "L"	Not Rated	7.01- 9.27	2,180	2	3¼x4	1,550	J.D.Her.	**Gas	10	0.490	0.292	39	10.42	1,549	1.061	9.82	0.625	0.000	196	62	100.0	10.42	1,549	1.061	9.82	0.625	0.000	196	62	9.28	1,550	0.946	9.81	0.626	0.000	193	61	1,644	1,471
314	Caterpillar Diesel D-8 [9]	....	73.43-95.99	32,925	6	5¾x8	850	Own	*F.Oil	20	8.168	5.505	43	109.64	850	7.029	15.60	0.449	0.000	165	58	...... [9]	109.64	850	7.029	15.60	0.449	0.000	165	58	96.46	850	6.083	15.86	0.442	0.000	165	57	931	839
	1939																																							
315	Int. TrTrTor(WT)TD-18 [10]	....	52.94-71.96	23,360	6	4¾x6½	1,200	Own	*F.Oil	30	6.081	4.682	72	80.32	1,199	5.215	15.40	0.455	0.000	187	69	.... [10]	80.32	1,199	5.215	15.40	0.455	0.000	187	69	72.56	1,199	4.739	15.31	0.458	0.000	183	71	1,246	1,187
316	Allis-Chalmers RC	Not Rated	12.09-16.16	4,005	4	3⅜x3½	1,500	Own	Dist.	30	1.963	0.771	51	18.21	1,500	1.510	12.06	0.574	0.000	179	70	99.1	18.14	1,501	1.480	12.26	0.564	0.000	163	65	16.24	1,500	1.374	11.82	0.586	0.000	179	67	1,669	1,421
317	M-M Twin City GT	Not Rated	36.27-48.93	9,445	4	4⅝x6	1,075	Own	**Gas	30	3.004	2.226	52	55.08	1,075	4.808	11.46	0.526	0.000	176	74	98.0	53.99	1,073	4.647	11.62	0.519	0.000	177	75	49.02	1,074	4.300	11.40	0.529	0.000	173	75	1,206	1,023
319	M-M Twin City UTU	Not Rated	28.32-38.48	6,773	4	4¼x5	1,275	Own	**Gas	30	2.211	1.889	55	42.71	1,275	4.108	10.40	0.590	0.000	190	78	97.2	41.51	1,274	3.731	11.13	0.551	0.000	190	78	38.77	1,275	3.662	10.59	0.579	0.000	186	80	1,367	1,233
320	Caterpillar R-2	Not Rated	18.80-26.41	6,835	4	3¾x5	1,525	Own	Gas	30	3.041	2.897	50	29.31	1,525	3.559	8.24	0.752	0.000	182	86	99.7	28.95	1,525	3.359	8.62	0.718	0.000	195	95	26.49	1,526	3.181	8.33	0.743	0.000	185	87	1,619	1,439
321	Caterpillar R-2	Not Rated	18.50-26.20	6,835	4	3¾x5	1,525	Own	Dist.	30	3.037	3.164	55	28.56	1,526	3.684	7.75	0.894	0.000	204	99	96.6	27.78	1,525	3.089	8.99	0.771	0.000	202	93	26.28	1,524	3.000	8.76	0.791	0.000	194	87	1,602	1,467
322	Caterpillar Diesel D-2 [11]	....	19.40-27.19	7,420	4	3¾x5	1,525	Own	*F.Oil	20	4.040	3.336	41	29.98	1,526	2.264	13.24	0.528	0.000	174	94	.... [11]	29.98	1,526	2.264	13.24	0.528	0.000	174	94	27.46	1,525	2.062	13.32	0.525	0.000	172	96	1,617	1,506
323	Cletrac General GG	Not Rated	10.46-	3,115	4	3 x4	1,400	Hercules	**Gas	30	1.503	1.245	42																											
323	(1700 r.p.m.)		-17.42				1,700							19.29	1,700	2.194	8.79	0.695	0.024	196	86	100.0	19.29	1,700	2.194	8.79	0.695	0.024	196	86	17.42	1,700	2.124	8.20	0.745	0.000	193	87	1,726	1,674
324	Cletrac HG	Not Rated	11.09-	3,510	4	3 x4	1,400	Hercules	**Gas	30	1.544	1.364	64																											
324	(1700 r.p.m.)		-17.49				1,700							19.36	1,700	2.187	8.85	0.690	0.005	199	86	100.0	19.36	1,700	2.187	8.85	0.690	0.005	199	86	17.59	1,700	1.969	8.93	0.684	0.049	188	79	1,760	1,700
325	Cletrac BD [12]	....	28.54-40.88	9,500	6	3¾x4½	1,400	Hercules	*F.Oil	30	3.815	2.677	45	45.37	1,400	3.459	13.12	0.533	0.000	195	88	.... [12]	45.37	1,400	3.459	13.12	0.533	0.000	195	88	40.67	1,399	2.963	13.73	0.509	0.000	186	86	1,507	1,374
326	Cletrac FD [13]	Not Rated	71.58-96.54	30,030	6	5 x6	1,300	Hercules	*F.Oil	30	5.904	5.271	46	107.25	1,300	7.660	14.00	0.499	0.042	208	91	.... [13]	107.25	1,300	7.660	14.00	0.499	0.042	208	91	96.73	1,300	6.567	14.73	0.475	0.078	188	88	1,393	1,276
327	Mc'Deering Farmall M	Not Rated	24.65-31.20	6,770	4	3⅞x5¼	1,450	Own	Dist.	30	2.143	2.120	57	34.82	1,447	2.800	12.44	0.557	0.000	197	81	98.5	34.16	1,452	2.733	12.50	0.554	0.000	196	82	31.38	1,450	2.580	12.16	0.570	0.000	201	82	1,536	1,433
328	Mc'Deering Farmall M	Not Rated	25.83-33.35	6,770	4	3⅞x5¼	1,450	Own	**Gas	20	2.090	1.722	53	36.66	1,451	2.996	12.24	0.502	0.000	198	95	98.9	36.07	1,451	2.966	12.16	0.505	0.000	200	100	33.46	1,451	2.834	11.81	0.520	0.000	203	97	1,530	1,426
329	Mc'Deering Farmall A	Not Rated	13.01-16.20	3,570	4	3 x4	1,400	Own	**Gas	20	1.681	1.560	37	18.34	1,400	1.569	11.69	0.528	0.067	199	69	92.8	16.86	1,400	1.408	11.97	0.515	0.031	200	75	16.34	1,399	1.404	11.64	0.530	0.209	201	78	1,526	1,339
330	Mc'Deering Farmall A	Not Rated	11.65-14.55	3,570	4	3 x4	1,400	Own	Dist	20	2.439	1.209	66	16.51	1,400	1.396	11.83	0.582	0.099	197	79	92.1	15.18	1,400	1.276	11.90	0.579	0.000	194	75	14.58	1,399	1.254	11.63	0.593	0.070	196	73	1,588	1,359
331	Mc'Deering Farmall B	Not Rated	12.98-16.34	3,740	4	3 x4	1,400	Own	**Gas	20	1.493	0.931	46	18.39	1,400	1.580	11.64	0.530	0.098	196	65	91.9	16.82	1,400	1.415	11.89	0.519	0.490	200	69	16.41	1,401	1.426	11.51	0.536	0.127	200	70	1,525	1,355
332	Mc'Deering Farmall B	Not Rated	11.47-14.20	3,740	4	3 x4	1,400	Own	Dist.	20	1.381	1.469	53	16.00	1,400	1.361	11.76	0.586	0.033	194	73	95.4	15.36	1,399	1.300	11.82	0.583	0.112	201	72	14.31	1,401	1.276	11.21	0.614	0.151	198	79	1,498	1,336
333	Mc'Deering Farmall H	Not Rated	19.13-23.72	5,550	4	3⅜x4¼	1,650	Own	**Gas	20	1.501	1.438	43	26.20	1,652	2.261	11.59	0.532	0.000	191	86	93.3	24.28	1,650	2.075	11.70	0.527	0.000	196	89	23.83	1,649	2.079	11.46	0.538	0.000	199	90	1,710	1,595
334	Mc'Deering Farmall H	Not Rated	16.99-20.69	5,550	4	3⅜x4¼	1,650	Own	Dist.	20	1.740	1.165	55	23.31	1,650	2.044	11.40	0.604	0.013	189	79	95.1	22.14	1,650	1.885	11.75	0.587	0.546	199	77	20.80	1,651	1.872	11.11	0.620	0.000	199	78	1,741	1,593
335	John Deere A	Not Rated	20.12-26.33	6,410	2	5½x6¾	975	Own	Dist.	30	3.305	1.844	50	29.59	975	2.809	10.53	0.655	0.019	209	79	96.9	28.93	975	2.561	11.30	0.611	0.049	207	70	26.36	975	2.397	11.00	0.627	0.079	207	69	1,046	936
336	Allis-Chalmers WK	Not Rated	41.24-54.37	13,340	4	5 x6½	1,050	Own	**Gas	30	4.746	3.467	36	62.22	1,050	6.160	10.10	0.597	0.000	138	56	99.2	62.15	1,050	6.027	10.31	0.585	0.000	129	49	54.54	1,050	5.247	10.39	0.580	0.000	143	69	1,167	1,000
337	Allis-Chalmers WS	Not Rated	52.83-72.90	20,330	4	5¾x6½	1,050	Own	**Gas	30	5.866	4.856	36	84.34	1,050	8.210	10.27	0.587	0.000	136	47	100.0	84.34	1,050	8.210	10.27	0.587	0.000	136	47	72.92	1,049	7.028	10.38	0.581	0.000	167	74	1,158	1,011
338	Allis-Chalmers L	Not Rated	68.65-93.63	26,105	6	5¼x6½	1,050	Own	**Gas	30	7.163	5.012	37	108.84	1,050	10.538	10.33	0.584	0.000	139	43	100.0	108.84	1,050	10.538	10.33	0.584	0.000	139	43	93.65	1,051	9.169	10.21	0.590	0.000	158	55	1,204	962
	1940																																							
339	Ford-Ferguson System 9N	Not Rated	12.68-	3,375	4	3-3/16x3¾	1,400	Own	**Gas	20	1.413	1.065	64																											
339	(2,000 r.p.m.)		-20.29				2,000							23.56	2,000	2.434	9.68	0.623	0.000	183	55	97.4	23.07	2,001	2.368	9.74	0.619	0.000	179	51	20.24	1,996	2.310	8.76	0.688	0.000	171	51	2,172	1,872
340	Case DC	Not Rated	25.56-32.83	7,010	4	3⅞x5½	1,100	Own	**Gas	20	3.038	2.316	78	37.28	1,100	3.401	10.96	0.550	0.000	186	63	95.3	35.51	1,100	2.927	12.13	0.497	0.000	187	61	32.94	1,099	2.728	12.07	0.499	0.000	183	64	1,236	1,070
341	M-M Twin City RTU	Not Rated	15.58-20.49	4,135	4	3⅝x4	1,400	Own	**Gas	30	1.686	1.552	55	23.22	1,400	2.394	9.70	0.626	0.000	171	62	92.9	21.63	1,401	2.133	10.14	0.599	0.000	169	59	20.52	1,400	2.058	9.97	0.609	0.000	170	63	1,496	1,376
342	McCormick-Deering W-4	Not Rated	16.87-21.14	5,690	4	3⅜x4¼	1,650	Own	Dist.	20	1.883	1.099	59	23.11	1,651	2.005	11.53	0.600	0.000	191	89	96.8	22.16	1,653	1.906	11.63	0.595	0.000	191	97	21.18	1,653	1.871	11.32	0.611	0.000	191	98	1,800	1,643
343	Int. TrTrTor(WT)TD-14 [14]	....	40.53-54.42	17,595	4	4¾x6½	1,350	Own	*F.Oil	30	4.213	3.039	51	61.56	1,350	3.918	15.71	0.444	0.000	194	70	.... [14]	61.56	1,350	3.918	15.71	0.444	0.000	194	70	54.61	1,350	3.501	15.60	0.448	0.000	192	75	1,434	1,326
344	Int. TrTrTor(WT)TD-9 [15]	....	29.16-39.02	10,955	4	4.4x5.5	1,400	Own	*F.Oil	20	2.518	2.660	28	43.93	1,399	2.973	14.78	0.472	0.000	200	73	.... [15]	43.93	1,399	2.973	14.78	0.472	0.000	200	73	39.15	1,400	2.652	14.76	0.473	0.000	200	74	1,488	1,372
345	Int. TrTrTor(WT)TD-6 [16]	....	22.12-30.80	7,950	4	3⅞x5¼	1,450	Own	*F.Oil	20	2.504	2.175	47	34.54	1,448	2.427	14.23	0.490	0.000	189	75	.... [16]	34.54	1,448	2.427	14.23	0.490	0.000	189	75	30.82	1,451	2.119	14.54	0.480	0.000	191	74	1,559	1,422
346	Int. TrTrTor(WT)T-6	Not Rated	24.69-33.12	7,680	4	3⅞x5¼	1,450	Own	**Gas	20	2.760	2.266	65	36.96	1,446	3.044	12.14	0.502	0.000	201	75	97.9	36.06	1,449	2.982	12.09	0.504	0.000	206	76	33.04	1,448	2.761	11.97	0.510	0.000	206	75	1,582	1,422
347	Int. TrTrTor(WT)T-6	Not Rated	23.72-31.18	7,680	4	3⅞x5¼	1,450	Own	Dist.	20	2.445	2.396	48	34.54	1,448	2.822	12.24	0.565	0.000	199	86	99.8	34.22	1,445	2.783	12.30	0.563	0.000	201	93	31.21	1,450	2.694	11.59	0.597	0.000	202	95	1,591	1,427
348	Case VC	Not Rated	15.00-	4,290	4	3 x4⅜	1,425	Cont'al	**Gas	20	1.182	0.979	52																											
348	(1650 r.p.m.)		-22.06				1,650							24.48	1,650	2.205	11.10	0.549	0.000	184	88	95.9	23.76	1,650	2.097	11.33	0.538	0.000	164	76	22.07	1,647	2.007	11.00	0.555	0.000	176	91	1,740	1,597
349	Case D	Not Rated	24.77-31.83	7,005	4	3⅞x5½	1,200	Own	Dist.	30	2.775	1.154	60	35.36	1,200	3.308	10.69	0.649	0.000	205	82	93.8	33.45	1,199	2.621	12.76	0.544	0.000	189	71	31.87	1,199	2.488	12.81	0.542	0.000	192	82	1,321	1,108
350	John Deere D	Not Rated	30.46-38.11	8,125	2	6¾x7	900	Own	Dist.	40	3.639	2.992	45	42.05	900	4.764	8.83	0.786	3.238	188	88	95.9	40.24	900	3.968	10.14	0.684	2.432	199	90	38.15	899	3.872	9.85	0.704	0.902	210	96	959	879
351	Oliver Row Crop 70 HC	Not Rated	22.64-28.37	6,770	6	3⅛x4⅜	1,500	Own	**Gas	30	1.216	1.145	49	31.52	1,500	2.954	10.67	0.580	0.000	171	78	96.4	30.37	1,499	2.693	11.28	0.549	0.000	180	79	28.46	1,500	2.532	11.24	0.551	0.000	193	81	1,667	1,467
352	M-M Twin City ZTU	Not Rated	20.98-27.95	5,025	4	3⅝x4½	1,500	Own	**Gas	30	1.702	1.615	56	31.14	1,500	2.909	10.70	0.578	0.000	189	83	97.8	30.55	1,500	2.792	10.94	0.566	0.000	184	81	27.94	1,499	2.777	10.06	0.615	0.000	184	87	1,590	1,460
353	McCormick-Deering W-4	Not Rated	19.25-23.71	5,850	4	3⅜x4¼	1,650	Own	**Gas	20	1.487	1.433	52	26.21	1,650	2.254	11.63	0.532	0.000	192	85	92.6	24.30	1,651	2.086	11.65	0.531	0.000	193	84	23.77	1,650	2.058	11.55	0.536	0.000	196	84	1,785	1,617
354	McCormick-Deering W-6	Not Rated	24.60-31.19	7,744	4	3⅞x5¼	1,450	Own	Dist.	20	1.975	1.868	51	34.23	1,450	2.805	12.20	0.569	0.000	198	98	100.0	34.23	1,450	2.805	12.20	0.569	0.000	198	98	31.23	1,450	2.584	12.09	0.574	0.000	200	90	1,557	1,436
355	McCormick-Deering W-6	Not Rated	25.36-32.93	7,610	4	3⅞x5¼	1,450	Own	**Gas	20	2.486	2.153	45	36.97	1,450	2.964	12.47	0.496	0.000	193	78	98.1	36.15	1,449	2.895	12.49	0.496	0.000	192	82	33.11	1,449	2.750	12.04	0.514	0.000	199	91	1,561	1,393
356	McCormick-Deering WD-6 [17]	....	23.54-30.92	7,995	4	3⅞x5¼	1,450	Own	*F.Oil	20	2.263	1.675	39	34.75	1,450	2.340	14.85	0.471	0.000	195	72	.... [17]	34.75	1,450	2.340	14.85	0.471	0.000	195	72	30.98	1,451	2.073	14.94	0.468	0.000	198	70	1,545	1,433
357	Caterpillar Diesel D-8 [18]	....	84.86-111.35	35,000	6	5¾x8	950	Own	*F.Oil	20	8.123	6.980	32	127.93	950	8.611	14.86	0.471	0.000	167	62	.... [18]	127.93	950	8.611	14.86	0.471	0.000	167	62	111.37	949	7.413	15.02	0.466	0.000	163	68	1,031	937
358	Caterpillar Diesel D-7 [19]	....	60.33-78.91	24,790	4	5¾x8	1,000	Own	*F.Oil	20	5.435	3.713	42	89.10	1,000	5.994	14.86	0.471	0.000	174	68	.... [19]	89.10	1,000	5.994	14.86	0.471	0.000	174	68	78.96	1,000	5.330	14.81	0.473	0.000	171	72	1,072	989
359	Massey-Harris "101" R Jr.	Not Rated	19.45-	4,825	4	3-3/16x4⅜	1,500	Cont'al	**Gas	10	1.103	1.101	40																											
359	(1800 r.p.m.)		-27.40				1,800							30.56	1,800	2.606	11.73	0.521	0.000	171	77	98.0	30.15	1,799	2.566	11.75	0.520	0.000	163	72	27.57	1,802	2.378	11.59	0.527	0.000	160	73	1,894	1,761
360	Allis-Chalmers HD-7W [20]	Not Rated	45.08-60.42	14,175	3	4¼x5	1,500	G.M.	*F.Oil	20	3.168	2.777	49	68.68	1,499	4.835	14.20	0.493	0.000	164	66	.... [20]	68.68	1,499	4.835	14.20	0.493	0.000	164	66	60.52	1,499	4.330	13.98	0.501	0.000	185	56	1,612	1,484
361	Allis-Chalmers HD-10W [21]	Not Rated	64.97-86.38	21,630	4	4¼x5	1,600	G.M.	*F.Oil	20	3.608	2.845	63	98.47	1,600	6.832	14.41	0.486	0.000	171	59	.... [21]	98.47	1,600	6.832	14.41	0.486	0.000	171	59	86.45	1,601	6.114	14.14	0.496	0.000	173	68	1,729	1,566
362	Allis Chalmers HD-14 [22]	Not Rated	99.14-127.91	28,750	6	4¼x5	1,500	G.M.	*F.Oil	20	4.827	4.204	62	145.39	1,501	9.575	15.18	0.462	0.000	175	59	.... [22]	145.39	1,501	9,575	15.18	0.462	0.000	175	59	128.17	1,498	8.441	15.18	0.462	0.000	178	58	1,688	1,485
363	Allis-Chalmers "C"	Not Rated	12.57-17.17	3,205	4	3⅜x3½	1,500	Own	Dist.	10	1.366	0.877	53	19.40	1,500	1.629	11.91	0.583	0.000	157	61	97.0	18.82	1,501	1.558	12.08	0.574	0.000	152	59	17.24	1,500	1.481	11.64	0.596	0.000	163	60	1,683	1,412
364	Allis-Chalmers "C"	Not Rated	14.06-20.24	3,205	4	3⅜x3½	1,500	Own	**Gas	10	1.253	0.848	45	23.30	1,499	2.033	11.46	0.533	0.000	152	55	93.2	21.83	1,499	1.841	11.86	0.515	0.000	152	51	20.27	1,499	1.722	11.77	0.519	0.000	171	51	1,767	1,408
365	Oliver Standard 80 HC	Not Rated	27.66-36.07	8,145	4	4¼x5¼	1,200	Own	**Gas	30	1.949	1.351	40	41.27	1,200	3.845	10.73	0.569	0.000	179	63	92.5	38.11	1,200	3.311	11.51	0.531	0.000	191	63	36.25	1,200	3.144	11.53	0.530	0.000	198	62	1,302	1,112

[1] Now called Oliver 90, formerly known as Oliver Hart Parr 28-44.
[2] Now called Bates Steel Mule 50.
[3] Now called Massey-Harris 25.
[4] Now called Caterpillar Diesel D-6, formerly called Caterpillar Diesel RD-6. Rating: Drawbar—52 H. P. maximum, 1000 r.p.m., second gear, standard conditions. Belt—not rated. One fuel pump setting to develop maximum specified by manufacturer was used throughout the test.
[5] Now called Caterpillar Diesel D-6, formerly called Caterpillar Diesel RD-6. Rating: Drawbar—44 H. P. maximum, 850 r.p.m., second gear, standard conditions. Belt—not rated. One fuel pump setting to develop maximum specified by manufacturer was used throughout the test.
[6] Now called McCormick-Deering W-40.
[7] Now called Caterpillar R-4.
[8] Now called Caterpillar Diesel D-4. Rating: Drawbar—35 H. P. maximum, 1400 r.p.m., second gear, standard conditions. Belt—not rated. One fuel pump setting to develop maximum specified by manufacturer was used throughout the test.
[9] Rating: Drawbar—97 H. P. maximum, 850 r.p.m., second gear, standard conditions. Belt—not rated. One fuel pump setting to develop maximum specified by manufacturer was used throughout the test.
[10] Rating: Drawbar—not rated. Belt—approximately 80 H. P. maximum observed. One fuel pump setting to develop maximum specified by manufacturer was used throughout the test.

SUMMARY OF RESULTS OF THE NEBRASKA TRACTOR TESTS (continued)

CONDUCTED BY THE UNIVERSITY OF NEBRASKA, DEPARTMENT OF AGRICULTURAL ENGINEERING, LINCOLN, NEBRASKA

INCLUDES ONLY TRACTORS ON THE MARKET JANUARY 1, 1941

DRAWBAR TESTS

Test No.	TRACTOR: NAME	Tires or Type of Lug	TEST F—100% MAXIMUM LOAD, Rated Gear: Draw Bar H. P.	Draw Bar Pull Lbs.	Miles per Hour	Slip of Drivers %	TEST G—OPERATING MAXIMUM LOAD, First Gear: Draw Bar H. P.	Draw Bar Pull Lbs.	Miles per Hour	Slip of Drivers %	Second Gear: Draw Bar H. P.	Draw Bar Pull Lbs.	Miles per Hour	Slip of Drivers %
	1	41	42	43	44	45	46	47	48	49	50	51	52	53
	1920-1935													
135	Huber Super Four	Spade	....	....	....	....	49.44	7,608	2.44	9.15	50.03	5,355	3.51	7.81
168	Huber Light Four	Spade	....	....	....	....	27.25	4,483	2.28	19.19	29.18	3,298	3.32	11.76
183	Oliver Hart Parr 3-5 Plow [1]	Spade	36.34	3,851	3.54	11.00	28.36	5,116	2.08	22.32	34.21	3,650	3.51	11.35
184	Eagle 6A	Spade	29.15	3,423	3.19	15.95	27.87	4,650	2.25	20.63	29.52	3,282	3.37	11.38
186	Bates Steel Mule 35	Cleats	41.83	5,147	3.05	0.97	43.73	8,375	1.96	4.43	41.83	5,147	3.05	0.97
187	Bates Steel Mule 45 [2]	Cleats	52.22	6,371	3.07	1.30	54.43	10,312	1.98	3.68	52.22	6,371	3.07	1.30
198	Mc'Deering Farmall F-30	Spade	25.73	3,151	3.06	5.78	24.85	4,157	2.24	9.70	23.93	2,953	3.04	6.54
210	McCormick-Deering W-30	Spade	24.35	2,454	3.72	5.59	24.29	3,118	2.92	6.58	23.36	2,359	3.71	5.59
216	Allis-Chalmers M	Cleats	27.46	3,256	3.16	2.29	29.65	5,166	2.15	5.00	27.28	3,248	3.15	2.48
219	Massey-Harris 3-4 Plow [3]	Spade	33.93	3,641	3.49	8.36	26.45	4,501	2.20	25.10	33.12	3,534	3.51	7.60
224	Caterpillar R-5	Cleats	51.14	7,041	2.72	1.90	48.93	10,384	1.77	5.96	49.44	6,778	2.74	1.73
226	Caterpillar Twenty-two	Cleats	23.72	3,499	2.54	2.23	23.43	4,534	1.94	2.84	22.32	3,294	2.54	1.75
228	Caterpillar Twenty-two	Cleats	25.33	3,723	2.55	1.86	25.26	4,900	1.93	3.29	25.21	3,705	2.55	1.69
237	Allis-Chalmers U	Spade	23.83	2,502	3.57	14.82	23.78	3,428	2.60	9.62	23.06	2,415	3.58	14.57
237	(With Rubber Tires)	Tires	....	....	....	....	13.40	2,389	2.11	20.72	20.49	2,452	3.13	15.07
238	Allis-Chalmers UC	Spade	23.67	2,297	3.86	4.07	24.17	3,422	2.65	6.87	22.53	2,176	3.86	4.15
238	(With Rubber Tires)	Tires	....	....	....	....	14.69	2,566	2.15	17.20	21.14	2,593	3.06	16.49
239	Allis-Chalmers M	Cleats	28.66	3,450	3.12	3.74	27.43	5,009	2.05	8.96	28.44	3,421	3.12	3.36
242	Caterpillar Diesel Forty [4]	Cleats	50.44	6,716	2.82	3.74	50.26	9,750	1.93	3.01	50.44	6,716	2.82	3.74
243	Caterpillar Diesel Forty [5]	Cleats	42.78	6,524	2.46	0.97	42.82	9,692	1.66	2.10	42.78	6,524	2.46	0.97
	1936													
251	Case RC	Spade	14.21	1,519	3.51	4.51	13.26	2,103	2.36	8.98	12.60	1,343	3.52	4.38
260	Cletrac AG	Cleats	24.44	3,584	2.56	3.12	22.74	5,254	1.62	10.09	24.58	3,586	2.57	2.65
261	Cletrac E	Cleats	20.47	2,940	2.61	6.61	19.01	3,867	1.84	13.28	20.20	2,887	2.62	6.37
262	Cletrac FG	Cleats	87.85	13,094	2.52	2.34	87.02	19,107	1.71	3.47	87.85	13,094	2.52	2.34
265	Massey-Harris Challenger	Spade	20.03	2,049	3.67	6.66	20.01	2,883	2.60	10.15	18.90	1,929	3.67	6.44
266	Massey-Harris Pacemaker	Spade	19.76	2,051	3.61	9.19	18.45	2,878	2.40	18.09	19.08	1,983	3.61	9.14
267	Oliver Row-Crop 70 KD	Spade	20.48	1,778	4.32	8.61	15.02	2,523	2.23	16.46	19.22	2,220	3.25	10.92
269	McCormick-Deering WK-40 [6]	Spade	35.22	3,873	3.41	6.72	34.86	4,982	2.62	7.77	33.23	3,656	3.41	6.86
271	Caterpillar Thirty [7]	Cleats	30.88	5,032	2.30	2.67	26.71	6,120	1.64	3.94	26.15	4,264	2.30	2.48
272	Caterpillar Thirty [7]	Cleats	35.05	5,775	2.28	3.44	30.99	7,211	1.61	5.39	31.48	5,186	2.28	3.33
273	Caterpillar Diesel RD-4 [8]	Cleats	35.36	5,811	2.28	3.27	33.48	7,852	1.60	6.17	35.36	5,811	2.28	3.27
274	Co-op No. 3	Tires	37.50	2,308	6.09	9.67	16.99	2,679	2.38	12.10	28.22	2,635	4.02	14.56
275	Co-op No. 2	Tires	29.67	1,839	6.05	9.57	12.23	1,980	2.32	13.81	20.13	1,863	4.05	13.18
276	Mc'Deering Farmall F-20	Spade	20.66	2,089	3.71	6.89	19.60	2,927	2.51	12.53	20.64	2,500	3.10	8.84
	1937													
283	Oliver Standard 70 HC	Spade	19.84	1,529	4.87	4.83	16.17	2,493	2.43	15.81	17.92	1,833	3.67	7.46
284	Oliver Standard 70 KD	Spade	19.83	1,535	4.85	5.71	15.85	2,634	2.26	22.15	19.16	1,937	3.71	6.42
289	Cletrac CG	Cleats	43.23	5,367	3.02	1.42	45.79	9,385	1.83	2.29	43.23	5,367	3.02	1.42
290	M-M Twin City ZT	Spade	20.53	2,270	3.39	4.51	19.82	3,262	2.28	8.31	20.37	2,691	2.84	5.14
291	Huber LC	Spade	30.47	2,823	4.05	5.98	28.55	3,990	2.68	9.06	29.49	2,696	4.10	4.83
291	(With Rubber Tires)	Tires	....	....	....	....	18.06	3,265	2.07	18.42	26.97	3,390	2.98	19.55
292	Huber B	Spade	20.72	2,164	3.59	2.13	19.05	3,059	2.34	8.04	20.72	2,164	3.59	2.13
292	(With Rubber Tires)	Tires	....	....	....	....	14.16	2,668	1.99	13.79	21.16	2,767	2.87	14.59
293	Massey-Harris TP-CH	Spade	25.52	2,672	3.58	8.17	24.34	3,432	2.66	7.99	23.69	2,449	3.63	7.49
293	(With Rubber Tires)	Tires	....	....	....	....	23.90	4,009	2.24	15.60	29.23	3,376	3.25	9.17
293	(1400 r.p.m.)													
294	Massey-Harris TP-PA	Spade	26.26	2,720	3.62	8.42	24.37	3,555	2.57	11.80	26.17	2,669	3.68	6.86
294	(With Rubber Tires)	Tires	....	....	....	....	19.23	3,236	2.23	14.37	25.83	3,164	3.06	13.45
294	(1400 r.p.m.)													
295	John Deere G	Spade	27.63	3,079	3.36	6.40	25.86	4,085	2.37	6.21	25.55	2,818	3.40	5.18

Test No.	NAME	TEST G, Third Gear: Draw Bar H. P.	Draw Bar Pull Lbs.	Miles per Hour	Slip of Drivers %	Fourth Gear: Draw Bar H. P.	Draw Bar Pull Lbs.	Miles per Hour	Slip of Drivers %	Fifth Gear: Draw Bar H. P.	Draw Bar Pull Lbs.	Miles per Hour	Slip of Drivers %	Sixth Gear: Draw Bar H. P.	Draw Bar Pull Lbs.	Miles per Hour	Slip of Drivers %
	1	54	55	56	57	58	59	60	61	62	63	64	65	66	67	68	69
	1920-1935																
135	Huber Super Four	....	....	....	....	....	....	....	....	....	....	....	....	....	....	....	....
168	Huber Light Four	....	....	....	....	....	....	....	....	....	....	....	....	....	....	....	....
183	Oliver Hart Parr 3-5 Plow [1]	31.18	2,412	4.85	7.75	....	....	....	....	....	....	....	....	....	....	....	....
184	Eagle 6A	25.75	2,120	4.55	10.59	....	....	....	....	....	....	....	....	....	....	....	....
186	Bates Steel Mule 35	39.19	3,475	4.23	0.00	....	....	....	....	....	....	....	....	....	....	....	....
187	Bates Steel Mule 45 [2]	49.63	4,547	4.09	0.79	....	....	....	....	....	....	....	....	....	....	....	....
198	Mc'Deering Farmall F-30	23.04	2,327	3.71	5.23	22.00	1,849	4.46	4.34	....	....	....	....	....	....	....	....
210	McCormick-Deering W-30	22.71	1,961	4.34	4.25	....	....	....	....	....	....	....	....	....	....	....	....
216	Allis-Chalmers M	25.18	2,295	4.11	1.73	23.03	1,484	5.82	1.22	....	....	....	....	....	....	....	....
219	Massey-Harris 3-4 Plow [3]	31.46	2,616	4.51	6.17	....	....	....	....	....	....	....	....	....	....	....	....
224	Caterpillar R-5	48.20	5,049	3.58	1.32	44.83	3,288	5.11	0.77	....	....	....	....	....	....	....	....
226	Caterpillar Twenty-two	20.88	2,214	3.54	1.50	....	....	....	....	....	....	....	....	....	....	....	....
228	Caterpillar Twenty-two	23.14	2,448	3.55	1.44	....	....	....	....	....	....	....	....	....	....	....	....
237	Allis-Chalmers U	19.58	1,333	5.51	12.86	....	....	....	....	....	....	....	....	....	....	....	....
237	(With Rubber Tires)	26.64	2,179	4.59	18.31	30.07	1,037	10.87	2.84	....	....	....	....	....	....	....	....
238	Allis-Chalmers UC	18.75	1,178	5.92	2.78	....	....	....	....	....	....	....	....	....	....	....	....
238	(With Rubber Tires)	28.44	2,121	5.03	17.26	28.85	995	10.87	2.10	....	....	....	....	....	....	....	....
239	Allis-Chalmers M	25.45	2,358	4.05	3.12	23.22	5,576	1.56	15.82	....	....	....	....	....	....	....	....
242	Caterpillar Diesel Forty [4]	49.64	4,991	3.73	2.53	47.91	3,348	5.37	1.22	....	....	....	....	....	....	....	....
243	Caterpillar Diesel Forty [5]	40.55	4,714	3.23	0.77	35.91	2,939	4.58	0.58	....	....	....	....	....	....	....	....
	1936																
251	Case RC	12.08	942	4.81	4.16	....	....	....	....	....	....	....	....	....	....	....	....
260	Cletrac AG	23.23	2,346	3.71	1.67	....	....	....	....	....	....	....	....	....	....	....	....
261	Cletrac E	19.08	1,853	3.86	3.31	....	....	....	....	....	....	....	....	....	....	....	....
262	Cletrac FG	73.67	6,551	4.22	2.27	....	....	....	....	....	....	....	....	....	....	....	....
265	Massey-Harris Challenger	18.11	1,452	4.68	4.09	11.72	439	10.01	1.05	....	....	....	....	....	....	....	....
266	Massey-Harris Pacemaker	18.63	1,499	4.66	5.68	10.61	394	10.10	1.81	....	....	....	....	....	....	....	....
267	Oliver Row-Crop 70 KD	19.97	1,726	4.34	8.34	17.56	1,071	6.15	4.65	....	....	....	....	....	....	....	....
269	McCormick-Deering WK-40 [6]	32.24	3,008	4.02	5.41	....	....	....	....	....	....	....	....	....	....	....	....
271	Caterpillar Thirty [7]	28.26	3,642	2.91	2.19	24.77	2,536	3.66	1.40	23.93	1,680	5.34	1.22	....	....	....	....
272	Caterpillar Thirty [7]	31.60	4,105	2.89	2.52	30.54	3,147	3.64	1.69	29.09	2,045	5.34	0.99	....	....	....	....
273	Caterpillar Diesel RD-4 [8]	35.06	4,541	2.90	2.44	33.71	3,471	3.64	1.90	31.76	2,230	5.34	1.11	....	....	....	....
274	Co-op No. 3	35.61	2,180	6.13	9.29	35.04	967	13.59	2.44	....	....	....	....	....	....	....	....
275	Co-op No. 2	27.67	1,662	6.24	6.82	26.57	741	13.45	2.80	....	....	....	....	....	....	....	....
276	Mc'Deering Farmall F-20	20.31	2,064	3.69	7.51	19.09	1,645	4.35	6.14	....	....	....	....	....	....	....	....
	1937																
283	Oliver Standard 70 HC	17.96	1,385	4.86	4.92	16.41	922	6.67	4.51	....	....	....	....	....	....	....	....
284	Oliver Standard 70 KD	18.31	1,414	4.86	5.84	16.91	957	6.63	5.77	....	....	....	....	....	....	....	....
289	Cletrac CG	36.24	3,092	4.40	1.40	....	....	....	....	....	....	....	....	....	....	....	....
290	M-M Twin City ZT	19.53	2,149	3.41	4.20	17.55	1,302	5.05	3.75	....	....	....	....	....	....	....	....
291	Huber LC	26.60	1,839	5.42	4.07	....	....	....	....	....	....	....	....	....	....	....	....
291	(With Rubber Tires)	31.35	2,824	4.16	14.34	....	....	....	....	....	....	....	....	....	....	....	....
292	Huber B	19.72	1,534	4.82	1.19	13.32	544	9.18	1.42	....	....	....	....	....	....	....	....
292	(With Rubber Tires)	22.47	2,078	4.06	9.04	22.94	1,039	8.28	2.95	....	....	....	....	....	....	....	....
293	Massey-Harris TP-CH	24.35	1,983	4.61	5.41	16.60	630	9.88	2.15	....	....	....	....	....	....	....	....
293	(With Rubber Tires)	29.83	2,691	4.16	6.52	27.10	1,138	8.93	3.29	....	....	....	....	....	....	....	....
293	(1400 r.p.m.)																
294	Massey-Harris TP-PA	25.10	2,025	4.65	5.59	18.77	700	10.06	1.65	....	....	....	....	....	....	....	....
294	(With Rubber Tires)	30.58	2,864	4.00	8.94	29.63	1,254	8.86	3.06	....	....	....	....	....	....	....	....
294	(1400 r.p.m.)																
295	John Deere G	22.13	1,839	4.51	3.36	20.28	1,171	6.49	2.34	....	....	....	....	....	....	....	....

Test No.	NAME	TEST H—RATED LOAD—TEN HOURS: Gear Used	Draw Bar H. P.	Draw Bar Pull Lbs.	Miles per Hour	Av. Eng. R.P.M.	Slip of Drivers %	Fuel: Gals. per Hour	D. B. H. P. Hours p'r Gal.	Lbs. per D. B. H.P.hr.	Water: Gals. per Hour	Temp. °F.: Rad.	Atm.	Test No.
	1	70	71	72	73	74	75	76	77	78	79	80	81	
	1920-1935													
135	Huber Super Four	High	40.26	4,442	3.40	1,113	11.44	6.121	6.58	0.934	0.14	184	84	135
168	Huber Light Four	High	20.35	2,191	3.48	1,153	7.06	4.035	5.04	1.225	0.00	166	91	168
183	Oliver Hart Parr 3-5 Plow [1]	Int.	28.44	2,940	3.63	1,128	8.39	3.974	7.16	0.953	0.113	185	67	183
184	Eagle 6A	Int.	22.20	2,422	3.44	1,418	9.80	4.235	5.24	1.181	0.132	186	48	184
186	Bates Steel Mule 35	2nd	35.43	4,368	3.04	1,496	0.89	5.556	6.38	0.947	0.000	190	78	186
187	Bates Steel Mule 45 [2]	2nd	45.12	5,482	3.09	1,504	1.07	6.691	6.74	0.896	0.067	181	75	187
198	Mc'Deering Farmall F-30	2nd	20.88	2,520	3.11	1.150	3.44	3.152	6.62	1.024	0.053	175	69	198
210	McCormick-Deering W-30	2nd	20.23	2,008	3.78	1,160	3.94	2.870	7.05	0.956	0.000	184	89	210
216	Allis-Chalmers M	2nd	22.98	2,714	3.18	1,202	2.06	3.065	7.50	0.805	0.145	198	60	216
219	Massey-Harris 3-4 Plow [3]	2nd	24.92	2,714	3.44	1,200	9.83	3.680	6.77	1.028	0.000	183	54	219
224	Caterpillar R-5	2nd	41.29	5,649	2.74	950	1.30	5.511	7.49	0.824	0.000	187	103	224
226	Caterpillar Twenty-two	2nd	19.08	2,806	2.55	1,250	1.75	2.355	8.10	0.858	0.000	188	68	226
228	Caterpillar Twenty-two	2nd	19.55	2,870	2.55	1,249	1.50	2.862	6.83	0.899	0.000	180	54	228
237	Allis-Chalmers U	2nd	19.35	1,891	3.84	1,202	3.29	2.825	6.85	1.039	0.053	192	91	237
237	(With Rubber Tires)	3rd	22.72	1,680	5.07	1,200	9.10	2.650	8.57	0.815	0.000	184	87	237
238	Allis-Chalmers UC	2nd	18.90	1,808	3.92	1,201	2.76	3.062	6.18	1.132	0.000	194	94	238
238	(With Rubber Tires)	3rd	24.57	1,770	5.21	1,199	5.36	2.743	8.96	0.780	0.000	190	96	238
239	Allis-Chalmers M	2nd	22.61	2,709	3.13	1,201	3.12	3.087	7.32	0.954	0.022	188	93	239
242	Caterpillar Diesel Forty [4]	2nd	39.83	5,173	2.89	1,000	1.32	3.384	11.77	0.594	0.000	157	70	242
243	Caterpillar Diesel Forty [5]	2nd	33.15	5.071	2.46	850	1.09	2.880	11.51	0.611	0.000	159	75	243
	1936													
251	Case RC	2nd	11.61	1,228	3.55	1,424	3.47	1.728	6.72	0.917	0.298	196	57	251
260	Cletrac AG	2nd	19.40	2,830	2.57	1,400	2.57	3.104	6.25	0.982	0.056	193	96	260
261	Cletrac E	2nd	16.48	2,435	2.54	1,301	5.50	3.183	5.18	1.186	0.000	188	93	261
262	Cletrac FG	2nd	69.55	10,325	2.53	1,119	1.65	11.077	6.28	0.981	0.103	195	96	262
265	Massey-Harris Challenger	2nd	16.29	1,638	3.73	1,200	5.00	2.450	6.65	1.038	0.028	196	99	265
266	Massey-Harris Pacemaker	2nd	16.21	1,658	3.67	1,199	7.82	2.439	6.65	1.038	0.030	182	87	266
267	Oliver Row-Crop 70 KD	3rd	16.45	1,363	4.53	1,500	4.40	2.269	7.25	0.953	0.065	184	77	267
269	McCormick-Deering WK-40 [6]	2nd	28.50	3,117	3.43	1,750	6.26	4.262	6.69	1.036	0.022	194	61	269
271	Caterpillar Thirty [7]	2nd	24.46	3,968	2.31	1,400	2.15	3.145	7.78	0.892	0.012	204	56	271
272	Caterpillar Thirty [7]	2nd	26.59	4,346	2.29	1,398	2.52	3.714	7.16	0.853	0.000	175	47	272
273	Caterpillar Diesel RD-4 [8]	2nd	26.80	4,374	2.30	1,400	2.63	2.281	11.75	0.595	0.000	173	55	273
274	Co-op No. 3	3rd	29.19	1,719	6.37	1,600	5.50	3.114	9.37	0.649	0.000	151	49	274
275	Co-op No. 2	3rd	22.77	1,344	6.35	1,500	5.03	2.451	9.29	0.654	0.000	153	47	275
276	Mc'Deering Farmall F-20	3rd	16.12	1,619	3.73	1,200	6.02	2.258	7.14	0.966	0.177	192	52	276
	1937													
283	Oliver Standard 70 HC	3rd	15.92	1,233	4.84	1,500	5.45	2.299	6.92	0.886	0.000	171	89	283
284	Oliver Standard 70 KD	3rd	15.92	1,230	4.85	1,502	5.39	2.438	6.53	1.063	0.163	186	85	284
289	Cletrac CG	2nd	33.98	4,198	3.04	1,564	0.99	5.151	6.60	0.933	0.011	171	76	289
290	M-M Twin City ZT	3rd	15.98	1,768	3.39	1,497	4.34	2.145	7.45	0.931	0.016	187	50	290
291	Huber LC	2nd	23.34	2,122	4.12	1,200	4.45	3.207	7.28	0.952	0.004	188	60	291
291	(With Rubber Tires)	3rd	24.90	2,086	4.48	1,200	7.89	2.705	9.21	0.753	0.001	189	43	291
292	Huber B	2nd	16.29	1,695	3.60	1,301	2.08	2.917	5.58	1.098	0.015	153	58	292
292	(With Rubber Tires)	3rd	18.62	1,647	4.24	1,300	4.85	2.636	7.06	0.868	0.000	123	36	292
293	Massey-Harris TP-CH	2nd	19.63	1,999	3.68	1,200	5.89	2.969	6.61	0.921	0.009	168	56	293
293	(With Rubber Tires)	3rd	24.69	2,202	4.20	1,200	5.41	2.902	8.51	0.707	0.000	164	29	293
293	(1400 r.p.m.)	....	....	....	....	....	....	....	....	....	....	....	....	293
294	Massey-Harris TP-PA	2nd	20.26	2,048	3.71	1,200	5.50	3.038	6.67	0.903	0.000	168	60	294
294	(With Rubber Tires)	3rd	24.01	2,173	4.14	1,201	5.62	2.871	8.36	0.720	0.000	167	50	294
294	(1400 r.p.m.)	....	....	....	....	....	....	....	....	....	....	....	....	294
295	John Deere G	2nd	20.75	2,252	3.46	975	3.70	2,822	7.35	0.942	0.101	199	23	295

DRAWBAR TESTS

Test No.	TRACTOR: NAME	Tires or Type of Lug	TEST F—100% MAXIMUM LOAD, Rated Gear: Draw Bar H. P.	Draw Bar Pull Lbs.	Miles per Hour	Slip of Drivers %	TEST G—OPERATING MAXIMUM LOAD, First Gear: Draw Bar H. P.	Draw Bar Pull Lbs.	Miles per Hour	Slip of Drivers %	Second Gear: Draw Bar H. P.	Draw Bar Pull Lbs.	Miles per Hour	Slip of Drivers %	Third Gear: Draw Bar H. P.	Draw Bar Pull Lbs.	Miles per Hour	Slip of Drivers %	Fourth Gear: Draw Bar H. P.	Draw Bar Pull Lbs.	Miles per Hour	Slip of Drivers %	Fifth Gear: Draw Bar H. P.	Draw Bar Pull Lbs.	Miles per Hour	Slip of Drivers %	Sixth Gear: Draw Bar H. P.	Draw Bar Pull Lbs.	Miles per Hour	Slip of Drivers %	TEST H—RATED LOAD—TEN HOURS: Gear Used	Draw Bar H. P.	Draw Bar Pull Lbs.	Miles per Hour	Av. Eng. R.P.M.	Slip of Drivers %	Fuel: Gals. per Hour	D. B. H. P. Hours p'r Gal.	Lbs. per D. B. H.P.hr.	Water: Gals. per Hour	Temp. °F.: Rad.	Atm.	Test No.
	1	41	42	43	44	45	46	47	48	49	50	51	52	53	54	55	56	57	58	59	60	61	62	63	64	65	66	67	68	69	70	71	72	73	74	75	76	77	78	79	80	81	
	1938																																										
296	Graham-Bradley 503-108	Tires	25.20	2,283	4.14	9.29	19.11	3,013	2.38	17.26	23.92	2,163	4.15	9.34	24.01	1,640	5.49	6.47	......	......	......	......	......	......	......	......	......	......	......	......	2nd	20.03	1,751	4.29	1,500	6.35	2.414	8.30	0.734	0.000	157	76	296
297	Mc'Deering Farmall F-14	Tires	14.84	1,920	2.90	8.16	13.24	2,369	2.10	12.66	14.24	1,841	2.90	8.09	14.76	1,445	3.83	5.87	......	......	......	......	......	......	......	......	......	......	......	......	2nd	11.57	1,451	2.99	1,650	5.25	1.352	8.56	0.807	0.064	195	74	297
300	Oliver Row Crop 80KD	Spade	29.92	3,300	3.40	5.91	27.13	3,785	2.69	7.87	26.99	2,984	3.39	5.77	25.01	2,101	4.46	4.40	......	......	......	......	......	......	......	......	......	......	......	......	2nd	23.32	2,558	3.42	1,200	5.29	2.912	8.01	0.863	0.000	188	75	300
301	Oliver Standard 80KD	Spade	28.55	2,911	3.68	4.83	27.97	3,596	2.92	6.54	26.11	2,648	3.70	4.27	24.19	1,890	4.80	3.70	......	......	......	......	......	......	......	......	......	......	......	......	2nd	22.13	2,246	3.70	1,199	4.25	2.845	7.78	0.888	0.010	188	68	301
302	Allis-Chalmers B	Tires	12.97	1,473	3.30	15.23	7.76	1,406	2.07	16.30	12.68	1,382	3.44	11.91	13.17	684	7.22	4.51	......	......	......	......	......	......	......	......	......	......	......	......	2nd	10.31	1,074	3.60	1,402	7.92	1.089	9.47	0.728	0.000	172	72	302
303	Allis-Chalmers WC	Spade	18.72	1,841	3.81	2.02	18.04	2,599	2.60	2.84	17.12	1,678	3.83	1.65	14.98	1,106	5.08	1.61	......	......	......	......	......	......	......	......	......	......	......	......	2nd	14.81	1,447	3.84	1,301	4.29	1.944	7.62	0.904	0.015	178	79	303
303	(With Rubber Tires)	Tires	......	......	......	......	17.27	3,054	2.12	17.04	20.16	2,222	3.40	8.02	20.41	1,644	4.65	5.12	18.76	768	9.22	1.92	......	......	......	......	......	......	......	......	3rd	17.90	1,439	4.67	1,299	4.87	1.926	9.29	0.741	0.000	178	80	303
304	Allis-Chalmers WC	Spade	22.29	2,198	3.80	4.94	19.77	2,941	2.52	8.09	21.16	2,071	3.83	4.25	20.13	1,480	5.10	4.08	......	......	......	......	......	......	......	......	......	......	......	......	2nd	17.67	1,727	3.84	1,300	0.28	2.173	8.13	0.758	0.000	155	80	304
304	(With Rubber Tires)	Tires	......	......	......	......	17.15	3,136	2.05	18.91	22.48	2,722	3.10	15.91	24.16	2,017	4.49	8.12	23.52	976	9.04	8.49	......	......	......	......	......	......	......	......	3rd	20.10	1,640	4.60	1,299	5.87	2.099	9.58	0.648	0.000	152	72	304
305	John Deere B	Spade	13.41	1,508	3.34	2.44	14.03	2,088	2.52	3.19	13.32	1,497	3.34	2.58	11.25	974	4.33	2.65	9.57	616	5.83	0.50	......	......	......	......	......	......	......	......	2nd	10.84	1,228	3.31	1,149	0.66	1.589	6.82	1.010	0.005	202	65	305
305	(With Rubber Tires)	Tires	......	......	......	......	15.92	2,690	2.22	11.27	16.25	2,005	3.04	7.80	16.44	1,518	4.06	4.96	16.05	1,105	5.45	3.53	......	......	......	......	......	......	......	......	3rd	13.12	1,195	4.12	1,149	3.70	1.451	9.04	0.762	0.011	196	70	305
306	Massey Harris "101"S	Tires	30.80	2,612	4.42	7.15	20.13	3,571	2.11	16.69	27.95	3,449	3.04	14.49	29.51	2,498	4.43	7.05	......	......	......	......	......	......	......	......	......	......	......	......	3rd	23.94	1,987	4.52	1,500	5.12	2.431	9.85	0.626	0.000	166	75	306
306	(With Steel Wheels)	Spade	......	......	......	......	19.92	3,076	2.43	13.78	22.85	2,380	3.60	8.49	23.11	1,703	5.09	3.88	......	......	......	......	......	......	......	......	......	......	......	......	2nd	18.29	1,862	3.68	1,505	6.79	2.453	7.46	0.829	0.000	175	84	306
306	(1800 r.p.m.)																														......	......	......	......	......	......	......	......	......	......	......	......	306
307	Massey Harris "101"R	Tires	31.50	2,487	4.75	5.34	20.16	3,199	2.36	11.72	29.20	3,233	3.39	9.39	30.36	2,392	4.76	5.14	......	......	......	......	......	......	......	......	......	......	......	......	3rd	24.79	1,936	4.80	1,500	4.21	2.489	9.96	0.619	0.000	167	71	307
307	(With Steel Wheels)	Spade	......	......	......	......	21.98	3,265	2.53	9.87	25.38	2,624	3.63	5.30	25.12	1,872	5.03	2.52	......	......	......	......	......	......	......	......	......	......	......	......	2nd	20.76	2,138	3.64	1,500	5.55	2.625	7.91	0.782	0.000	160	62	307
307	(1800 r.p.m.)																														......	......	......	......	......	......	......	......	......	......	......	......	307
308	Case "R"	Tires	18.23	1,979	3.45	6.45	15.58	2,574	2.27	12.59	17.02	1,832	3.48	5.62	16.53	1,264	4.90	3.59	......	......	......	......	......	......	......	......	......	......	......	......	2nd	14.19	1,512	3.52	1,425	4.63	1.660	8.55	0.719	0.000	154	61	308
309	Case "L"	Tires	40.80	3,635	4.21	8.79	36.81	4,474	3.09	14.79	39.90	3,506	4.27	7.68	39.73	2,736	5.45	5.50	......	......	......	......	......	......	......	......	......	......	......	......	2nd	32.14	2,768	4.35	1,100	5.77	3.198	10.05	0.688	0.014	197	74	309
309	(With Steel Wheels)	Spade	......	......	......	......	31.64	4,472	2.65	11.75	31.94	3,394	3.53	6.68	29.41	2,470	4.46	5.00	......	......	......	......	......	......	......	......	......	......	......	......	2nd	25.74	2,657	3.63	1,098	3.62	3.240	7.94	0.870	0.053	191	76	309
310	M-M Twin City UTS	Tires	39.00	3,285	4.45	6.63	30.02	4,959	2.27	16.51	37.40	4,481	3.13	10.86	38.12	3,195	4.47	6.24	38.22	2,367	6.06	4.38	......	......	......	......	......	......	......	......	3rd	30.90	2,559	4.53	1,275	4.96	3.513	8.80	0.698	0.011	160	64	310
311	M-M Twin City UTS	Tires	33.29	2,781	4.49	5.11	28.94	4,773	2.27	16.01	31.67	3,713	3.20	8.14	31.86	2,647	4.51	4.74	31.18	1,924	6.07	3.49	......	......	......	......	......	......	......	......	3rd	24.51	2,018	4.55	1,275	3.68	2.669	9.18	0.754	0.072	180	24	311
312	John Deere "H"	Tires	12.48	1,386	3.38	6.72	11.67	1,839	2.38	10.14	11.96	1,323	3.39	6.35	11.90	759	5.88	2.99	......	......	......	......	......	......	......	......	......	......	......	......	2nd	9.77	1,065	3.44	1,400	4.98	0.962	10.16	0.680	0.013	195	51	312
313	John Deere "L"	Tires	9.06	908	3.74	7.65	8.40	1,235	2.55	16.05	9.06	908	3.74	7.65	9.06	571	5.95	4.07	......	......	......	......	......	......	......	......	......	......	......	......	2nd	7.06	688	3.85	1,549	4.98	0.872	8.10	0.758	0.006	190	58	313
314	Caterpillar Diesel D-8 [9]	Cleats	96.37	18,740	1.93	2.59	95.27	26,111	1.37	4.58	96.37	18,740	1.93	2.59	93.81	15,580	2.26	2.32	93.85	13,344	2.64	1.86	90.61	10,713	3.17	1.15	86.87	7,524	4.33	0.66	2nd	73.71	14,249	1.94	850	2.06	5.525	13.34	0.525	0.000	158	37	314
	1939																																										
315	Int. TrTrTor (WT) TD-18 [10]	Cleats	70.66	10,561	2.51	1.38	72.38	18,973	1.43	3.27	71.97	13,357	2.02	1.79	70.66	10,561	2.51	1.38	69.45	7,827	3.33	0.71	63.85	5,157	4.64	0.18	58.67	3,833	5.74	0.42	3rd	53.22	7,943	2.51	1,199	0.99	4.214	12.63	0.555	0.000	185	69	315
316	Allis-Chalmers RC	Tires	15.25	1,551	3.69	5.12	12.60	2,840	1.66	17.38	14.88	2,067	2.70	7.75	14.84	1,509	3.69	5.09	13.93	716	7.30	2.02	......	......	......	......	......	......	......	......	3rd	12.11	1,221	3.72	1,500	4.29	1.282	9.45	0.732	0.000	186	77	316
317	M-M Twin City GT	Tires	44.99	4,964	3.40	13.69	31.60	5,068	2.34	16.42	44.31	4,866	3.41	13.54	47.85	3,186	5.63	5.16	46.24	1,793	9.67	2.89	......	......	......	......	......	......	......	......	2nd	36.96	3,849	3.60	1,074	8.63	3.918	9.43	0.639	0.000	163	64	317
319	M-M Twin City UTU	Tires	35.16	3,582	3.68	15.41	24.65	4,496	2.06	17.52	32.58	4,585	2.66	17.00	35.30	3,384	3.91	9.99	36.40	2,498	5.46	5.61	......	......	......	......	......	......	......	......	3rd	28.32	2,620	4.05	1,276	6.80	3.255	8.70	0.705	0.024	191	99	319
320	Caterpillar R-2	Cleats	23.46	3,655	2.41	3.23	23.84	5,676	1.58	6.59	22.70	3,544	2.40	3.51	22.49	2,891	2.92	2.31	22.23	2,401	3.47	2.87	20.22	1,506	5.04	1.34	......	......	......	......	2nd	18.86	2,922	2.42	1,525	2.76	2.971	6.35	0.975	0.033	182	89	320
321	Caterpillar R-2	Cleats	22.98	3,596	2.40	3.53	22.49	5,379	1.57	6.89	22.26	3,455	2.42	2.72	21.36	2,722	2.94	1.36	21.31	2,258	3.54	0.64	19.35	1,423	5.10	0.06	......	......	......	......	2nd	18.95	2,898	2.45	1,524	1.19	2.801	6.77	1.024	0.013	192	93	321
322	Caterpillar Diesel D-2 [11]	Cleats	24.12	3,798	2.38	4.65	25.15	5,903	1.60	5.71	24.12	3,798	2.38	4.65	24.25	3,069	2.96	1.13	23.53	2,485	3.55	0.83	21.47	1,585	5.08	0.83	......	......	......	......	2nd	19.45	2,993	2.44	1,525	2.36	1.880	10.35	0.676	0.000	163	85	322
323	Cletrac General GG	Tires	13.29	1,661	3.00	18.09	8.44	1,699	1.86	19.65	13.29	1,661	3.00	18.09	14.26	942	5.68	5.91	......	......	......	......	......	......	......	......	......	......	......	......	2nd	10.38	1,172	3.32	1,401	9.52	1.443	7.19	0.849	0.037	199	83	323
323	(1700 r.p.m.)																																										323
324	Cletrac HG	Cleats	14.01	1,748	3.00	2.74	13.44	2,800	1.80	7.65	14.01	1,748	3.00	2.74	11.63	883	4.94	2.91	......	......	......	......	......	......	......	......	......	......	......	......	2nd	11.14	1,415	2.95	1,400	4.42	1.530	7.28	0.839	0.072	201	80	324
324	(1700 r.p.m.)																																										324
325	Cletrac BD [12]	Cleats	36.07	5,295	2.55	4.23	31.15	7,530	1.55	15.18	36.07	5,295	2.55	4.23	35.85	3,916	3.43	1.59	......	......	......	......	......	......	......	......	......	......	......	......	2nd	28.83	4,136	2.61	1,399	1.81	2.625	10.98	0.636	0.022	170	82	325
326	Cletrac FD [13]	Cleats	90.38	12,459	2.72	0.99	91.16	21,831	1.57	2.93	90.38	12,459	2.72	0.99	83.97	8,584	3.67	0.66	72.05	5,468	4.94	0.04	......	......	......	......	......	......	......	......	2nd	72.03	9,917	2.72	1,299	0.93	6.186	11.64	0.600	0.056	187	90	326
327	Mc'Deering Farmall M	Tires	30.62	3,653	3.14	7.87	25.49	4,365	2.19	14.89	30.36	3,634	3.13	8.21	30.52	2,951	3.88	7.01	30.65	2,396	4.80	5.08	......	......	......	......	......	......	......	......	2nd	24.89	2,941	3.17	1,450	6.80	2.769	8.99	0.771	0.019	190	94	327
328	Mc'Deering Farmall M	Tires	33.05	4,060	3.05	10.46	24.49	4,233	2.17	15.57	32.52	3,963	3.08	10.17	31.70	3,120	3.81	8.69	32.02	2,503	4.80	5.54	......	......	......	......	......	......	......	......	2nd	26.23	3,102	3.17	1,450	7.22	2.495	10.51	0.584	0.000	190	89	328
329	Mc'Deering Farmall A	Tires	16.32	1,827	3.35	8.98	12.27	2,387	1.93	17.49	14.79	1,622	3.42	7.20	14.38	1,165	4.63	4.92	13.88	528	9.86	2.17	......	......	......	......	......	......	......	......	2nd	13.11	1,434	3.43	1,401	7.03	1.365	9.60	0.642	0.054	200	81	329
330	Mc'Deering Farmall A	Tires	15.17	1,648	3.45	7.73	12.42	2,360	1.97	16.01	14.76	1,599	3.46	7.53	14.27	1,148	4.66	5.14	13.48	505	10.01	2.04	......	......	......	......	......	......	......	......	2nd	11.37	1,211	3.52	1,400	5.77	1.225	9.28	0.742	0.041	198	78	330
331	Mc'Deering Farmall B	Tires	16.21	1,794	3.39	8.99	12.06	2,377	1.90	18.87	14.86	1,633	3.41	7.82	14.86	1,201	4.64	4.91	13.02	494	9.88	2.31	......	......	......	......	......	......	......	......	2nd	13.04	1,420	3.44	1,400	7.05	1.361	9.58	0.644	0.038	198	85	331
332	Mc'Deering Farmall B	Tires	14.73	1,598	3.46	6.91	12.02	2,463	1.83	21.68	13.95	1,505	3.48	6.38	13.97	1,119	4.68	4.18	13.01	493	9.90	2.31	......	......	......	......	......	......	......	......	2nd	11.55	1,242	3.49	1,399	6.05	1.203	9.60	0.717	0.037	201	50	332
333	Mc'Deering Farmall H	Tires	24.17	2,841	3.19	6.09	19.84	3,603	2.06	17.86	22.65	2,634	3.23	5.59	22.85	2,211	3.88	4.65	22.03	1,673	4.94	3.47	......	......	......	......	......	......	......	......	2nd	19.14	2,211	3.25	1,650	4.83	2.020	9.48	0.651	0.006	193	62	333
334	Mc'Deering Farmall H	Tires	21.37	2,484	3.23	4.83	19.38	3,169	2.29	7.75	20.22	2,362	3.21	5.00	19.72	1,903	3.89	3.68	19.70	1,495	4.94	2.74	......	......	......	......	......	......	......	......	2nd	17.07	1,968	3.25	1,651	4.01	1.683	10.14	0.679	0.041	199	48	334
335	John Deere A	Tires	26.20	2,339	4.20	4.96	24.64	4,110	2.25	13.03	25.53	3,053	3.14	7.41	25.47	2,269	4.21	4.82	25.13	1,682	5.60	3.33	......	......	......	......	......	......	......	......	3rd	20.48	1,815	4.23	975	4.21	2.140	9.57	0.721	0.040	204	59	335
336	Allis-Chalmers WK	Cleats	53.60	7,813	2.57	1.63	51.85	11,785	1.65	4.32	52.89	7,727	2.57	1.61	53.08	6,165	3.23	1.07	44.60	2,815	5.94	0.62	......	......	......	......	......	......	......	......	2nd	41.50	6,063	2.57	1,050	1.52	4.630	8.96	0.673	0.009	170	38	336
337	Allis-Chalmers WS	Cleats	68.86	11,367	2.27	2.67	68.72	17,843	1.44	5.00	68.86	11,367	2.27	2.67	67.44	7,900	3.20	2.08	62.20	5,152	4.53	1.03	55.14	3,245	6.37	0.71	......	......	......	......	2nd	53.80	8,828	2.29	1,050	2.10	6.677	8.06	0.748	0.043	165	46	337
338	Allis-Chalmers L	Cleats	91.99	18,410	1.87	3.44	90.42	23,851	1.42	4.32	91.99	18,410	1.87	3.44	90.67	12,919	2.63	2.86	90.12	9,770	3.46	1.32	84.14	6,448	4.89	0.71	77.06	4,520	6.39	0.48	2nd	69.34	13,727	1.89	1,050	2.40	8.300	8.35	0.722	0.022	177	42	338
	1940																																										
339	Ford-Ferguson System 9N	Tires	16.31	2,146	2.85	13.55	12.61	2,236	2.11	17.33	15.92	2,101	2.84	14.46	17.02	872	7.32	4.45	......	......	......	......	......	......	......	......	......	......	......	......	2nd	12.80	1,568	3.06	1,399	6.70	1,610	7.95	0.758	0.000	178	60	339
339	(2,000 r.p.m.)																		......	......	......	......	......	......	......	......	......	......	......	......													339
340	Case DC	Tires	33.06	3,616	3.43	8.58	24.39	4,128	2.22	15.88	31.30	3,413	3.44	8.12	31.34	2,395	4.91	5.06	28.70	1,058	10.17	2.51	......	......	......	......	......	......	......	......	2nd	25.74	2,730	3.54	1,099	5.46	2.578	9.98	0.604	0.000	192	67	340
341	M-M Twin City RTU	Tires	20.01	1,938	3.87	7.58	13.47	2,666	1.89	17.47	18.11	2,330	2.91	10.75	18.72	1,804	3.89	7.06	16.40	531	11.58	1.96	......	......	......	......	......	......	......	......	3rd	15.66	1,477	3.98	1,403	5.32	1.926	8.13	0.746	0.015	172	67	341
342	McCormick-Deering W-4	Tires	21.38	2,573	3.11	6.40	19.46	3,297	2.21	9.44	19.89	2,377	3.14	5.68	19.74	1,844	4.01	4.56	19.77	1,455	5.10	3.59	......	......	......	......	......	......	......	......	2nd	16.92	2,004	3.17	1,648	4.80	1.733	9.76	0.709	0.020	191	79	342
343	Int. TrTrTor (WT) TD-14 [14]	Cleats	51.71	7,919	2.45	6.07	51.43	13,426	1.44	7.18	51.82	9,645	2.01	6.58	51.71	7,919	2.45	6.07	51.28	5,683	3.38	4.85	48.21	3,824	4.73	4.49	44.67	2,925	5.73	4.45	3rd	40.81	6,178	2.48	1,351	5.16	3.106	13.14	0.531	0.047	190	74	343
344	Int. TrTrTor (WT) TD-9 [15]	Cleats	37.21	6,637	2.10	4.60	34.28	9,014	1.43	8.07	37.21	6,637	2.10	4.60	36.66	4,368	3.15	2.36	36.16	3,551	3.82	1.46	32.21	2,304	5.24	0.93	......	......	......	......	2nd	28.97	4,971	2.19	1,401	0.81	2.359	12.28	0.568	0.026	197	74	344
345	Int. TrTrTor (WT) TD-6 [16]	Cleats	28.14	4,929	2.14	3.27	26.75	7,160	1.40	8.24	28.14	4,929	2.14	3.27	27.29	3,368	3.04	3.40	26.35	2,641	3.74	2.53	23.84	1,661	5.38	1.22	......	......	......	......	2nd	22.28	3,818	2.19	1,452	1.77	1.824	12.21	0.571	0.038	188	82	345
346	Int. TrTrTor (WT) T-6	Cleats	30.85	5,400	2.14	3.74	28.52	7,652	1.40	8.59	29.55	5,215	2.13	4.56	29.38	3,579	3.08	2.06	28.08	2,767	3.81	1.32	25.24	1,756	5.39	1.05	......	......	......	......	2nd	24.63	4,266	2.16	1,450	2.72	2.636	9.34	0.653	0.047	195	86	346
347	Int. TrTrTor (WT) T-6	Cleats	29.53	5,146	2.15	3.49	28.45	7,434	1.44	6.02	28.88	5,042	2.15	3.44	28.03	3,433	3.06	2.82	26.99	2,700	3.75	2.78	24.40	1,712	5.35	2.06	......	......	......	......	2nd	23.58	4,098	2.16	1,450	2.99	2.431	9.70	0.713	0.051	198	84	347
348	Case VC	Tires	18.55	2,034	3.42	7.05	17.50	2,798	2.35	12.28	18.18	1,990	3.43	7.05	18.70	1,567	4.48	5.11	17.86	682	9.82	2.76	......	......	......	......	......	......	......	......	2nd	15.07	1,627	3.47	1,424	5.50	1.624	9.28	0.657	0.000	168	84	348
348	(1650 r.p.m.)																		......	......	......	......	......	......	......	......	......	......	......	......													348
349	Case D	Tires	30.67	3,091	3.72	6.84	26.78	4,192	2.40	14.88	28.10	2,807	3.75	6.00	27.57	1,948	5.31	3.53	25.84	885	10.95	1.42	......	......	......	......	......	......	......	......	2nd	24.86	2,465	3.78	1,200	5.34	2.197	11.32	0.613	0.000	198	88	349
350	John Deere D	Tires	38.02	3,641	3.92	8.61	34.50	4,830	2.68	14.97	37.12	3,537	3.94	8.11	37.02	2,661	5.22	5.94	......	......	......	......	......	......	......	......	......	......	......	......	2nd	30.77	2,907	3.97	900	7.30	3.505	8.78	0.790	1.198	196	87	350
351	Oliver Row Crop 70 HC	Tires	28.63	2,528	4.25	5.82	24.79	4,170	2.23	12.23	26.14	3,040	3.23	7.17	27.01	2,376	4.26	5.30	27.43	1,737	5.92	3.66	25.82	1,324	7.31	2.67	......	......	......	......	3rd	22.72	1,986	4.29	1,500	4.76	2.317	9.81	0.631	0.000	194	79	351
352	M-M Twin City ZTU	Tires	26.39	2,258	4.38	8.94	16.16	3,254	1.86	18.47	19.18	3,168	2.27	17.40	25.78	2,869	3.37	11.97	26.10	2,233	4.38	8.89	......	......	......	......	......	......	......	......	4th	21.04	1,748	4.51	1,500	4.05	2.467	8.53	0.726	0.000	192	92	352
353	McCormick-Deering W-4	Tires	23.97	2,957	3.04	7.24	21.12	3,671	2.16	10.20	22.17	2,693	3.09	5.80	21.91	2,087	3.94	4.76	22.00	1,643	5.02	3.47	......	......	......	......	......	......	......	......	2nd	19.43	2,362	3.09	1,650	5.75	1.963	9.90	0.625	0.000	193	89	353
354	McCormick-Deering W-6	Tires	30.74	3,963	2.91	8.71	24.84	4,755	1.96	18.51	30.74	3,963	2.91	8.71	31.12	2,987	3.91	5.96	31.40	2,451	4.80	4.60	......	......	......	......	......	......	......	......	2nd	24.77	3,149	2.95	1,448	7.24	2.389	10.37	0.669	0.000	199	87	354
355	McCormick-Deering W-6	Tires	32.48	4,282	2.84	10.84	25.13	4,777	1.97	17.23	31.44	4,173	2.83	11.44	32.33	3,156	3.84	7.66	32.14	2,545	4.74	6.12	......	......	......	......	......	......	......	......	2nd	25.48	3,238	2.95	1,450	6.84	2.525	10.09	0.613	0.000	199	63	355
356	McCormick-Deering WD-6 [17]	Tires	30.58	3,997	2.87	9.78	25.25	4,806	1.97	18.03	30.58	3,997	2.87	9.78	31.02	2,980	3.90	5.78	30.96	2,414	4.81	4.43	......	......	......	......	......	......	......	......	2nd	23.43	2,948	2.98	1,450	6.23	1.799	13.02	0.537	0.000	195	65	356
357	Caterpillar Diesel D-8 [18]	Cleats	110.06	19,537	2.11	4.25	104.12	26,208	1.49	6.58	110.06	19,537	2.11	4.25	108.91	15,973	2.56	1.11	108.41	13,707	2.97	1.03	106.58	11,266	3.55	0.91	102.92	7,995	4.83	0.62	2nd	85.33	14,728	2.17	951	1.67	6.541	13.05	0.537	0.000	162	56	357
358	Caterpillar Diesel D-7 [19]	Cleats	78.48	13,454	2.19	1.98	75.83	21,351	1.33	4.80	78.48	13,454	2.19	1.98	76.99	9,090	3.18	0.71	73.16	5,994	4.58	0.48	72.57	4,550	5.98	0.46	......	......	......	......	2nd	60.63	10,335	2.20	1,000	1.63	4.734	12.81	0.547	0.000	171	68	358
359	Massey-Harris "101" R Jr.	Tires	24.65	2,118	4.36	6.14	18.30	3,278	2.09	16.18	23.54	2,840	3.11	10.59	24.53	2,105	4.37	6.00	......	......	......	......	......	......	......	......	......	......	......	......	3rd	19.44	1,648	4.42	1,500	4.94	1.903	10.22	0.598	0.000	164	70	359
359	(1800 r.p.m.)																		......	......	......	......	......	......	......	......	......	......	......	......													359
360	Allis-Chalmers HD-7W [20]	Cleats	57.29	8,570	2.51	1.86	57.31	12,171	1.77	4.43	57.29	8,570	2.51	1.86	55.56	6,167	3.38	2.36	51.19	3,311	5.80	0.81	......	......	......	......	......	......	......	......	2nd	45.45	6,753	2.52	1,500	1.24	4.015	11.32	0.619	0.000	170	71	360
361	Allis-Chalmers HD-10W [21]	Cleats	82.19	15,507	1.99	3.68	81.25	19,002	1.60	4.83	82.19	15,507	1.99	3.68	80.47	11,421	2.64	2.13	78.24	7,867	3.73	1.88	75.78	6,190	4.59	1.21	66.69	4,157	6.02	1.01	2nd	65.55	12,159	2.02	1,601	2.13	5.640	11.62	0.603	0.000	168	72	361
362	Allis Chalmers HD-14 [22]	Cleats	126.98	22,699	2.10	4.14	122.09	28,019	1.63	5.64	126.98	22,699	2.10	4.14	125.15	17,265	2.72	2.27	126.28	13,769	3.44	2.23	115.27	10,074	4.29	1.92	103.43	5,579	6.95	1.17	2nd	99.39	17,568	2.12	1,501	3.29	7.898	12.58	0.557	0.000	189	40	362
363	Allis-Chalmers "C"	Tires	15.96	1,878	3.19	13.15	11.83	2,368	1.87	19.24	15.97	1,818	3.29	10.47	16.10	890	6.78	4.69	......	......	......	......	......	......	......	......	......	......	......	......	2nd	12.68	1,407	3.38	1,500	8.00	1.349	9.40	0.739	0.000	181	53	363
364	Allis-Chalmers "C"	Tires	18.43	2,206	3.13	15.95	12.10	2,352	1.93	17.90	18.04	2,108	3.21	13.34	18.31	1,008	6.81	5.34	......	......	......	......	......	......	......	......	......	......	......	......	2nd	14.15	1,557	3.41	1,500	8.21	1.460	9.69	0.631	0.000	187	47	364
365	Oliver Standard 80 HC	Tires	35.91	3,982	3.38	8.59	32.67	5,079	2.41	13.07	33.70	3,706	3.41	7.60	33.42	2,786	4.50	5.66	32.38	1,976	6.14	3.92	......	......	......	......	......	......	......	......	2nd	27.65	3,002	3.45	1,200	6.31	2.912	9.50	0.643	0.000	195	58	365

[11] Rating: Drawbar—approximately 25.5 H. P. maximum, 1525 r.p.m., second gear, standard conditions. Belt—not rated. One fuel pump setting to develop maximum specified by manufacturer was used throughout the test.

[12] The fuel pumps as used throughout the complete test were set to develop approximately 48 maximum belt horsepower, standard conditions.

[13] The fuel pumps were used throughout the test as set and sealed by the manufacturer.

[14] The fuel pumps as used throughout the complete test were set to develop approximately 61.5 observed maximum belt horsepower.

[15] The fuel pumps as used throughout the complete test were set to develop approximately 44 observed maximum belt horsepower.

[16] The fuel pumps as used throughout the complete test were set to develop approximately 34.4 observed maximum belt horsepower.

[17] The fuel pumps as used throughout the complete test were set to develop approximately 34.5 observed maximum belt horsepower.

[18] Rating: Drawbar—26,200 pounds maximum pull, first gear, observed. Belt—not rated. One fuel pump setting to develop maximum specified by manufacturer was used throughout the test.

[19] Rating: Drawbar—80 H. P. maximum, 1000 r.p.m., second gear, corrected conditions. Belt—not rated. One fuel pump setting to develop maximum specified by manufacturer was used throughout the test.

[20] W indicates wide tread model. The fuel pumps were used throughout the test as set and sealed by the manufacturer.

[21] W indicates wide tread model. The fuel pumps were used throughout the test as set and sealed by the manufacturer.

[22] The fuel pumps were used throughout the test as set and sealed by the manufacturer.

* A fuel oil was used conforming to specifications of the manufacturer.

** 70 octane (regular grade) gasoline was used.

Some 1941 Models

Case SC

Case LA

Caterpillar D6

Deere LA

with disk-type turning brakes and tree-saving shields over the rear wheels and at the cowl. Their heavy Model LA (page 45) with a 4- or 5-bottom moldboard plow capacity appeared.

The Caterpillar Tractor Co. added Model D6 (page 45) to its line. This tractor was powered by a 6-cylinder vertical engine and apparently replaced the D6 -3 cylinder job, Nebraska tested in 1936, and introduced the year before.

The Huber Manufacturing Co. discontinued manufacturing tractors.

The Deere Co. announced its model LA tractor (page 45), which like the "L", had vertical 2-cylinder engine. The "LA" had slightly more power, capable of pulling one 16-inch plow compared to a 12-inch for the "L", and was fitted with a power take-off.

In the World War II period, beginning late in 1941, because of the acute manpower shortage, experimental work on tractors was curtailed and few new models appeared. Many tractor plants were converted into munition factories and plants suitable for making other war materials. Because of war needs, materials for civilian use were becoming scarce, and as the war progressed, steel and certain other metals for building tractors and farm machinery became more and more restricted. In many instances where possible, substitute materials were used. For example, steel back bearings replaced solid bronze, steel replaced copper where possible and the composition of steel and iron alloys was altered - molybdenum replaced chromium. To offset this, in part, emphasis was being placed on preventative maintenance and repairs of agricultural equipment to prolong its usefulness. Accordingly, better care was taken of all such equipment and much was being repaired that in normal times would have been discarded.

At the close of the war, 1946, a year or so was found necessary to reconvert the plants to the manufacture of tractor and related equipment. A large backlog of orders had already piled up, and because of the need of new and additional tractors and the fact that the farmer's income had become relatively high, tractor demands became staggering.

When the Marshall Plan started up, there developed a huge demand for all sorts of tractors in Europe, particularly of the larger sizes - both crawler and wheel type. Accordingly, export quotas were placed on regular manufacturers to permit a gradual filling of the domestic backlog and yet permit a reasonable number of tractors to be exported. As a result of this restriction, many new tractor companies sprung up overnight and offered their products for sale - immediate delivery - and many tractors of inferior design and quality were turned loose in foreign countries. Considerable amounts of needed material were therefore dumped abroad of little use to anyone. This situation fortunately did not last long. The fly-by-night manufacturers passed out with the increasing ability of the domestic manufacturers to supply the demands.

1942

Because of the war, the Nebraska tests were suspended during the years 1942, 1943, 1944 and 1945.

The Minneapolis-Moline Power Implement Co. put out its GTA tractor (page 46) of 4- to 5-plow capacity. This was a 4-wheel rubber-tired job of more or less conventional design. The engine was available with a high compression head, standard, and optional with low compression head.

The Keck-Gonnerman Co., Mt. Vernon, Ind., discontinued the manufacture of tractors.

The J.I. Case Co. added the models VAC (page 46), VA, VAI, and VAO to its line. The VAC was a 2-plow 4-wheel tricycle type with tread of rear wheels adjustable and powered with a 4 cylinder vertical engine. Hydraulic lift control was provided and 4 forward speeds from 2.32 m.p.h. to 8.40.

1943

The Deere Co. introduced its Model GM (page 46), a 3-plow tricycle-type tractor. This was powered with the company's own 2-cylinder horizontal engine. The tread of the rear wheels was adjustable.

B. F. Avery & Sons Co., Louisville, Ky. started production on their Model A tractor (page 46) of the 3-wheel tricycle type with adjustable rear axle and two 12 in. plow capacity.

1944

On October 4, the Oliver Farm Equipment Co. changed its corporate name to The Oliver Corporation and took over the Cleveland Tractor Co.

The Deere Co. introduced its Model HNH 1-plow tractor of the single front-wheel high-clearance type, the Model HWH 1-plow adjustable front axle high-clearance type tractor (page 46), the BNH, 2-plow size with single front wheel and high clearance, the BWH 2-plow tractor with adjustable front axle and high clearance, the ANH 3-plow tractor with single front wheel and high clearance and the AWH same as ANH except provided with an adjustable front axle.

Some 1942-1944 Models

1942 - Minneapolis-Moline GTA

1942 - Case VAC

1943 - Deere GM

1943 - B. F. Avery A

1944 - Deere HWH

1945

The International Harvester Co. introduced two new orchard and grove models, the Os-4 (page 47) and OS-6. These tractors were the same tractors as the O-4 and O-6, put out in 1940 except that the stream-lined fenders were replaced with plain fenders. Streamlining is not necessary in many apple, cherry, prune, pecan and other orchards, where trees are pruned high and are not so closely spaced as in citrus groves. These OS tractors are intended for use in any such orchards or in the larger vineyards, where tractor power can often be used to advantage. The WR-9 and WDR-9 (page 47) both 4-plow tractors, were also introduced for the special use of the rice growers.

The Willys-Overland Co., Toledo, Ohio introduced its Universal Jeep (page 47) which was produced by altering the design of the Jeep developed for the Army. It differed from the latter in that it had a lower low-speed, more radiator capacity, a lower hitch and a slightly larger transmission and rear axle ratio. The Jeep was provided with a 4- or 2-wheel drive which could be selected as desired by the use of a transfer lever.

The Minneapolis-Moline Co. came out with the UTC cane tractor (page 47). This was a special model of the standard U, a 3-4 plow tractor, with added features to meet the needs of modern cane farming. A high overall clearance was provided to allow longer crop care and cultivation without damage to the cane.

Some 1945 Models

Willys-Overland Universal Jeep

Minneapolis Moline UTC Cane

International Harvester OS4

International Harvester WDR9

1946

The Empire Tractor Corp., Philadelphia, Pa. announced its 40 hp. Empire tractor Model 88. This was powered with a 4-cylinder 40 hp. Willys-Overland engine and fitted with a transmission transfer case claimed to be the same as was used in the jeep. The frame was built on the conventional lines of an earlier period when channel iron and companion construction were common. In 1948, Model 90 (page 48) came into production. This had the same general appearance as the Model 88 and had the same power but made use of lower transmission speeds. Relatively few tractors were sold and the company soon encountered financial difficulties sufficient to cause it to close its doors a couple years later.

The B. F. Avery Co., Louisville, Ky. added the Model V (page 48) one plow tractor to its line. This was powered with a 4-cylinder high compression engine mounted on a channel frame fitted with a front cross rod, serving as a hitching point for various tools and also as a bumper.

At the close of this year Harry Ferguson, Inc. and the Ford Motor Co. terminated the oral agreement made by Harry Ferguson and Henry Ford for the manufacture of tractors embodying the Ferguson system. This resulted in the formation of a new company, the Dearborn Motors Corp., Birmingham, Mich. which distributed the Ford tractor while Harry

Ferguson, Inc. continued to distribute the Ferguson. The Ford Motor Co. powered its tractors with its own 4-cylinder engine, and made use of the 3-point suspension and the hydraulic system, in principle, the same as the original Ferguson system. The treads of both front and rear wheels were adjustable. Under a written agreement the manufacture of the Ford tractor with Ferguson system by the Ford Co. for Harry Ferguson was continued until June 30, 1947.

The Earthmaster Farm Equipment Co., Hollydale, Calif. put out a Model C series 1 to 2 plow general-purpose tractor. The outstanding feature was the "Duomatic" hydraulic control or so-called effortless implement handling. It was claimed that 9 speeds forward were possible with the company's 3 transmission ratios and 3 throttle settings. The Model C had 20 in. clearance while Model CH (page 48) had 24 1/2 in. Adjustments for altering clearance, wheel base and tread were also claimed.

The Leader Tractor Mfg. Co., Chagrin Falls, Ohio produced its Leader (page 48) which was powered with a Hercules 35 hp. engine. This tractor featured a hydraulic lift and power take-off. A clearance of 21 in. was provided and forward speeds up to 13 mi. per hour.

Empire 90

Earthmaster CH

Some 1946 Models

Minneapolis-Moline-Avery V

Leader

1947

The Cockshutt Plow Co. of Brantford, Ontario, Canada developed for their tractors a continuous running power take-off which continued to operate when the clutch was released. This was an important innovation as heretofore machinery operated by the regular p.t.o. would stop upon release of the clutch, such as sprayers and small combines so driven. Because of the importance of this development it was not long until many U. S. manufacturers fitted their tractors with a similar mechanism.

The John Deere Co. presented its Model M (page 49) a general-purpose tractor powered with a 2-cylinder vertical engine — a departure from their customary 2-cylinder twin horizontal engines. The M was of two plow capacity and featured "quik-tach" to give ease in attaching drawn implements. The tread could be varied from 36 to 42 in. and 4 speeds from 1-5/8 to 12 mi. per hr. were provided. This same year another new feature "Roll-O-Matic" was introduced on the tricycle model and consisted of a mechanism which equalized the load on both front wheels and cut in half the up-and-down movement on the front end of the tractor. Either wheel would ride over an obstruction while the other hugged the ground. The slightest up-and-down movement of one wheel was immediately transferred through gears to the other wheel thereby automatically equalizing the load on both front wheels. It was claimed this made steering and riding much easier on the operator.

The new Ford Tractor Model 8N (page 49) was introduced. New features included a mounted implement control system as a part of the hydraulic controls, a 4-speed constant mesh helical gear transmission, new brakes and steering gear, and a new safety starter. Horsepower as tested at Nebraska in 1948 was 22.76 - 26.42.

The Oliver Corporation's Cletrac Model HG track-layer (page 49) for use in row crops appeared. Four tread widths were available — standard 42 in. and specials 31, 60 and 68 in. and a 20 in. clearance. This was powered by a 4-cylinder engine delivering 22 belt hp. and 18 drawbar hp. and had 3 forward speeds - 2.01 - 5.24 mph. Steering was by planetary gears.

Production started on the Allis-Chalmers HD 19 (page 49) crawler tractor which featured the use of a torque converter. This was a great step forward in performance although it was not accomplished by other manufacturers for many years. It was claimed this was the world's largest and most powerful tractor and was powered with a 6-cylinder 2-cycle diesel engine. As tested at Nebraska the weight was 40,395 lbs. and the maximum drawbar horsepower was 118. The tread was 84 in. with a 106 in. length of track on the ground.

The Massey-Harris Co. put out six new models in five power ranges — Model 20 of 1 plow size and powered with a 4-cylinder engine, designed for auxiliary power on the larger farms; Model 30 with a 4-cylinder engine but of 2 plow capacity and adaptable for row crops; Model 44 also with 4-cylinders but of 3 plow capacity; Model 44-6 a 3 plow row-crop tractor; Model 55 a heavier tractor with 4-5 plow capacity and designed for big jobs; and the Pony (page 50) a 1 plow tractor with high compression (6.5 to 1) engine and adjustable treads for both front and rear wheels. All row crop models were tricycle, two wheels in front close together. All models except Model 20 were factory equipped with "Velvet Ride", a seat designed to do "for the tractor what the shock absorber did for the automobile".

The International Harvester Co. added to their line a row crop tractor — the Farmall Super A (see page 000) which had either a 4 cylinder gasoline or distillate engine and adjustable front and rear axles. It was fitted with "touch control" which permitted raising or lowering implements by a touch of the control levers.

The smallest tractor of the International Harvester Co. line, the Cub, (page 50) of the culti-vision type was offered to the trade. Capable of pulling one 12 in. plow it was fitted with other equipment such as cultivator, planters, etc. to round out its usefulness. It was fitted with a high tension magneto and electric starting and lighting equipment was special.

The International Harvester Co. added the TD-24 diesel crawler (page 50) to its line. It was powered with a 6-cylinder engine and as Nebraska tested it weighed 42,211 lbs., with maximum drawbar h.p. 161, being still larger than the Allis-Chalmers HD 19 recently announced. By means of a suitable mechanism the motor was started as a gasoline engine with an electric starter, when the compression was released and spark plugs in an auxiliary combustion chamber were energized. After having attained a speed of about 100 r.p.m. the changeover lever was moved to running position. This caused the auxiliary combustion chamber to close, spark ignition to be shut off, compression to be increased to 15 to 1 ratio and diesel fuel injected for normal diesel operation.

Some 1947 Models

Deere M

Ford 8N

Oliver HG

Allis-Chalmers HD 19

Massey Harris Pony

International Harvester Farmall Super A

International Harvester Farmall Cub

International Diesel TD-24

1948

Tractor production this year reached an all time peak when 753,623 units were produced - 529,587 wheel type, 184,624 garden tractors, both riding and walking, and 39,412 track type. Of this total production some 15,000 units, or about 2 percent were exported to ECA countries.

The Allis-Chalmers Co. at the Gladsden, Ala. plant offered its Model G tractor (page 52) with one 12 in. plow capacity. The 4 cylinder engine was located in the rear and provisions made for attaching the implements ahead of the operator. The tubular open frame construction served as a mounting place for the master tool carrier to which the various tools were attached. The high arch of the frame provided the necessary clearance for attaching these tools; a bucket seat was provided between front and rear wheels.

The Allis-Chalmers Model WD (page 51) with several innovations was introduced: the power of the engine was used to adjust the rear wheel treads by spiral bars on the rim to 10 different spacings; two clutches were provided which permitted continuous operation of the power take-off and hydraulic system, independent of tractor motion; and hydraulic control of mounted and pulled implements. The tractor was of the general-purpose type built around a frame of straight steel channels. When desirable to change the front end, the adjustable front axle can be converted into a single or dual front wheel style.

The first Ferguson tractor, Model TO 20, (page 51) was produced at the new Ferguson Park plant in Detroit. It was powered with a 4-cylinder Continental engine having a 6.1 to 1 compression ratio with 16.33 - 22.53 hp. as Nebraska tested. The wheel treads were adjustable and 4 forward speeds from 2.48 to 13.13 m.p.h. were provided. A safety feature was incorporated such that the starter would not operate when the machine was in gear.

The General Tractor Co. of Seattle, Wash. entered the tractor field with its Westrac track type tractor (page 51). It was powered with a 24 B.H.P. Waukesha or 29 B.H.P. Hercules engine. Final drive was effected with a heavy duty worm, and planetary type steering was used.

The Intercontinental Mfg. Co., Grand Prairie, (near Dallas) Texas also entered the field with the Model C-26 (page 52) of 3 plow capacity. The unit was powered with a 4-cylinder Continental Red Seal engine and was of the tricycle type, adjustable rear wheel tread and two narrow spaced wheels in front. Hydraulic controls, p.t.o., and electric starting and lighting were provided.

The Cimco, Model PH 31 (formerly PH 18) tractor was being manufactured by the newly established consolidated Implement Mfg. Co., Ogden, Utah. It was a short coupled machine with all four wheels drivers - 6 inches between front and back tires, and was powered by a Continental 4-cylinder engine having a maximum horsepower of 42 at 2200 r.p.m. Four speeds forward from 2.5 to 9 m.p.h. were provided and a power take-off front and rear.

The Jumbo Steel Products Co., Azusa, Calif. announced its Simpson Jumbo (page 51) 3 plow tractor said to be designed especially for the extra rugged soil and climatic conditions of the west. This machine was available as a standard 4 wheel job (Model C) and 4 wheel tricycle (Model B) and was powered with a 6-cylinder 35 drawbar hp. Chrysler engine. Five forward speeds up to 22.4 m.p.h. and hydraulic brakes were provided.

The Minneapolis-Moline Co. started manufacturing its Model ZA (page 52) 2-3 plow tractor. Five forward speeds ranging from 2.4 to 13.1 m.p.h. were provided. Four types of this model were produced featuring different front end designs; ZAU with 2 front wheels close together, but which could be reversed for wider spacing ZAN with single front wheel for narrow spaced row crops; ZAE with adjustable front and rear axles and ZAS a standard tread type model.

The Oliver Corporation as a part of its centennial celebration announced a new and improved line of tractors - the Oliver 66 (page 51), Oliver 77, and Oliver 88, unique in having continuous running power take-offs (direct engine driven p.t.o.) - the first wheel-type tractor to be thus equipped in the United States. With the small Model 66 (1-2 plow size) there was a choice of a 4-cylinder so-called high compression, with 129 cu. in. piston displacement, a diesel of the same size or a 145 cu. in. engine with lower compression for operation on kerosene or distillate. The 77 (2-3 plow) was provided with a 6-cylinder high compression or diesel engine of 194 cu. in. or with a 216 cu. in. engine for kerosene or distillate. The 88 (3-4 plow) was powered with a 6-cylinder 231 cu. in. high compression or diesel engine or a 265 cu. in. engine for kerosene or distillate. The standard, row crop with close spaced front wheels and row crop with adjustable front axles were provided for each model as were also standard treads. Magneto or battery ignition was optional on all 3 models, but battery ignition and electric starting were standard equipment.

The Farmall C (page 52) was added to the International Harvester Co. line. It was fitted with a gasoline or distillate engine of tricycle design with double front wheels and adjustable rear wheel tread. In addition, it was fitted with "touch control" which permitted raising or lowering of the drawn implements by a touch of the small control levers.

Some 1948 Models

Oliver 66

General Tractor Co. Westrak

Ferguson TO-20

Allis Chalmers WD

Jumbo

Intercontinental C-26

International Harvester Farmall C

Minneapolis-Moline ZAS

Allis-Chalmers G

1949

The Detroit Tractor Corp., Detroit, Mich. offered a 4-wheel, all wheel drive tractor in 3 sizes, 16, 35 and 45 hp., (page 54) with magneto ignition or battery ignition, electric starting and lighting, and powered with either a 4-cylinder vertical Waukesha or Continental engine. Steering was similar to that in crawler tractors and the top speed was 17 mi. per hr.

The Long Manufacturing Co., Tarboro, N. C. started production on the Long tractor (page 54) powered with a 4-cylinder vertical 26-29 hp. engine. This tractor was provided with electric starting and lighting and a hydraulic lift and had 4 forward speeds from 2.45 to 18 mi. per hr.

The National Farm Machinery Cooperative Inc., Bellevue, Ohio introduced its CO-OP E3 tractor (page 53) of 2-3 plow capacity, in 2 types, standard and row crop. Hydraulic control and a "live" power take-off were provided, and eight forward speeds 1 1/2 to 10 mph. When use was made of the "creeper" speed especially helpful with transplanters, an auxiliary set of reduction gears was used.

The Massey-Harris Co., Racine, Wis. during the year came out with five models; models 22 RT (page 53) and STD tricycle row crop and standard respectively, powered with a 4-cylinder vertical Continental Red Seal engine (Nebraska test 17.93 - 26.85 hp.). Depth-O-Matic hydraulic control was provided, power take-off and electric starting and lighting. This model succeeded the company's light 2-plow 20 placed on the market last year: Models 44 diesel standard and 44 K kerosene standard - the diesel with a 15 to 1 compression ratio (Nebraska test rating 29.61 - 36.58 hp.) and the kerosene with 4.65 to 1. (Nebraska test rating 27.64 - 33.53 hp.); and the 55 K standard with a Nebraska rating of 37.12 - 46.32 hp. This was a standard type and used kerosene fuel.

The Fate-Root-Heath Co., Plymouth, Ohio announced the production of a new Silver King (page 54) 3-wheel row crop tractor, powered with a 4-cylinder vertical Continental engine having a compression ratio of 6.23 to 1. The Nebraska test rating was 22.35 - 29.35 hp.

Deere and Co., Moline, Ill. introduced several new models to the trade; the Model R, (page 54) the first diesel built by this company, was powered by a 2-cylinder twin horizontal engine with 16 to 1 compression ratio. Starting was accomplished by first releasing the engine compression and starting with a 2-cylinder auxiliary engine in turn started by an electric starter, then releasing compression release and engaging starting engine pinion with fly-wheel ring. Power-Trol was provided - hydraulic control by independent clutch for continuous operation; the Model AR, powered with a 2-cylinder twin horizontal gasoline engine with a 5.57 to 1 compression ratio and a Nebraska rating of 26.16 - 33.24 hp. Hydraulic control was provided and 6 forward speeds 1 1/2 to 11 mph; the model MC (page 54), the first track-type tractor produced by the company was powered by the same 2-cylinder engine (20 b.h.p.) used in models M (produced in 1947). Twelve-inch tracks were standard equipment but 10 and 14 in. tracks were available and various treads from 36 to 46 in. Four forward speeds were available from 1.2 to 6 mi per hour. The MT (2 plow) which was designed mainly for extremely hilly farms, orchards and vineyards was a 1-row wheel tractor with a capacity to pull 2 12 in. bottoms. Touch-o-matic (hydraulic) control was provided, power take-off and belt pulley.

The Dodge Division of the Chrysler Corp., Detroit, Mich. announced the production of a 4-wheel drive Power Wagon (page 54) powered with a 94 hp. engine, and fitted with a 4-speed transmission, 2-speed transfer case, dual power take-off which delivered 536 r.p.m. at tail shaft or 3124 r.p.m. at belt pulley. A front mounted winch was also provided. This Power Wagon was orginally designed for the military.

The Love Tractor, Inc., Eau Claire, Wis., announced the production of a 3-plow Love tractor (page 55) powered by a 6-cylinder Chrysler engine. Standard equipment included a built-in three point hydraulic lift implement hitch, electric starting and lighting system, belt pulley and power take-off. The front axle was adjustable for use in row crops. Ten forward speeds were provided, the highest being variable between 21.2 and 42.3 m.p.h. according to the engine speed which could be altered by 250 r.p.m. increments from 100 r.p.m to 2000 r.p.m.

The R. H. Sheppard Co., Hannover, Pa. introduced 3 models of diesel powered tractors; model Sheppard SD air cooled engine of 1-12 in. plow capacity (3.6-5.4 hp.) adjustable front and rear wheel spacing and 3 forward speeds 3-10 m.p.h.; Sheppard SD-2 (page 55) with 24 hp. 2-cylinder engine (2 plow capacity). Four standard speeds forward 2.76-10.7 m.p.h. or with auxiliary transmission, 7 speeds forward 2.7 to 14.3 m.p.h. 12 and 24 volt electric starting and lighting systems were available; Sheppard SD-3, with a 29.4-32 hp. (3 plow capacity) 3-cylinder engine and 4 forward tractor speeds slightly greater than the SD-2, with auxiliary overdrive top speed was 15 m.p.h. This company manufactured its first tractors in 1940 when they were put in the field for test purposes, but due to the war emergency, was unable to market the tractors until 1949.

The Metal Parts Corp., Racine, Wis. announced the Haas "Atomic" tractor (page 56) unique in that it was powered with a 1-cylinder 12 1/2 hp. airplane type engine. It was of the conventional 4-wheel type and had 6 speeds forward.

The Brockway Tractor Co., Chagrin Falls, Ohio entered the market with three 4-wheel tractors with adjustable treads front and rear. Forward speeds were from 1.75 to 10 m.p.h. The Brockway 49G burning gasoline (page 54) had 24-28 hp. capacity, the Brockway 49D was powered with a diesel engine and capable of pulling 2 bottoms, and the 49K had same specifications as the 49G but fitted with a kerosene engine. Hydraulic lift and combination power take-off and belt pulley were optional with either model.

The Farmaster Corp. with executive offices in New York City offered its Farmaster in 2 models, the FD-33 (page 55) and the FG-33. The FD-33 was powered by a 4-cylinder Buda diesel engine and had a 27-32 hp. rating and the FG-33 was powered by a 4-cylinder Buda gasoline engine and had a 28.43-32.95 hp. rating. Both engines had identical piston displacement. The tractors were of the conventional 4-wheel type with the master touch hydraulic control, 2 speed power take-off and auxiliary transmission gears providing 8 forward speeds 1.65 to 12 mi. per hr.

The Earthmaster Farm Equipment Co., Hollydale, Calif. added the Earthmaster D, (page 54) a 2-plow tractor to its line. It was a 4-wheel tractor with tread adjustments from 56 to 84 in. It was provided with a Speedi-Hitch hydraulic control.

The Friday Tractor Co., Hartford, Mich. entered the tractor field with its 3-plow Friday (page 55). It was powered with a 46 hp. 6-cylinder Chrysler engine and with a 2 speed rear axle and had 10 forward speeds from 1.9 to 32.4 m.p.h. Optional were power take-off, hydraulic control and variable speed governor.

The Laughlin Tractor., Inc., Marshall, Texas entered the field with its model C-27, 2 to 3 plow size, 4-wheel tricycle tractor (page 55). It was rated as 27-31 hp. and was powered with a 4-cylinder vertical Continental Red Seal engine. Battery ignition and electric starting and lighting were provided and 4 speed transmission with forward speeds 1.64 to 6.50 m.p.h. Hydrualic lift was optional.

The Gibson Corp., Longmont, Colo. started production on 2 tractor models the I, 2-3 plow and H 2 plow size (page 55). Both were of the 4-wheel tricycle type and were equipped with power take-off and hydraulic lift. The I was powered with a 4-cylinder vertical Hercules engine and the H with a 6-cylinder Hercules engine.

The Custom Manufacturing Corp., Shelbyville, Ind. offered its 4-wheel standard tractor, C (page 55) and its 4-wheel tricycle tractor B, each powered with a 25 hp. 6-cylinder Chrysler engine. Five forward speeds were provided, 2.32 to 21.14 m.p.h. for B and 1.96 to 17.82 m.p.h. for C.

Some 1949 Models

Coop E3 (National Machine Corp.)

Massey-Harris 22RT

Detroit

Long A

Silver King (Fate-Root-Heath)

Dodge Power Wagon

Deere Diesel R

Earthmaster D

Brockway 49G

Deere MC (Crawler)

Love (Chrysler)

Sheppard Diesel SD2

Friday

Laughlin C-27

Farmaster FD-33

Gibson H-2

Custom C

Haas Atomic

1950

This year brought the total number of tractors on American farms to nearly 4 million and roughly marked the end of half a century of unparalleled progress in power farming. During this period a number of factors contributed such as research undertaken by industry, state and federal agencies, high cost of production and labor scarcity mainly brought about by two world wars. The first tractors produced half a century ago were large and cumbersome and were suited mainly for plowing and threshing. Then came the motor cultivator, another step toward mechanization, and finally the general-purpose tractor with suitable attachments to perform major farm operations. To this latter in turn was added: the power take-off, the continuous running p.t.o. and the hydraulic controls for both attached and drawn implements. Rubber tires, when they first appeared in the early thirties, were rapidly accepted. Electric starting and lighting systems became common and the efficiency of diesel engine-powered tractors was recognized as evidenced by increasing numbers of manufacturers of such tractors. Tractors using liquefied petroleum gas (LP-gas) for use in certain areas were coming more and more into the picture.

At the turn of the century a wide variety of the then called gasoline traction engines were in the making. They were crude affairs, barked loudly, frightened horses, were difficult to start and were heavy and inefficient. Their acceptance was slow, the steam boys ridiculed the gasoline boys' efforts and many clashes occurred. Gradually and finally the gas engine boys won out, and 50 years later, the steam engine traction engine became as much of a curiosity as the gasoline traction engine was in 1900, the so-called "iron horse" of that period.

The year 1950 marked the continued improvement in tractors from the standpoints of design and efficiency of the products of all tractor manufacturers and related accessories. The performance chart on page 57 furnished by the University of Nebraska Tractor Testing Lab., Lincoln, Nebr. shows the average performance of wheel-type tractors by years.

The Oliver Corp. announced the development of diesel engines for all models which made available tractors powered with Oliver engines in 3 distinct fuel types.

The Global Trading Corp., Washington, D.C. became the distributors of the Global or Huber B tractor (page 59) said to be built by the Huber Co. It was powered with a 29.16 - 38.13 hp. (Nebraska test) Continental diesel engine and provided with an adjustable front axle and 5 speed transmission for speeds from 1.9 to 15.3 m.p.h. It was designed for the purpose of selling to the export trade which did not materialize, so the company did not go into production of this model.

The National Implement Co., Stockton, Calif. entered the tractor field with a Harris Power Horse 4 wheel drive tractor (page 60) which was powered with a 6 cylinder 96 hp. Chrysler engine. Independent axle for each wheel was provided and roller chains united each front wheel with the corresponding rear wheel - the pair moving in tandem.

The Minneapolis-Moline Co. converted its Model G tractor for the use of LP-gas fuel. This company began in 1941 to release tractors equipped to utilize the LP or bottled gas.

The B. F. Avery Co., Lousville, Ky., introduced its Model BF tractor (page 59). This 2-plow tractor was fitted with hydraulic controls and adjustable rear tread and featured three interchangeable front end types - the BFD with dual front wheels; the BFS with single front wheel and the BFW with wide, adjustable front axle. The Minneapolis-Moline Co. acquired the B. F. Avery Co. in 1951 and continued to make these models.

The American Tractor Corp., Cherubusco. Ind. started production on its Model GT-25 Terratrac (page 60) crawler tractor. It was powered with a 20-25 hp. 4-cylinder Continental engine and was provided with a hydraulic lift and power take-off. The clearance was 20 in. and 6 track gauges were available from 36 to 72 in. for various row crop spacings when used for row-crop work.

The General Tractor Corp., Detroit, Mich., came into the picture with its 4-wheel drive Powerbilt (page 60) tractor powered with a 36 hp. 4-cylinder gasoline engine with diesel power available. Steering was accomplished by the use of steering brakes assisted as necessary with a drawbar which could be moved laterally under hydraulic control of the operator, a so-called auxiliary steering unit. In turning left, for instance, with a load attached, the drawbar would be shifted to the left thereby tending to force the front end of the tractor to the left. A three-point suspension lift or hydraulic control was interchangeably mounted front or rear for the manipulation of various implements.

Table 4 gives the number of tractors on farms by states at 5 and 10 year intervals from 1920 to 1050 inclusive.

Table 5 gives the number of tractor manufacturers, tractors manufactured by type and the number of horses and mules on farms by years from 1931 to 1950 inclusive.

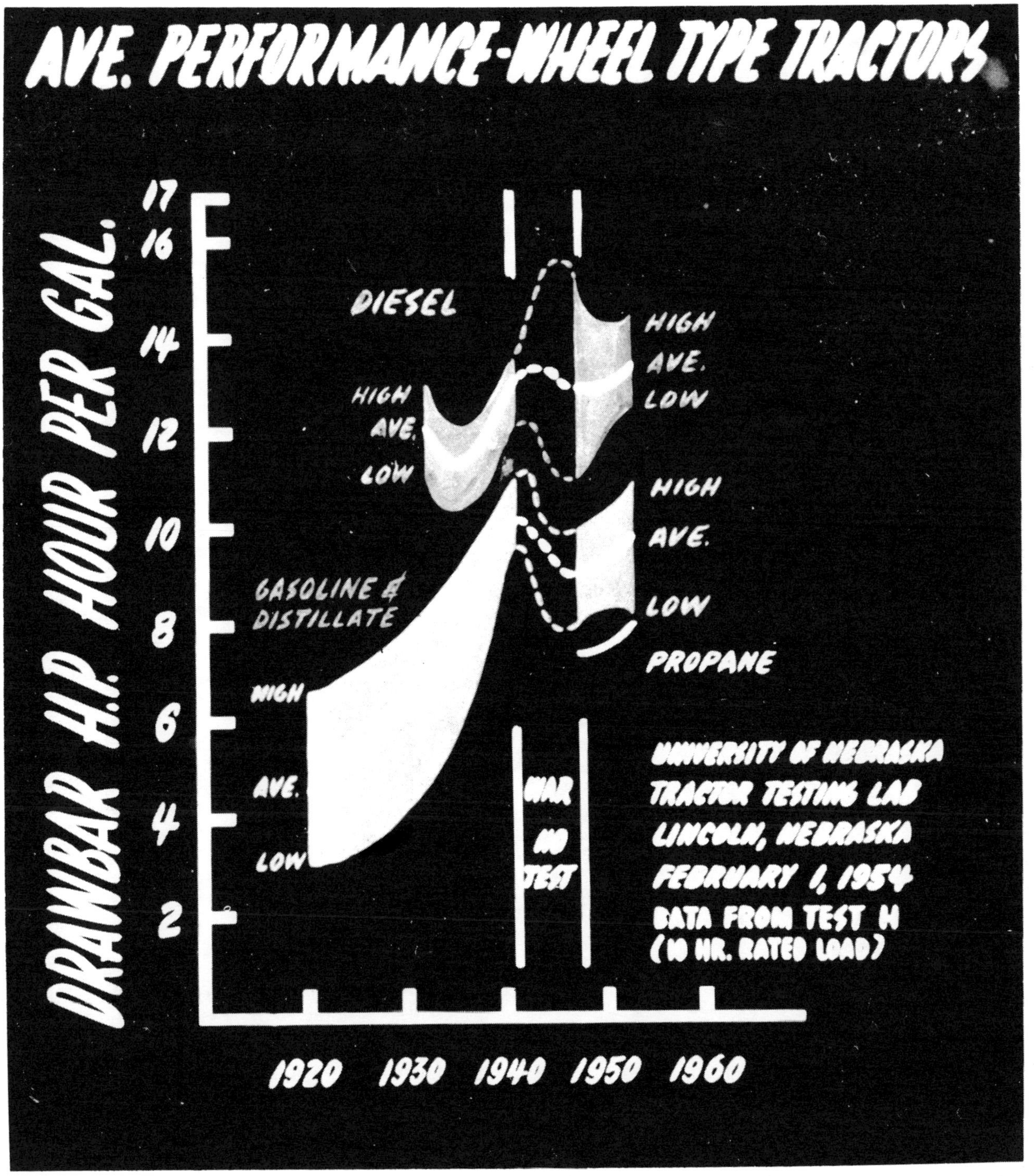
AVE. PERFORMANCE-WHEEL TYPE TRACTORS
DRAWBAR H.P. HOUR PER GAL.
17
16
14
12
10
8
6
4
2
DIESEL
HIGH
AVE.
LOW
HIGH
AVE.
LOW
HIGH
AVE.
LOW
PROPANE
GASOLINE &
DISTILLATE
HIGH
AVE.
LOW
WAR
NO
TEST
UNIVERSITY OF NEBRASKA
TRACTOR TESTING LAB
LINCOLN, NEBRASKA
FEBRUARY 1, 1954
DATA FROM TEST H
(10 HR. RATED LOAD)
1920
1930
1940
1950
1960

Table 4. U. S. Census of Tractors on Farms

State	April 1 1950	Jan. 1 1945	April 1 1940	April 1 1930	Jan. 1 1925	Jan. 1 1920
Alabama	45,751	17,060	7,638	4,664	2,465	811
Arizona	9,746	6,372	4,129	2,558	1,239	930
Arkansas	60,308	26,537	12,564	5,684	3,476	1,822
California	124,582	79,839	55,191	44,437	29,948	13,852
Colorado	53,117	32,766	21,423	13,334	6,693	4,990
Connecticut	11,954	9,740	5,349	2,667	1,441	440
Delaware	6,125	4,604	2,661	1,600	694	239
Florida	22,018	12,812	7,703	5,244	2,777	680
Georgia	60,269	24,648	9,327	5,870	4,145	2,252
Idaho	39,398	20,299	11,103	4,691	1,927	1,587
Illinois	234,789	174,270	126,069	69,628	43,325	23,102
Indiana	153,980	105,263	73,221	41,979	23,567	9,230
Iowa	240,941	181,049	128,516	66,258	37,230	20,270
Kansas	146,266	116,651	95,139	66,275	31,171	17,177
Kentucky	58,818	24,409	11,927	7,322	4,994	2,029
Louisiana	35,735	17,630	9,476	5,016	3,482	2,812
Maine	16,665	14,794	8,093	3,410	1,774	635
Maryland	29,905	18,789	10,335	7,208	4,026	1,525
Massachusetts	14,697	14,026	7,335	3,921	2,212	592
Michigan	149,377	110,120	66,524	34,579	19,217	5,884
Minnesota	204,200	152,555	105,075	48,457	26,739	15,503
Mississippi	51,698	21,077	10,577	5,542	1,871	667
Missouri	125,536	76,110	45,155	24,999	12,745	7,889
Montana	44,890	31,725	22,587	19,031	6,602	7,647
Nebraska	127,154	96,203	70,761	40,729	18,765	11,100
Nevada	3,213	1,827	681	360	221	210
New Hampshire	6,946	6,094	3,129	1,096	531	207
New Jersey	26,039	19,809	12,920	8,088	4,419	946
New Mexico	15,780	10,735	5,832	2,497	1,080	491
New York	119,302	93,292	58,906	40,369	25,681	7,497
North Carolina	73,497	31,189	12,756	11,426	7,909	2,277
North Dakota	98,676	73,984	49,361	37,605	17,483	13,006
Ohio	182,481	130,486	89,999	52,974	30,905	10,469
Oklahoma	93,705	70,395	45,369	25,962	10,950	6,210
Oregon	52,426	28,646	17,077	9,838	5,768	3,070
Pennsylvania	125,851	92,638	54,842	33,513	19,764	5,697
Rhode Island	1,944	1,962	1,008	589	318	79
South Carolina	30,282	12,447	4,791	3,462	2,906	1,304
South Dakota	88,274	62,772	44,154	33,837	17,426	12,939
Tennessee	59,798	24,052	11,817	6,865	4,767	1,872
Texas	232,328	162,381	98,923	37,348	16,780	9,048
Utah	15,987	6,876	3,041	1,426	850	583
Vermont	10,662	7,327	3,566	2,426	1,568	444
Virginia	48,133	23,418	11,951	9,757	6,750	2,379
Washington	55,061	32,050	18,019	8,388	4,490	2,635
West Virginia	13,934	6,863	3,656	2,792	1,860	572
Wisconsin	171,433	123,280	81,195	50,173	29,651	9,407
Wyoming	15,610	9,855	6,534	4,110	1,323	1,075
U. S. Total	3,609,281	2,421,747	1,567,430	920,021	505,933	246,083

From the 1950 U. S. Census of Agriculture, U. S. Dept. of Commerce. Of the tractors on farms, the following were reported as garden tractors:

215,521 in 1950; 69,745 in 1945; not reported for previous years.

Table 5. Number of tractor manufacturers, number of tractors produced, and number of horses and mules on farms

			Number of Tractor Manufacturers				Horses and Mules on Farms
Year	Total	Garden	Wheel Type	Tracklaying	Garden	Total	(000 omitted)
1931		11	61,940	7,089	2,675	71,704	18,468
1932							17,812
1933		Not available					17,337
1934							16,997
1935		16	138,084	18,774	4,273	161,131	16,683
1936		13	193,947	27,299	5,939	227,185	16,226
1937		13	237,837	34,602	10,716	283,155	15,802
1938	42	15	172,437	16,837	9,949	199,223	15,245
1939		16	185,558	20,127	9,777	215,462	14,792
1940		14	249,434	24,762	9,350	283,546	14,478
1941		15	313,432	28,661	16,427	358,520	14,104
1942		16	172,123	29,578	13,373	215,074	13,655
1943		20	105,248	29,453	8,788	143,489	13,231
1944	32	17	249,131	44,860	16,999	310,990	12,613
1945	35	20	244,430	44,872	27,966	317,268	11,950
1946	45	31	258,274	25,902	118,237	402,413	11,108
1947	67	45	433,334	37,295	172,938	643,567	10,129
1948	104	71	529,587	39,412	184,624	753,623	9,279
1949	141	90	555,523	44,613	126,839	726,975	8,498
1950	137	92	498,768	43,680	151,198	693,646	7,781

From Agricultural Statistics, 1952, and the Bureau of Census, U. S. Department of Commerce. The total number of tractor manufacturers is shown when available from Census reports.

Some 1950 Models

Global or Huber B

Minneapolis-Moline BF

Terratrac GT 25 (American Tractor Corp.)

General Powerbilt

Harris Power Horse 53

Index to Manufacturers

Brand Name Index

This index identifies those tractors sold by brand name rather than the manufacturer's cororate name. For example, all references to Farmall, a tractor produced by the International Harvester Company are listed below. References to Farmall can also be found under International Harvester Company in the Index to Manufacturers. Tractors produced by J.I. Case, with a few exceptions, are listed in the Index to Manufacturers. In some instances, the author did not list a manufacturer. However, in most cases, a brand name can be easily cross-referenced to the Index of Manufacturers.